Critical Realism, Environmental Learning and Social-Ecological Change

Southern Africa, where most of these book chapters originate, has been identified as one of regions of the world most at risk of the consequences of environmental degradation and climate change. At the same time, it is still seeking ways to overcome the century-long ravages of colonial and apartheid impositions of structural and epistemic violence. Research deliberations and applied research case studies in environmental education and activism from this region provide an emerging contextualized engagement that is related to a wider internationally articulated quest to achieve social-ecological justice, resilience and sustainability through educational interventions.

This book introduces a decade of mainly southern African critical realist environmental education research and thinking that asks the question: 'How can we facilitate learning processes that will lead to the flourishing of the Earth's people and ecosystems in more socially just ways?' The environmental education research topics represented in this book are wide-ranging. However, they all exhibit the common theme of social justice and wanting to create change towards a better future. All the authors have used critical realist or critical realist-influenced research methodologies. Offering contributions from a small but growing community of researchers working with critical realism in the global South, this book will be of interest to students, scholars and practitioners in the areas of environmental education, sustainability, development and the philosophy of critical realism in general.

Leigh Price is a member of the Centre for Critical Realism, an Associate of the International Centre for Critical Realism, a Research Associate of Rhodes University, South Africa, and a Visiting Research Associate of the Institute of Education, University College London, UK.

Heila Lotz-Sisitka holds the Murray and Roberts Chair of Environmental Education and Sustainability, and is also part time Director of Postgraduate Studies at Rhodes University, South Africa. She is editor of the *Southern African Journal of Environmental Education.*

Ontological explorations

Other titles in this series:

From One 'Empire' to the Next
Radha D'Souza

Science for Humanism
The recovery of human agency
Charles R. Varela

Philosophical Problems of Sustainability
Taking sustainability forward with a critical realist approach
Jenneth Parker

Dialectic and Difference
Dialectical critical realism and the grounds of justice
Alan Norrie

Interdisciplinarity and Climate Change
Transforming knowledge and practice for our global future
Edited by Roy Bhaskar, Cheryl Frank, Karl Georg Høyer, Petter Naess and Jenneth Parker

Conversations about Reflexivity
Edited by Margaret S. Archer

Relational Sociology
A new paradigm for the social sciences
Pierpaolo Donati

Sociological Realism
Edited by Andrea M. Maccarini, Emmanuele Morandi and Riccardo Prandini

The Economics of Science: A Critical Realist Overview
Volume 1: Illustrations and philosophical preliminaries
David Tyfield

The Economics of Science: A Critical Realist Overview
Volume 2: Towards a synthesis of political economy and science and technology studies
David Tyfield

Ontology Revisited
Metaphysics in social and political philosophy
Ruth Groff

Childhoods, Real and Imagined
Volume 1: An introduction to critical realism and childhood studies
Priscilla Alderson

Naturalizing Critical Realist Social Ontology
Tuukka Kaidesoja

What's Critical about Critical Realism?
Essays in reconstructive social theory
Frederic Vandenberghe

Integrating Knowledge through Interdisciplinary Research
Problems of theory and practice
Dominic Holland

Post-Secularism, Realism and Utopia
Transcendence and immanence from Hegel to Bloch
Jolyon Agar

Critical Realism, Somalia and the Diaspora Community
Abdullahi Haji-Abdi

Reality and Self Realisation
Bhaskar's metaphilosophical journey toward non-dual emancipation
Seo MinGyu

The Contradictions of Love
Towards a feminist-realist ontology of sociosexuality
Lena Gunnarsson

Capitalism, Citizenship and the Arts of Thinking
A Marxian-Aristotelian linguistic account
Kathryn Dean

Understanding Mental Health
A critical realist exploration
David Pilgrim

Indigenist Critical Realism
Human rights and first Australians' wellbeing
Gracelyn Smallwood

The Denial of Nature
Environmental philosophy in the era of global capitalism
Arne Johan Vetlesen

The Politics of Childhoods Real and Imagined
Volume 2: Practical application of critical realism and childhood studies
Priscilla Alderson

Critical Realism, Environmental Learning and Social-Ecological Change
Edited by Leigh Price and Heila Lotz-Sisitka

Metatheory for the 21st Century
Critical realism and integral theory in dialogue
Edited by Roy Bhaskar, Sean Esbjörn-Hargens, Nicholas Hedlund and Mervyn Hartwig

Critical Realism, Environmental Learning and Social-Ecological Change

Edited by Leigh Price and
Heila Lotz-Sisitka

Routledge
Taylor & Francis Group
LONDON AND NEW YORK

First published 2016 by Routledge

2 Park Square, Milton Park, Abingdon, Oxfordshire OX14 4RN
52 Vanderbilt Avenue, New York, NY 10017

Routledge is an imprint of the Taylor & Francis Group, an informa business

First issued in paperback 2020

British Library Cataloguing in Publication Data
A catalogue record for this book is available from the British Library

Library of Congress Cataloging in Publication Data
Critical realism, environmental learning, and social-ecological change / edited by Leigh Price and Heila Lotz-Sisitka.
pages cm. – (Ontological explorations)
ISBN 978-1-138-02519-6 (hardback) – ISBN 978-1-315-66089-9 (e-book) 1. Critical realism. 2. Environmental education–Africa, Southern. 3. Human ecology–Africa, Southern. I. Price, Leigh, editor.
B835.C745 2015
149'.2–dc23 2015022104

ISBN: 978-1-138-02519-6 (hbk)
ISBN: 978-0-367-59768-9 (pbk)

Typeset in Baskerville
by Wearset Ltd, Boldon, Tyne and Wear

Contents

Figures

Tables

Contributors

Adesuwa Vanessa Agbedahin has a Bachelor's Degree in Education of the Hearing Impaired from University of Jos, Nigeria. She obtained a Master's degree from Rhodes University, using Cultural Historical Activity Theory (CHAT) and its related expansive learning theory to facilitate learning and change in university dining halls, instigating food waste reduction within the Food Services sector. Her PhD research at Rhodes University focuses on the mainstreaming of Education for Sustainable Development (ESD) in African Higher Education institutions, with a novel focus on the position-practice system of academics vis-à-vis their participation in an international change-oriented professional development programme on Education for Sustainable Development.

Million Belay Ali is a founder and Director of MELCA-Ethiopia, an indigenous NGO working on issues of agro-ecology, intergenerational learning, advocacy and livelihood improvement of local and indigenous peoples. He also co-ordinates the Alliance for Food Sovereignty for Africa. He has been working for over two decades on the issues of intergenerational learning of biocultural diversity, sustainable agriculture, forests and the right of local communities to seed and food sovereignty. His main interest is currently advocacy on food sovereignty, learning among generations, knowledge dialogues and the use of participatory mapping for social learning, identity building and mobilization of memory for resilience. He has an MSc in tourism and conservation and a PhD in environmental learning from Rhodes University.

Aristides Baloi was born in Mozambique. He has a degree in Geography from the Ukraine and a Masters degree in Water Management from the Netherlands. He is a teacher and researcher in the Geography Department at the University of Eduardo Mondlane in Mozambique. His research interests include learning practice and processes in social-ecological systems. He is currently a PhD scholar at Rhodes University in the Environmental Learning Research Centre.

Jane Caroline Burt is a researcher-educator in the environmental/development sector with a deep interest in learning as emancipatory action.

She is currently a research associate of the Rhodes University Environmental Learning Research Centre and is working on a PhD. Her chapter is based on work for the Association of Water and Rural Development, a South African-based NGO.

Heila Lotz-Sisitka is Murray and Roberts Chair of Environmental Education and Sustainability, and Director of the Centre for Postgraduate Studies at Rhodes University. She has been involved in national and international education and environmental education research and policy development since 1992. In the past 15 years she has co-ordinated the masters and PhD programmes in environmental education at Rhodes University. She has served on a number of international scientific committees, including as scientific chair of the World Environmental Education Congress in 2007, and as scientific committee member of the PhD Graduate School for Education for Sustainability in Sweden, and as scientific advisor to the Global Universities Partnership for Environment and Sustainability. She is editor of the *Southern African Journal of Environmental Education.* Her research interests include education and agency in contexts of risk and vulnerability, participation in education, and critical research methodologies.

Justin Lupele works for FHI 360 as a Chief of Party for the USAID-funded Schools Promoting Learning Achievement through Sanitation and Hygiene (SPLASH) in Zambia. He is a distinguished Rhodes and Commonwealth scholar and has made several contributions to refereed journals. His professional interests are in research; curriculum development; climate change; water, sanitation and hygiene education; participatory educational materials development and institutional/organisational capacity building. He has a PhD in Environmental Education from Rhodes University in South Africa.

Mutizwa Mukute is a Research and Development Consultant and Rhodes University Research Associate with over 25 years experience of working in the agriculture and natural resources management sector. He briefly worked for Zimbabwe's Parks and Wildlife Department before joining the civil society sector where he served as an environmental researcher/writer, college co-ordinator and director. He received a Certificate of Outstanding Leadership from the Participatory Ecological Land Use Management Association (PELUM) in 2005 and a Certificate of Outstanding Contribution from the Eastern and Southern African Farmers' Forum (ESAFF) in 2014. He holds a PhD, masters and Certificate in Environmental Education from Rhodes University, funded by scholarships from the South African Qualification Authority (SAQA) and the MacArthur Foundation respectively.

Victor Munnik is an associate at the Society, Work and Development (SWOP) Institute at the University of the Witwatersrand, and a research associate of the Institute for Water Research (IWR) at Rhodes University. His activist work, within the South African environmental justice movement for the past 30 years, has focused on extending the human right to water, the right to live in a healthy environment and the defence of fenceline communities, the protection of water resources and ecosystems, social justice and environmental sustainability, and building a strong participatory democracy in South Africa. His academic work brings together action research, political ecology and discourse analysis in support of environmental activism. He holds honours degrees in Journalism (Stellenbosch), Comparative Literature (Wits) and Geography (Wits), and a PhD in Geography and Environmental Studies (Wits).

Rob O'Donoghue is an Associate Professor and Director of the Environmental Learning Research Centre (ELRC) at Rhodes University. In his research on environmental education processes of learning-led change, he has given close attention to indigenous knowledge practices, social theory and environmental learning in post-colonial curriculum and community contexts. In his recent work with critical realism, he has probed generative research and transformative social learning to inform the Fundisa [Teaching] for Change teacher professional development programme and the Amanzi [Water] for Food initiative on local food production with rainwater harvesting to mitigate the emerging patterns of climate change in eastern southern Africa.

Lausanne Laura Olvitt has a background as an English language teacher but has been working in environmental education since 2002. Currently, she is a senior lecturer in the Environmental Learning Research Centre (ELRC) at Rhodes University where she supervises masters and doctoral research in environmental education and co-ordinates the BEd Honours programme. She is also co-editor of the *Southern African Journal of Environmental Education* and is involved in the ELRC's national Green Skills research programme. She has a strong interest in understanding the historical, socio-cultural and socio-material dimensions of environmental learning, especially in relation to values and ethical deliberations.

Tichaona Pesanayi is currently a PhD researcher in the Environmental Learning Research Centre at Rhodes University. For the past few years he has been manager of the Southern African Development Community's Regional Environmental Education Programme, a programme that works across 14 southern African countries. Tichaona has a long history of working in environmental education, especially with a focus on education and community-based agriculture. He has also served as secretary general of the Environmental Education Association of Southern Africa. His research interests include boundary crossing, social learning, farmers' knowledge and social change in post-colonial social, agriculture and food production contexts.

Leigh Price recently completed a three-year Postdoctoral Research Fellowship at the Environmental Learning Research Centre, Education Department, Rhodes University. She is also a member of the Centre for Critical Realism, an Associate of the International Centre for Critical Realism, a Research Associate of Rhodes University and a Visiting Research Associate of the Institute of Education, University College London. Leigh's (2007) PhD on environmental education from Rhodes University was funded by a scholarship from the MacArthur Foundation for Peace and Justice. In 2010, Leigh joined Roy Bhaskar at the Institute of Education, University College London, where she worked with him during much of her three-year post-doctoral fellowship.

Presha Ramsarup is currently completing a PhD thesis at Rhodes University, focusing on a dialectical critical realist approach to the emergence of change-orientated learning pathways and sustainable development, with a focus on scarce skills. She has worked as an educator, professional development facilitator and researcher. She has served as an environmental education technical advisor to both provincial and national government to support the development of environmental and sustainability education within South Africa's education and training transformation process. She is currently a part-time researcher at the University of Witwatersrand Researching Education, Work and Labour Centre, and a part-time researcher at the Rhodes University Environmental Learning Research Centre, where she is involved in research and co-ordination of a national Green Skills programme supporting system capacity development, and employer-led demand analysis to identify green jobs and opportunities for green jobs.

Daniel Sabai has worked as an Environmental Educator and Communicator for over 15 years. He currently works as a lecturer at the Dar es Salaam University College of Education, a Constituent College of University of Dar es Salaam, where he is researching experiential learning processes in the coastal and marine context with an emphasis on social learning. His PhD thesis, which he completed at Rhodes University, focused on analysing learning at the interface of scientific and traditional ecological knowledge. The study created a participatory learning process, which enabled fishers and mangrove restorers in the Eastern Coast of Tanzania to develop indicators for monitoring mangroves and fisheries.

Ingrid Joan Schudel is a Senior Lecturer at Rhodes University and is involved in pre-service and in-service courses at Certificate and Honours levels with a focus on environmental education and Natural Science teaching in schools. She is also a lecturer and co-ordinator of the Masters in Education (Environmental Education) degree programme which supports environmental education in school, higher education, community, NGO and government contexts. She is involved in a number of research programmes, particularly as co-ordinator of the national Fundisa [Teaching] for Change Network and the Citizen Science research programme at the Environmental Learning Research Centre. She supervises masters and doctoral students. Her research interests include transformative learning, teacher professional development and citizen science.

Muchaiteyi Togo holds a PhD in Environmental Education from Rhodes University (South Africa). Her PhD explored the use of systems approaches in mainstreaming sustainable development in universities. She currently works for the University of South Africa (UNISA) as a senior lecturer in Environmental Science in the College of Agriculture and Environmental Sciences. With a background in Geography and Environmental Science, Muchaiteyi works with an interdisciplinary approach to environmental and sustainability research. Her research interests are in education for sustainable development, sustainability mainstreaming in higher education, environmental management, climate change, green economy and green skills.

Acknowledgements

There are several institutions and people that we would like to thank. First and foremost, we would like to thank Rhodes University's *Sandisa Imbewu* project, for its generous support in allocating funding to the Environmental Learning Research Centre for a three-year post-doctoral fellowship for Leigh Price, which allowed us to bring this volume into being. While the fellowship was enabling, Leigh's contributions stretch way beyond the fellowship's boundaries. We also thank the many funders, partners of the Environmental Learning Research Centre, and community members that supported the range of PhD studies that are reported on in this volume. We also thank the Rhodes University Faculty of Education for their support of staff in the Environmental Learning Research Centre who have supervised studies, and supported the work in this volume.

Thanks are also due to the Department of Curriculum, Pedagogy and Assessment, Institute of Education, University College London who enabled institutional collaboration by making Leigh Price a Visiting Research Associate. This provided a multi-institutional and reflexive platform for the exploration of many of the ideas in this book, through its Post-Graduate Reading Group sessions and its International Centre for Critical Realism Discussion Series. We would also like to thank Routledge for being willing to publish our work.

Second, we would like to thank the authors themselves who have persevered through numerous revisions and resubmissions, patiently fine-tuning already good papers. Their deeply ethical commitment to social change and transformative praxis in Africa provides the foundational narrative of this book.

Third, we would like to thank specific people who worked on the technical side of the preparation of the manuscript. In particular, we thank Kim Ward who provided kind, responsive and valuable final assistance with a final proofread and cross-checking of references, and helped us to get some of the diagrams into a professional format. Emma Chappell at Routledge was always ready to offer advice and skilfully guided us through the publishing process. Carol Thomson ably assisted us in the early stages with advice on how to improve our academic writing.

Last, but certainly not least, we would like to thank the late Roy Bhaskar, who encouraged us to submit our book proposal to Routledge and whose ideas provide the core of the book itself.

1 Why critical realism, environmental learning and social-ecological change?

Introducing the book

Heila Lotz-Sisitka and Leigh Price

The contemporary social-ecological condition is characterized by powerful changes in the way that we relate to each other and to the environment. This has led to increased ecological vulnerability, which is also accompanied by ongoing, and increased societal vulnerability. Nevertheless there remain opportunities for developing new social-ecological relations, and for social-ecological learning and change. This would seem to require a strong project of recovering ontology, and a challenging and broadening of dominant ways of knowing (Mignolo, 2000) that also tend to commit what Bhaskar describes as the 'epistemic fallacy', or the 'the analysis or reduction of being to knowledge of being' (Bhaskar, 2010, p. 1). In response, Bhaskar (ibid.) suggests critical realism as an alternative that embodies a 'compatibility of ontological realism, epistemological relativism and judgmental rationality'. This includes a 're-vindication of ontology' and the possibility of recognizing and accounting for structure, difference and change in the world in ways that escape ontological actualism and ontological monovalence or 'the generation of a purely positive account of reality' (ibid., p. 15).

Addressing the social-ecological concerns of our times would seem to require intentional human agency of all 'planetary subjects'[1] (Spivak, 1999, p. 46), to achieve transformation towards sustainability in ways that ensure the flourishing of all current and future generations. Bhaskar (2008 [1993], p. 9) describes agency as: '*radically transformed transformative praxis*, oriented to rationably groundable projects'. This implies a role for education and social learning, as transformative praxis involves knowledge, values, skills, beliefs and motives that are learned in various sociocultural and educational contexts, both formal and informal. Bhaskar (2010), arguing from the standpoint of concrete utopianism, suggests that a key role for intellectuals and educators is be involved in the envisaging of alternative possible futures for humanity. The field of environmental education has been engaging with these ideas and possibilities for the past 40 or more years, but has experienced difficulties which are ascribed to underlying commitments to untenable philosophical positions of either positivism, participation or postmodernism (Price, 2007). In this book we explore how critical realism can offer possibilities that overcome these difficulties. In doing this, we recognize that our efforts are necessarily fallible, partial and incomplete since all of our efforts take place in open systems.

Southern Africa, where most of these book chapters originate, has been identified as one of the regions of the world most at risk of the consequences of environmental degradation and climate change (UNEP, 2006; IPCC, 2014). At the same time, it is still seeking ways to overcome the century-long ravages of colonial and apartheid impositions, structural and epistemic violence (Neocosmos, 2012). Research deliberations and applied research case studies from this region provide an emerging contextualized engagement that is related to a wider internationally articulated quest to achieve social-ecological justice, resilience and sustainability through educational interventions.

This book introduces a decade of mainly southern African critical realist environmental education research and thinking that asks the question: 'How can we facilitate learning processes that will lead to the flourishing of the Earth's people and ecosystems in more socially just ways?' Many of the contributions in this book arise from academic work supervised and supported by the Rhodes University Environmental Learning Research Centre where explorations of critical realist scholarship emerged following the work of Price (2007). The environmental education research topics represented in this book are wide-ranging. However, they all exhibit the common theme of social justice and wanting to create change towards a better future. All the authors have used critical realist or critical realist-influenced research methodologies. They represent a small but growing community of researchers.

Perhaps unsurprisingly, many southern African researchers decided to use critical realism as a result of similar frustrations to those that originally motivated the founder of critical realism, Roy Bhaskar (1944–2014).[2,3] Essentially, Bhaskar wanted to contribute to making the world a better place. However, he was frustrated to discover that he was not allowed to talk about the world, only about theories of the world. Against the flow of contemporary philosophy of science at the time, he countered this prohibition by arguing that to change the world, and to change our beliefs about the world, it is necessary to have a conception of the world. If one says that a belief about the world is inadequate then there must be something which remains the same between the inadequate belief and the more adequate belief that one wants to replace it with (Bhaskar, 2014). Therefore, the starting point of critical realism was that to coherently critique beliefs, we need a notion of the (real) world.

In addition to making emancipatory research a possibility, critical realism also has important implications for education.[4] In discussing Bhaskar's naturalism in the context of educational research, David Scott (2010) explains that social objects, though they are real, change constantly.

He suggests that researchers need to focus on the changing object, as it is this that is 'relatively enduring'. Scott suggests that it is possible to see that objects can change to such an extent that they can become barely recognizable in relation to their former selves; it *is* possible to re-imagine education (Kanu, 2003). For example, unlike 50 years ago when curricula and education largely ignored issues of the environment, there is today a worldwide engagement with the processes of integrating environment and sustainability concepts, knowledge, pedagogical practices and values into national curricula, and into a range of community and public education programmes, including higher education (UNESCO, 2014). This project is not uncontested, and remains an emergent 'unfinished project' as can be seen from the chapters in this book.

This points to another key tenet of critical realism, which recognizes that knowledge is fallible, because it may be refuted. What counted as legitimate knowledge 50 years ago is not the same today. In interpreting this change, environmental education researchers need to take account of changes at deeper levels of social-ecological reality, where generative mechanisms and deep-seated structures are at work. Indeed, these have not only generated significant changes in educational thinking and practice the world over, but they have also generated a change within the field of environmental education research itself, where environmental education research and practice activities are nowadays often framed within, replaced by, or linked to the often conflicting discourses of Education for Sustainable Development (ESD), social-ecological justice, political ecology, deep ecology and/or conservation discourses.

In the past 40 or more years environment and development concerns have become more closely coupled, as environmental justice movements have gained ground, and as deeper understandings of the relationship between political economy and political ecology have emerged. In a global South context, this has also been accompanied by a greater focus on decolonization (Mignolo, 2000; Dussel, 1998) and framing of a de-colonial 'ecological consciousness' as counter-narrative to mainstream neo-liberal capitalism (which can include some forms of sustainable development thinking) (Leff, 2009). In this regard, Dussel (1998) suggests that the 'end' to the system of Western epistemology is rapidly being constituted by three limits: (1) the ecological destruction of the planet emerging from a concept that nature is an 'exploitable object' for serving capital's interests; (2) poverty, emerging from a conception that unequal distributions of wealth are limitless; and (3) the impossibility of subsuming peoples and cultures through homogenizing modernization processes (see also Neocosmos, 2012).

In ways that are not unrelated to this, Martinez-Alier (2002, p. 14) identifies what he calls 'three clusters' of environmental concern and activism that are emerging, and describes them as:

1 the 'cult of wilderness', concerned with the preservation of wild Nature, without anything to say on industry and urbanization, indifferent or opposed to economic growth, most worried by population growth, backed up scientifically by conservation biology;
2 the 'gospel of eco-efficiency', concerned with the sustainable management or 'wise use' of natural resources and with the control of pollution not only in industrial contexts, but also in agriculture, fisheries and forestry, resting on a belief in new technologies and the 'internalization of externalities' as instruments for ecological modernization, backed up by industrial ecology and environmental economics;
3 the environmental justice movement, popular environmentalism, the environmentalism of the poor, livelihood ecology, and liberation ecology, grown out of local, regional, national and global ecological distribution conflicts caused by economic growth and social inequalities. Examples are conflicts over water use, over access to forests, over the burdens of pollution and over ecologically unequal exchange, which are studied by political ecology.

Interesting for a text on education are the disciplinary roots that partly shape these forms of environmentalism vis-à-vis conservation biology, industrial ecology, economics and political ecology (amongst others). De-colonial theorists have much to say about how modern (Western-centric) education system disciplines have come to 'discipline reason' in particular ways that exclude and/or marginalize much of what there is to know or how we come to know reality. They also describe how the mainly Western epistemological disciplines tend to ignore the ontologies of a wider range of more complex forms of epistemology and reason (Gordon, 2014; Mignolo, 2000; Spivak, 1999; Kanu, 2003), a point that is reflected in a number of the chapters in this book, partly also motivating their use of critical realism.

Significant from an ontological and epistemological perspective for environmental education, Martinez-Alier (2002) argues that there are 'points of contact' and 'points of disagreement' amongst these varieties of environmentalism. He ultimately suggests that the points of contact are ontological, as the common ground that unites all environmentalists (despite epistemologically diverse perspectives) are the deep-seated structural conditions and mechanisms that shape and hold a (globally) powerful, colonially inspired anti-environmental lobby in place, which he suggests has more powerful effects in the global South than in the global North. This lobby tends to use epistemological strategies such as anti-development discourse to marginalize environmental action, a strategy that creates a tension-laden space for transformative environmental education praxis in the global South. As can be seen in the chapters of this book, this appears to give rise to a need for environmental educators to more clearly examine and understand the deep-seated and ontologically grounded contradictions that shape environmental practices and activism in a context where political ecology and livelihood ecology often show anti-development discourse to be filled with false assumptions (i.e. that environmental concerns can be separated out from development concerns). This is an issue that is considered by a number of the authors in this book, and appears to be partly motivating their use of critical realism. In Chapter 16 for example, Victor Munnik makes a clear argument as to how critical realism can advance environmental justice work in South Africa. As pointed out by Leigh Price in Chapter 2, such forms of critical engagement require humility, reflexivity and criticality.

As shown across the chapters in the book, critical realism provides explanatory tools and forms of reasoning that allow for making the complexities found in our contexts more visible and open for dialogue, engagement, learning and reflexivity. Most of the influences for the educational work shared in this book appear to be located at the intersections of sustainable development and environmental justice orientations outlined above, with many leaning in the direction of environmental justice, which is also where the emancipatory interests in environmentalism and associated forms of scholarship are most pronounced. There are also various intersectional forms of engagement with decolonization, especially in attempts to assert cognitive justice, indigenous knowledge and epistemology, and to open new spaces for learning engagements with absences. These various emancipatory/decolonization educational engagements tend to largely intersect with environmental justice discourse. This is a research arena that could potentially be further advanced in environmental education scholarship in southern Africa.

Being aware that our educational research activities and practices are an outcome of deep-seated generative mechanisms and emergence over time (morphogenesis) allows us to theorize our agency as educators. This, however, requires reflexivity as indicated by Price (Chapter 2). This agency is a pre-condition for an effective educational practice that will contribute to our emancipation from social and environmental ills (Bhaskar, 2008 [1993], pp. 163–173) (see also Chapters 2 and 17). As stated by Scott (2010, p. 6):

> Critical realists designate the relation between structure and agency as the key framing device at the ontological level; and furthermore, understand all observational or experiential statements as framed by a specific set of conceptual relations, that is, all observational or theoretical statements are in some sense theory-laden. As a consequence, any description of the world is both explanatory within a particular set of conceptual relations and potentially transformative of those relations. In short, educational processes take place in open systems.

That social researchers must engage with the systemic openness of the social world differentiates them from many (although not all) natural scientists who have the luxury of being able to close their system of enquiry. Social science thus requires the 'use of methods and strategies that fit with systemic openness' (Scott, 2010, p. 5). Scott suggests that such methods should include 'inferential judgements from the analysis of indirect evidence', that is *retroduction*, as described in the approach to science advocated by Bhaskar (2010) called the DREI(C).[5]

This book chronicles how environmental educators in southern Africa have tentatively explored what environmental education might look like if it were to assume that our research describes an open system context. As suggested by Scott (2010), several of the authors have made use of the DREI(C) schema or its relative, the MELD.[6] For examples of the use of the DREI(C), see Chapters 9 and 10, by Rob O'Donoghue and Daniel Sabai respectively. For examples of the use of the MELD, see Chapters 11, 12 and 14, by Mutizwa Mukute, Aristides Baloi and Ingrid Schudel respectively. Presha Ramsurup (Chapter 7), Victor Munnik (Chapter 16) and Heila Lotz-Sisitka (Chapter 17) specifically engage with absences, which are that aspect of the MELD associated with its Second Edge (E) and the possibilities for enabling learning as a passage through the dialectic (Bhaskar, 2008 [1993]).

Although social scientists prior to the postmodern movement often did retroductively theorize generative mechanisms (thus Marx was able to arrive at his retroductive theory of wage labour),[7] nevertheless, after postmodernism, intellectuals were dissuaded from such activity by the charge that to do so would be to commit the sin of 'grand theorizing'. The postmoderns were justified in their suspicion because, without the concept of emergence, most theories of structure were reified. For example, Jean-François Lyotard (1924–1998) sees the breakdown of the grand narratives (theories) that formerly legitimized the status quo as an important aspect of the postmodern condition. The postmoderns were right to avoid grand theories of the sort that stopped any possibility of debate and that legitimized oppression (for example, theories about the evolution of society along the lines of the survival of the fittest that were used to justify a competitive, marketized, imperialist society). However, then to throw out all theories of structures and mechanisms was mistaken and played into the hands of those who would keep the status quo since it is through our theories that we develop social critique. For example, Marx's theories of wage labour underpin a critique of society that has led to political action and social change (morphogenesis). Lyotard himself, whilst he critiques metaphysical theories of structure, at the same time admits there can be no politics without them. He writes (1997, p. 108): 'Can you act and think in politics without being captivated by these metaphysical motifs? Aren't they born with politics itself? Far away as you try to escape from them, it seems they have to return.' Such political action, based, as it must be, on theories of social structure, may lead to unintended consequences, but perhaps such consequences are a risk that we must be prepared to take in the face of the environmental crisis. Fortunately, we might argue that consciously engaging with ontologically layered explanatory theory is less likely to result in unintended consequences, and will allow a quicker response to such consequences, than the use of shallow theories, or protolaws (Bhaskar, 2008 [1975], pp. 163, 164). Such protolaws – or constant conjunctions of events found in statistical correlations that are presumed to be causal – are the preferred substitute when deep theoretical analyses are disallowed, but they are highly questionable as guides to social agency in the open system context of the world. Not only are they shallow and often incorrect,[8] but they also result in lengthy time delays, whilst the policy-analysts wait to allow correlations to develop and embark on the process of carrying out statistical research and its publication (which can take years).

However, only one instance of an unexpected consequence analysed from the perspective of a depth theory can justify action because case studies in this conception are instantiations of the whole society and can mirror society to us in a similar way to the way that the tiniest piece of a hologram contains all the information of the original undivided hologram. As environmental educators, this means that we can justify acting despite having only one piece of evidence that a generative mechanism is problematic and even if we cannot subject our claim to hypothetico-deductivist experimental research. For example, we can justify acting to reduce climate change because of the extensive evidence at hand that it is caused by human activity, even though we cannot prove anthropogenic climate change through hypothetico-deductive methods. We can justify a stance against nuclear power despite having only a few instances of its dangers, such as the 2011 Fukushima Daiichi nuclear disaster. Indeed, we could justify a stand against nuclear power simply based on the transcendental truth of the mechanism of nuclear power and its real *potential* for danger, as Einstein himself did in his 1939 letter to President Roosevelt. Leigh Price (Chapter 18), Victor Munnik (Chapter 16) and Heila Lotz-Sisitka (Chapter 17) explore the potential for critical realism to provide critique and guide action in this way.

Another aspect of critical realist action (Price, 2007, p. 249) involves the idea that, at any particular time, we can only do what we are potentially able to do, with the resources that we have. This echoes some of the work of Lev Vygotsky (1896–1934) and is reminiscent of aspects of his concept of the 'Zone of Proximal Development' (ZPD)[9] (Engeström and Sannino, 2010). Vygotsky is often thought of as a developmental psychologist whose 'general genetic law of cultural Development' suggested that learning has the power and potential to lead human development when he suggested that 'every function of the child's cultural development appears twice: first on the social level, and later, on the individual level; first between people (interpsychological), and then inside the child (intrapsychological)' (Vygotsky, 1978, p. 57). However, Vygotsky is also responsible for suggesting that social learning can result in social transformation; an idea that has been popularized by certain educators, notably Michael Cole and Yrjo Engeström (1993). They call their approach Cultural Historical Activity Theory (CHAT) and argue that as the learner changes, so culture, and ultimately society changes. Cole (in Barab *et al.*, 2004, p. 201) has thus described the ZPD as the zone where culture and cognition create each other. Therefore, we find within Vygotsky's work a mechanism for the emergent process by which individuals transform or reproduce society and society influences individuals, which mirrors the critical realist Transformational Model of Social Activity.

Relatedly, Nunez (2013) has argued that post-Vygotskian research has an 'implicit ontology' that is close to critical realism. Most research associated with Vygotsky's work has tended to focus on the epistemological and semiotic dimensions of meaning making. However, viewed from a critical realist perspective, the Zone of Proximal Development is an instance of a causally efficacious absence: the student's skill is absent, but there remains the potential for the absence to be absented if the student is given some assistance, where that assistance is judiciously designed to appreciate the potential skills already available to the student, but as yet not fully actualized. This is discussed further in Chapter 17 by Heila Lotz-Sisitka who probes potential dimensions of absenting absence in the Zone of Proximal Development, exploring the possibilities of this for emancipatory environmental education and learning. Meaning-making is central to the absenting of absence in emancipatory praxis, as explained by Newman and Holzman (1993, p. 86) as follows:

> human beings are never fundamentally changed (i.e. never develop) except insofar as, by our revolutionary activity, we change the totality of our continued historical existence. This we accomplish not by the humanly impossible act of materially altering all the elements of history, but by the uniquely human activity of materially reorganizing what there is to create a new meaning for everything. It is the child's meaning-making by which 'instruction … impels … a whole series of functions … in the zone of proximal development' (Vygotsky, 1987, p. 212). Without this revolutionary activity of meaning-making, i.e. when the human capacity for changing totalities is not able to be expressed (due to various forms of societal coercion), when toolmakers become mere tool users, when we engage in behaviors (even of incredible complexity and sophistication) and no longer activity, then learning ceases to lead development; instead, it replaces development.

Vygotsky's work is consistent with Bhaskar's (2002, xxx) claim for education that it should be focused on allowing learners to unfold that which they already potentially know, or that which is possible to know, and that which leads to agency and praxis, which provides a thesis of possibility similar to that proposed by Vygotsky's concept of the ZPD. Furthermore,

Vygotsky was well acquainted with the concept of the dialectic, which is a concept that Bhaskar (2008 [1993]) has expanded and refined. It is therefore not surprising that many of the southern African researchers have looked to Vygotsky's work to assist them with their commitment to an emancipatory environmental education. In this volume, Chapters 11, 12 and 17, by Mutizwa Mukute, Aristides Baloi and Heila Lotz-Sisitka respectively, explore Cultural Historical Activity Theory (CHAT); Million Belay Ali (Chapter 3) and Daniel Sabai (Chapter 10), amongst others, explore the idea of 'communities of practice'; and Rob O'Donoghue (Chapter 9) uses the idea of the 'germ cell', all of which are inspired in part by Vygotsky. Southern African authors working within cultural historical activity theory and socio-cultural learning theory traditions are also exploring how theories of cognitive justice, and transgressive learning found mainly in postcolonial and decolonization theory can extend the foundational learning process insights of Vygotsky and Bhaskar, as opened up in the chapter by Heila Lotz-Sisitka (Chapter 17), and as pointed to in the chapters by Million Belay Ali (Chapter 3), Justin Lupele (Chapter 4), Mutizwa Mukute (Chapter 11), Tichaona Pesanayi (Chapter 13) and Daniel Sabai (Chapter 10).

The work of Iskra Nunez (2013, 2014) and her immanent critique of CHAT is also pertinent to the research presented in this book as it draws attention to the empiricism and at times naïve realism in some formulations of CHAT and explains how critical realism, with its dialectical concept of absenting, can help to deepen the ontological dynamics of CHAT. Her work points to the need to be more reflexive of implicit ontology in CHAT research, a point with which the papers in this book seek to engage, as can be seen in Chapters 11, 12 and 13 by Mutizwa Mukute, Aristides Baloi and Tichaona Pesanayi respectively. In the case studies of CHAT research presented in this book, we see examples of where researchers have sought to show how the formative interventionist approach that is fostered by the expansive learning processes of cultural historical activity theory (Engeström, 1987; Engeström and Sannino, 2010) can be considered by seeing learning as a passage through the dialectic in ways that are inclusive of individual, social, material and psycho-social dynamics of the structure agency dynamic as proposed by Bhaskar's (2008 [1993]) transformative model of social action (TMSA, or four planar being) and his view of agency as being 'embodied *agentive* agency' that is emergent in open systems (ibid.). This is not unlike Engeström's (1987) proposal for expanding learning out of socio-culturally and historically situated contexts; and CHAT also takes account of and helps to explain the emergence of transformative agency (Engeström and Sannino, 2010). Closer engagement with Bhaskar's four planar being, and the post-Vygotskian work on transformative agency as proposed by Engeström and Sannino (2010) provides a potentially interesting area for further research and engagement.

Another possibly fruitful place of interaction in the environmental field is between systems theory, complexity theory and critical realism, because many environmentalists have turned to complexity theory and systems theory to articulate complex and emergent social-ecological systems dynamics (including their unpredictability and adaptive management strategies) (Folke *et al.*, 2005). Some argue that systems/complexity theory is a 'new name' for irrealism in sociology (Archer, 2012). However, others working with more mainstream approaches to systems theory suggest that there are powerful similarities between systems/complexity theory and critical realism and that they can both benefit from mutual interaction (Mingers, 2010). In this volume, Mucha Togo (Chapter 5) demonstrates how systems theory and critical realism (including Archer's social realist theory of emergence and morphogenesis) can be used to explain an educational research project focusing on the systemic dynamics of a university and how it responds to sustainability concerns.

As outlined above, developing an understanding of an emergent change in educational thinking and practice over time requires a commitment to the ontological depth of social reality. Social objects, such as our environmental education programmes and curricula, are 'the real manifestations of the idealized types used in discourse' (Scott, 2010, p. 5). According to Scott (2010), these provide a focus for any enquiry. As can be seen across the chapters of this book such enquiries include studies of learning programmes, education and training interventions, environmental education concepts and so on. These educational programmes and activities are all structured in various ways, and because of this, they posses powers. Critical realism proposes that such powers can be actualized, possessed and/or exercised (Brown *et al.*, 2002). Possessed powers may be held but not exercised. Exercised powers are actively having effects on open systems, and actualized powers are already generating effects through interacting with other exercised powers, and/or through possessed powers, potentially triggering changes and emergence. Examining these interacting powers provides for a generative-productive way of producing causal explanations (ibid). Sabai (Chapter 10) considers the consequences of unexercised powers for the efficaciousness of community natural resources management along the coastal areas in Tanzania, and Pesanayi (Chapter 13) considers a similar issue in community gardens in the Eastern Cape of South Africa.

In terms of where this book is situated in the context of environmental education, it has to be seen in relation to the emergence of the environmental movement over the past few decades which has gradually become a connected, internationally active or 'planetary' social movement, characterized by differing orientations as outlined by the work of Martinez-Alier (2002) referred to above. Since the publication of Rachel Carson's *Silent Spring* in 1962, which appears as a global 'marker' of the emergence of the modern environmental movement,[10] a number of institutional changes have taken place, and expanded numbers consisting of thousands and thousands of new organizations, programmes, NGOs, community-based organizations have emerged in this time to respond to ongoing and increased environmental degradation in what can be considered to be a relatively short period in human history (Hawkin, 2008). While it is difficult to quantify these, some indicators exist which show the expansion of this global social movement and its emergence over time. These can be found in the proliferation of environmental reporting, in the establishment and extension of major global organizations, the expansion of environmental technology development and many others. Overall, these represent a major shift in society, even if this shift is troubled by the sometimes contradictory and contested orientations referred to above, combined with the ongoing pressures of neo-liberal globalization, hegemonic states, and the continuities of (neo-)colonial intrusions.

Environmental education, and environmental education research has also emerged over this time, with varied attempts to 'mainstream' into national and international policy and practice narratives (UNESCO, 2014), with some levels of success (e.g. the Eco-Schools movement has been identified as the largest teacher and children's network in the world) (ibid.). Critical educational theory has, in general, resisted the cultural reproduction argument for education, and instead has sought to position education as a transformative social process, and it is in this space that this book is positioned. Such a view of education sees education not only as culturally reproductive, but it *also* (*in addition to, and often in tension with*) sees education as potentially transformative of the status quo. There has been a long and extended debate in educational circles, and in environmental education, about how such critical and transformative forms of education are to be constituted, and what their relation is to the mainstream (Stevenson *et al.*, 2013). In this regard, critical realism can provide metatheoretical tools that can be used to produce new ways of imagining emancipatory education.

The chapters in this book are not only applied studies, but are also deliberative of some of the key concepts in critical realism, as these may be applied to the field of environmental education. In Chapter 2, we provide an overview of critical realist concepts that have particular relevance to environmental education, and also outline some of the common 'mistakes' that are often made by novice critical realist researchers. We do this, not to criticize our own and other's work, but rather to provide tools for reflexivity, and to draw attention to the processes of sense-making involved in working with the sometimes complex but specifically framed concepts in Bhaskar's expanded critical realism *œuvre*, which expanded from basic critical realism, to dialectical critical realism and later meta-reality.

In Chapter 3, Million Belay Ali describes how Margaret Archer's morphogenesis (an expanded view of Roy Bhaskar's Transformational Model of Social Activity) helps to explain the process of social transformation that he has encountered as an activist educator in Ethiopia, where he works with communities in recovering the historical relations between biological diversity and cultural diversity and sustainable natural resource management. Justin Lupele (Chapter 4) uses critical realism as an underlabourer for a study on how social networking can enable or constrain institutionalization of environmental education courses in universities. Both Mucha Togo (Chapter 5) and Jane Burt (Chapter 6) use a critical realist approach to interdisciplinarity – specifically the seven laminations of scale (Bhaskar, 2010, pp. 9, 10) – to provide an in-depth explanation of their case studies. Togo uses the laminations to analyze Rhodes University's attempt to make environmental issues part of their day-to-day business practice, whilst Burt uses the laminations as a framework for reviewing the international literature on collective action. Presha Ramsurup in Chapter 7 also uses the laminations framework to identify absences in a laminated skills development system, arguing that these provide possibilities for transformative praxis in the emergence of learning pathways. Lausanne Olvitt in Chapter 8 refers to the laminations in framing and testing a critical realist discourse analysis tool with potential to explain emergence.

In Chapter 9, Rob O'Donoghue explains how the previous hegemony of Western empiricism and positivism has resulted in the dismissal of much indigenous knowledge as mere superstition. However, contemporary perspectives, such as poststructuralism and hermeneutics, whilst giving validity to indigenous knowledge, have nevertheless unfortunately resulted in exemplifying indigenous knowledge as different and opposing Western science. He argues that a critical realist perspective can overcome this issue and illustrates his argument with examples from his experience of teaching science in a South African university context. Chapter 10, written by Daniel Sabai, continues with the theme of indigenous knowledge. Sabai describes the epistemological processes of the artisanal fisher men and women of the mangroves of Eastern Tanzania. He argues that indigenous knowledge is entirely commensurable with Western scientific knowledge provided that one acknowledges the role of retroduction in scientific methodology. This shows possibilities for levelling the power gradient in knowledge co-production within the scientific-indigenous knowledge border zone where new syntheses are possible (Mignolo, 2000).

Mutizwa Mukute (Chapter 11) uses dialectical critical realism to conduct expanded learning processes based on Cultural Historical Activity Theory (CHAT) in sustainable agriculture workplace contexts. His study was conducted in three case study sites in Lesotho, South Africa and Zimbabwe and he concludes that sustainability in agriculture requires structural and epistemic transformations in government, business and civil society sectors, beyond the farm itself. Similarly to Mukute, Aristides Baloi (Chapter 12) uses dialectical critical realism and CHAT to support a community of farmers involved in an irrigation scheme in Mozambique to develop potential solutions to their water supply problems. Tichaona

Pesanayi (Chapter 13), also working with CHAT, develops historical insight into the changing 'object' of water supply and demand to farmers in the Lloyd Village communal garden in the Eastern Cape, showing the enduring deep-seated influence of generative mechanisms of poverty and disenfranchisement under colonialism and apartheid, influencing events and empirical experiences and the possibilities for dialectically framed transformative praxis from the level of 1M.

Ingrid Schudel (Chapter 14) reflects on the ways that both basic and dialectical critical realism offer insights into the transformative intent that dominates many perceptions of environmental learning. She uses the concept of ecological footprinting to illustrate her arguments. Also interested in change, Adesawu Agbedahin (Chapter 15) explains how an analysis of the critical realist position-practice system in universities can help to account for changes towards sustainability. Victor Munnik (Chapter 16) explores how dialectical critical realism (DCR) can strengthen activism towards environmental justice, specifically in relation to community resistance to pollution in the Steel Valley in South Africa's industrial heartland, the Vaal Triangle. Heila Lotz-Sisitka (Chapter 17) uses the context of waste management in a school to argue that a more careful engagement with the critical realist concept of absence and negation may help to respond to the problem of relativism and/or disillusionment in environmental education.

In the final chapter (Chapter 18), Leigh Price describes how Bhaskar's (2002) concept of metaReality is useful for environmental educators because it allows them to take seriously the knowledge and ethics of certain spiritual traditions associated with the environmental movement, such as indigenous knowledge, deep ecology, ecofeminism and the transcendentalism of the earlier forms of the movement. This points to the need for embracing a wider range of perspectives in our attempts to seek out transformative environmental education processes and praxis than those that have been explored to date, an argument also put forward by decolonial theorists such as Gordon (2014), Mignolo (2000) and De Sousa Santos (2007) and by environmental educators in Latin America (Leff, 2009).

Together these chapters provide insight into critical realist inspired explorations in the field of environmental education research, situated in a part of the world where emancipatory theorizing and praxis have come to be viewed as critically important for addressing the complex and historically constituted contradictions that arise at the intersection of colonialism, neo-liberal economics, ongoing environmental degradation, poverty and other ills that affect the well-being and livelihoods of people in the region, but also more widely in the global South. In this region, shaped by the oppressive forces of colonialism and apartheid, education has traditionally played mainly a reproductive role. This book has sought to mark out and share some exploratory research and thinking on how education may also play a transformative role in society. It is our hope that this volume will also help to make critical realism more accessible to others via the sharing of our experience of our engagement with it over the last ten years.

Postscript and acknowledgements

David Graber (December 2014), in his obituary to Roy Bhaskar, noted that '[t]he unifying purpose of his [Roy Bhaskar's] work was to establish that the pursuit of philosophical knowledge necessarily implied social transformation; the struggle for freedom and the quest for knowledge were ultimately the same.' We are grateful for Roy's initial generosity in encouraging us to produce this book based on a set of our conference presentations at the 2012 Critical Realism Conference hosted at Rhodes University in South Africa. We are also grateful to Rhodes University's *Sandisa Imbewu* grant programme for funding that helped us to compile and edit this book.

Notes

1 Spivak (1999, p. 46) suggests that

> if we imagine ourselves as planetary subjects rather than global agents, planetary creatures rather than global entities, alterity remains underived from us, it is not our dialectical negation, it contains us as much as it flings us away – and thus to think of it is already to transgress.

2 Researchers described these frustrations at an informal group interview held at the IACR 2012 annual conference at Rhodes University, South Africa.

3 The origins of this form of critical realism can be traced back to the publication of Roy Bhaskar's *A Realist Theory of Science* in 1975.

4 In this introductory chapter we assume a basic knowledge of critical realism. Readers who are unfamiliar with critical realism may find it useful to first turn to Chapter 2 for an outline of the most commonly used critical realist ideas.

5 For a description of the DREI(C), see Chapter 2. D = Description; R = Retroduction; E = Elimination (of competing theories); I = Identification (the knowledge is confirmed by testing it or using it); C = Correction.

6

> M – Moment signifies something finished, behind us, determinate – a product: transfactual (structural) causality, pertaining to NON-IDENTITY; first is for founding. E – Edge speaks of the point of transition or becoming, the exercise of causal powers in rhythmic (processual) causality, pertaining to NEGATIVITY. L – Level announces an emergent whole with its own specific determinations, capable of reacting back on the materials from which it is formed – process-in-product: holistic causality, pertaining to TOTALITY. D – Dimension singles out a geo-historically recent form of causality – product in-process: human intentional causality, transformative AGENCY or praxis.
>
> (Bhaskar, 2008 [1993], pp. xiii–xiv)

7 Marx and Freud were both accused of being unscientific because their theories, though based on empirical evidence, were nevertheless not submitted to empiricist-style methods such as deductive-hypothetico experimental validation (Popper, 2002 [1963], p. 44).

8 In the open system of our world, correlations do not necessarily indicate causation. This is because other extenuating circumstances can interfere with the expression of the correlations. The open system can lead to non-causative correlations, or suppress otherwise causative correlations.

9 The Zone of Proximal Development as described by Vygotsky (1978) is often described in terms of expansion of individual meaning making and higher-order thinking skills. Post-Vygotskian researchers have expanded the meaning of Zone of Proximal Development by considering it in cultural terms (i.e. cultural expansions) and in activity development terms (i.e. expansions of human activity) (Engeström and Sannino, 2010).

10 It is acknowledged here that earlier environmental movements and/or nature-culture relations are not included in this 'seminal date' that is said to mark the start of the modern environmental movement, and that naming this as the 'start' of a modern environmental movement may be partly a result of the 'epistemicides' (De Sousa Santos, 2007) that have characterized the emergence of (Western) modernity.

References

Archer, M. (2012). *Complexity Theory: A new challenge for Realism?* Presentation for the Post-Graduate Reading Seminar, International Centre for Critical Realism, Institute of Education, University of London, 9 October 2012. http://internationalcriticalrealism.com/2013/06/10/audio-margaret-archer-complexity-theory.

Barab, S., Evans, M., and Baek, E. (2004). *Constructivism: Implications for the design and delivery of instruction.* In D. H. Jonassen (Ed.), *Handbook of research on educational communications and technology.* London: Taylor & Francis.

Bhaskar, R. (1975). *A realist theory of science.* Leeds: Leeds Books.

Bhaskar, R. (1998). [1979]. *The possibility of naturalism: A philosophical critique of the contemporary human sciences* (3rd edn). London: Routledge.

Bhaskar, R. (2002). *Meta-reality: The philosophy of meta-reality. Volume 1: Creativity, love and freedom.* London: Sage.

Bhaskar, R. (2008) [1975]. *A realist theory of science.* London: Routledge.

Bhaskar, R. (2008) [1993]. *Dialectic: The pulse of freedom.* London: Routledge.

Bhaskar, R. (2010). Contexts of interdisciplinarity: Interdisciplinarity and climate change. In R. Bhaskar, C. Frank, K. G. Høyer, P. Naess, and J. Parker (Eds), *Interdisciplinarity and climate change: Transforming knowledge and practice for our global future* (pp. 1–24). London: Routledge.

Bhaskar, R. (2014). Critical realism as a resource for understanding and changing the world. Key note address at 17th Annual Conference of the International Association of Critical Realism. Institute of Education, University of London, 18 July 2014.

Brown, A., Fleetwood, S., and Roberts, J. (2002). *Critical Realism and Marxism.* London and New York: Routledge.

Carson, R. (1962). *Silent spring.* Boston: Houghton Mifflin.

Engeström, Y., and Cole, M. (1993). A cultural historic approach to distributed cognition. In G. Salomon (Ed.), *Distributed cognitions: Psychological and educational considerations* (pp. 1–46). Cambridge, UK: Cambridge University Press.

De Sousa Santos, B. (Ed.). (2007). *Another knowledge is possible.* London: Verso.

Dussel, E. (1998). Beyond Eurocentrism: The world-system and the limits of modernity. In F. Jameson and M. Myoshi (Eds), *The cultures of globalization* (pp. 3–31). Durham, NC: Duke University Press.

Engeström, Y. (1987). *Learning by expanding: An activity theoretical approach to developmental research.* Helsinki: Orienta-Konsultit.

Engeström, Y., and Sannino, A. (2010). Studies of expansive learning: Foundations, findings and future challenges. *Educational Research Review.* doi:10.1016/j.edurv.200212.002.

Folke, C., Hahn, T., Olsson, P., and Norberg, J. (2005). Adaptive governance of social–ecological systems. *Annual Review of Environment and Resources, 30,* 441–473.

Gordon, L. (2014). Disciplinary decadence and the decolonisation of knowledge. *Africa Development, 39*(1), 81–92.

Graber, D. (2014). Roy Bhaskar Obituary. *The Guardian,* 4 December 2014. www.theguardian.com/world/2014/dec/04/roy-bhaskar.

Hawkin, P. (2008). *Blessed unrest: How the largest social movement in history is restoring grace, justice, and beauty to the world.* New York: Penguin Books.

IPCC (Intergovernmental Panel on Climate Change). (2014). *Climate Change 2014: Impacts, adaptation, and vulnerability. Contribution of working group II to the fifth assessment report of the Intergovernmental Panel on Climate Change.* (C. B. Field, V. R. Barros, D. J. Dokken, K. J. Mach, M. D. Mastrandrea, T. E. Bilir,... L. L. White, Eds). Cambridge: Cambridge University Press.

Kanu, Y. (2003). Curriculum as cultural practice: Postcolonial imagination. *Journal of the Canadian Association for Curriculum Studies, 1*(1), 67–81.

Leff, E. (2009). *Latin American environmental thought.* Paper presented at Sixth Latin American Congress of Environmental Education. San Clemente de Tuyu, Argentina, 19 September 2009.

Lyotard, J.-F. (1997). *Postmodern fables.* Minnesota: University of Minnesota Press.

Martinez-Alier, J. (2002). *The environmentalism of the poor: A study of the ecological conflicts and valuation.* Cheltenham: Edward Elgar.

Mignolo, W. (2000). The geopolitics of knowledge and the colonial difference. *South Atlantic Quarterly, 101*(1), 57–97.

Mingers, J. (2010). *The contribution of systemic thought to critical realism.* Kent Business School Working Paper Series (232).

Newman, F., and Holzman, L. (1993). *Lev Vygotsky: Revolutionary scientist.* New York: Routledge.

Neocosmos, M. (2012). Are those-who-do-not-count capable of reason? Thinking political subjectivity in the (Neo-)Colonial World and the limits of history. *Journal of Asian and African Studies, 47*(5), 530–547. doi:10.1177/0021909612452701.

Nunez, I. (2013). Transcending the dualisms of activity theory. *Journal of Critical Realism, 12*(2), 141–165.

Nunez, I. (2014). *Critical Realist Activity Theory: An engagement with critical realism and cultural-historical activity theory.* London: Routledge.

Popper, K. (2002) [1963]. *Conjectures and refutations: The growth of scientific knowledge.* London: Psychology Press.

Price, L. (2007). *A transdisciplinary explanatory critique of environmental education.* Unpublished PhD thesis, Rhodes University, Grahamstown.

Scott, D. (2010). *Education, epistemology and critical realism.* London and New York: Routledge.

Spivak, G. (1999). *Imperatives to re-imagine the planet. Imperative zur Neuerfindung des Planeten.* Vienna: Passagem.

Stevenson, R., Brody, M., Dillon, J., and Wals, A. (Eds) (2013). *International handbook of research on environmental education.* New York: AERA/Routledge.

UNEP. (2006). *Africa environment outlook report 2: Our environment, our wealth.* Nairobi: UNEP (Division of Early Warning and Assessment).
UNESCO. (2014). *Shaping the future we want: UN Decade of Education for Sustainable Development (2005–2014) – Final Report.* Paris: UNESCO.
Vygotsky, L.S. (1978). *Mind in society.* Cambridge, MA: Harvard University Press.

2 Key critical realist concepts for environmental educators

Leigh Price

In this chapter, I provide a brief outline of aspects of critical realism that I have found useful for environmental education. Specifically, I describe: critical realism as underlabourer; the real, the actual and the empirical ontological domains; the epistemic and ontic fallacies; the transformational model of social activity; axiology and ethics; the seven laminations of scale; the concrete singular and concrete universal; and dialectical critical realism. Critical realism is notoriously difficult to understand and as a consequence many people new to its ideas inadvertently misread them. This is true of both those people sympathetic to its ideas and those people antagonistic to them. I therefore also draw attention to some of the more frequent misunderstandings about critical realism.

Critical realism as underlabourer

The primary business of critical realism is essentially 'philosophical underlabouring'. John Locke (1979 [1689], pp. 9–10) described *underlabouring* in this way:

> the commonwealth of learning is not at the this time without master-builders, whose mighty designs, in advancing the sciences will leave lasting monuments to the admiration of posterity: but everyone must not hope to be a Boyle or a Sydenham; and in an age that produced such masters as the great Huygenius and the incomparable Mr Newton, with some others of that strain, it is ambition enough to be employed as an under-labourer in clearing the ground a little, and removing some of the rubbish that lies in the way to knowledge.

Critical realism acts an underlabourer for the sciences, especially the human sciences and the practices of human emancipation. It attempts to 'clear' the ideological ground to allow the full expression of the sciences, which currently exist partially and incompletely. The objective of these sciences is not only to interpret but to change the world. Nevertheless, to change the world for the better, the sciences must achieve the condition that they interpret the world aright (Bhaskar, 2011 [1989], p. xii).

The real, the actual and the empirical ontological domains

According to Bhaskar (2009 [1986], p. 285), the Humean assumption that correlations (constant conjunctions) are the only way to know causation is a 'philosophical hallucination'. Critical realism offers environmental educators and researchers an alternative to thinking of causation in such flat, inadequate terms by describing the layered, deep nature of reality (the real, actual and empirical). It explains how higher levels of being are emergent from lower levels, therefore society is emergent from the activities of people, and our weather is emergent from the activities of the molecules of the atmosphere, amongst other things. Critical realists thus consider society to be 'a real thing', emergent from, but irreducible to, the people who daily reproduce it. Climate is also a 'real thing' and not reducible to the molecules which daily reproduce it. We thus assume that we cannot fully explain society by reducing it to the collective activities of empirical individual agents; neither can we fully explain climate by reducing it to the activities of the empirical molecules of the atmosphere (Figure 2.1). Furthermore, instead of trying to explain climate or society in terms of constant conjunctions of events it is useful to have law-like theories to explain them. Significantly, however, there are ideological advantages to

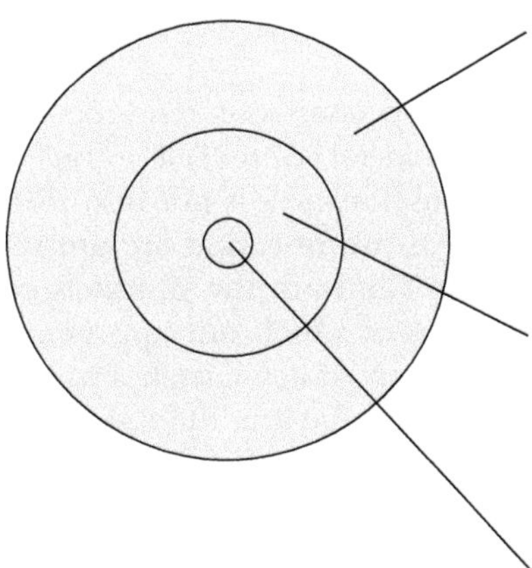

Figure 2.1 The real, actual and empirical domains of anthropogenic climate change (source: developed by author).

keeping explanations to the shallow empirical domain since this avoids placing responsibility for the change on our human activity (Bhaskar, 2009 [1986], p. 300).

An example of the usefulness of perceiving the world in terms of the real, actual and empirical domains is that it allows us to pull the ideological carpet from under a key argument used by climate change deniers, which is that climate change is 'unproven' and that it is 'not based on proper science' (Bratby, 2010; Meredith, 2014). Essentially, the scientific method – as it is taught in schools and undergraduate programmes – is actualist and empiricist. That is, it denies or at least ignores the domain of the real (Bhaskar, 2008 [1975]). Nevertheless, despite not including the real in their scientific method, scientists tacitly presume its existence. Climate scientists thus refer to the ontologically real with their theories about anthropogenic climate change whilst at the same time they contradict themselves by denying the existence of the real. The climate change deniers thus exploit the gap between 'scientific method as taught' and 'scientific method as actually practised' when they legitimately claim, by contemporary standards, that climate science is not true science because it is not possible to carry out experiments to disprove the theory of anthropogenic climate change. The deniers would not be able to use this argument if climate scientists had a more adequate version of science that more completely described their true scientific practice and that included the domain of the real (such a model is suggested in Figure 2.1). By keeping the discussion of climate change to the domain of the empirical – confining it within the bounds of positivist scientific method – the climate change deniers, who are also apologists for the current modes of capitalist consumption and production, effectively avoid critique.

The epistemic and ontic fallacies

Bhaskar (2009 [1986], pp. 285, 286) calls our spontaneous tendency to assume the naturalness of a fact (to disregard its social context) a 'mirage'. He suggests that to stop at this spontaneous consciousness is not only mistaken but also damaging for science, since 'facts are not what are apprehended in sense experience (things, events etc.) but are the products of the conceptual and cognitive structures in terms of which our apprehension of things is organised'. A common, related pair of key mistakes which have been used philosophically to underpin this mirage are: the epistemic fallacy, which is the reduction of being to knowledge; and the ontic fallacy, which is the reduction of knowledge to being. As an antidote to these fallacies, Bhaskar (2009 [1986], pp. 232–234) differentiates between the transitive realm and the intransitive realm.[1] Intransitive objects exist whether or not there are humans to see them. Transitive objects are 'artificial objects fashioned into items of knowledge by the science of the day' (Bhaskar, 2008 [1975], p. 21).

One of the ways that powerful actors in the arena of world policy can questionably insist on their agenda is to capitalize on the occlusion of truth that is the consequence of the epistemic and ontic fallacies. Critical realism gives environmental educators tools to uncover these practices. An example of how the occlusion of truth due to the ontic fallacy can be used strategically is the way that positivism can reify facts that supposedly 'tell' the policy-makers what to do. It occludes the truth that facts are not compulsively determined by being (ontic fallacy), but that there is a social component to facts (transitive realm): facts are construed by policy-makers to imply or suggest ways forward that are not immediately and obviously determined by the intransitive domain of those facts. Therefore, different ways of construing the facts are possible (transitive aspect of reality) and leading from this, different actions are possible, nevertheless still based on the facts. Thus the facts do not 'tell' policy-makers what to do, but rather the policy-makers manipulate the transitive processes of knowledge production in ways that benefit their own agendas. However, the ontic fallacy conveniently denies the possibility of the social aspect of knowledge, making action based on 'evidence' supposedly automatically objective (unbiased), thus hiding the process of manipulation and removing the possibility of contestation. As Bhaskar (1998 [1979], p. 263) explains, philosophical inconsistencies provide ideological opportunities. The ideological role of positivistic, reified facts are further explained in the section below on axiology.

Axiology and ethics

Environmental education is concerned with ethical behaviour and thus questions of ethics are at the core of our profession. Before critical realism, it seemed that our ethical options were all wanting in some way. Specifically, these ethical options were, and often still are, perceived to be as follows.

1 Absolutist ethics based on fideism (rejecting reason, associated with religious faith)

These ethics can fail to identify relevant-to-today behaviours and are associated with destructive fundamentalist positions. For example, absolutist ethics might claim that we 'have to' subordinate women to men, or one race to another, because it is the position of a particular group of people from a particular religion or culture.

2 Absolutist ethics based on positivism (using reified facts)

These ethics, often based on what is called 'evidence-based policy', can be commandeered by the powerful to insist on any ethical position that they deem would be in their best interests. For example, absolutist ethics of this type might claim that we 'ought to' increase taxes because of the presence of a correlation between countries with high taxation rates and high levels of happiness. This is the position of the UK happiness guru and advisor to David Cameron, Professor Lord Richard Layard (2011). However, such a correlation does not necessarily lead to the conclusion that we have to increase taxes. Not only may these correlations not be causal (always a question in open system correlative research) but even if they are causal, there may be a number of ways to react to them. Thus, if wealth does not necessarily lead to happiness, perhaps reducing the working week, rather than increasing tax, is a preferable response to the findings.

> the apologist can wean contingent from given facts and then … surmise the givenness of things, the naturalness of the status quo, insinuating human impotence in the face of it and the advisability of accepting it, at least grudgingly if not gracefully.
>
> (Bhaskar, 2009 [1986], p. 275)

When the social aspect of facts – the transitive dimension – are opaque, there is opportunity for corrupt action since the facts can be construed in a prejudiced way but since the prejudice is hidden, there is little space for resistance. By contrast, in a more democratic world, communities working along the lines of a non-relativist, i.e. a reformed, Habermasian-style ideal speech situation or a Vygotskian change laboratory, can arrive at ethical ways forward through a process of collective factualizing leading to democratic decisions. It might be useful here to look at how Bhaskar (2009 [1986], p. 281) describes a fact. At the level of the real, a fact is a potentiality, that is, it pre-exists its discovery. At the level of the actual, a fact is actualized in its discovery. At the level of the empirical, a fact is sustained in discursive and non-discursive activity and is objectified in sense experience. It is the way that a fact is sustained with discursive and non-discursive activity that has social consequences and requires social responsibility, transparency and accountability. When we get a chance to discuss the facts on an even playing field, we collectively sustain them and that makes them available in a way that is empowering, rather than disempowering. Note the difference here between the social constructionists, for whom facts are fully constructed by human activity, and the critical realists, for whom facts are sustained by human activity, but nevertheless have a relationship with reality beyond human activity.

3 Relativist ethics based on Humean empiricism

This position takes as a given the original Humean idea that we cannot arrive at values from facts. It might reasonably be assumed that this embargo is due to the epistemological issue of the social interpretation of facts, assuming the necessity of social mediation from fact to value, which would make it similar to the critical realist position. However, Hume's argument was ontological which makes his embargo far more profound (Price, 2014a). He believed that there are no deep ontological connections between things (he had an atomist view of the world) and therefore constant conjunction-type facts (correlations) are merely arbitrary and cannot be considered causal except psychologically.[2] Since one cannot know causality, one cannot use facts to indicate towards action, as there is supposedly no logical reason for why an action should cause an expected outcome. By contrast, critical realists suggest that there are deep causal connections between things, at an ontological level, and thus we can use facts to guide our meaning making, and our axiological process. Nevertheless, critical realists agree that epistemologically there can be many different interpretations of facts – leading to epistemic relativism and resulting in a wide variety of adequate ethical responses to the facts. Of course, in practice, we do use facts to inform our action but what is important in terms of emancipation is that we generally acknowledge that we do this; which removes the possibility that ideological positions can be defended using relativist or absolutist arguments. What is problematic is that ethically relativist empiricists – for whom science has no right to tell people what to do since this is a question of values and for whom there are no deep ontological connections and therefore no way to determine causation – can at any time transmogrify into positivist absolutists and insist that their particular position is an evidence-based imperative. They can do this simply when it suits them and without social sanction or questioning of the contradiction. For example, compare the following two scenarios.

> Scenario 1: A positive correlation is found between a diet high in vegetables and reduced chance of cancer (Graham *et al.*, 1978). People are told that 'according to the facts' they must eat more fruit and vegetables (amongst other things, protecting the interests of the food industry). This was despite there being, at the time, no well-established, clear reason for why this correlation might exist.[3]
>
> Scenario 2: A positive correlation is found between the presence of benzene in drinking water and the incidence of cancer. People are *not* told that 'according to the facts' we must reduce benzene in drinking water (Lombardi, 2014). Instead, the empirical correlation is contradictorily dismissed as irrelevant since, supposedly, we cannot be certain that the correlation is causal (protecting the interests of the petrochemical industry that releases benzene into the ground water). This is despite a generally accepted acknowledgement that at the level of real structures and mechanisms, benzene is a known cause of cancer, especially leukaemia (Snyder, 2002).

Here we have two positive correlations, construed contradictorily by policymakers to protect their own agendas, but both approaches are portrayed as good science.

4 Relativist ethics based on interpretivism (phenomenology and hermeneutics) and postmodernism

These ethics result in making ethical behaviour simply a question of personal or cultural preference. The postmodern version of this relativism is particularly ethically licentious as it tends to be associated with individualist relativism. As Bookchin (1999, p. 226) explained:

> self-styled postmodernist intellectuals have nourished an entirely new ensemble of cultural conventions that stem from a corrosive social, political, and moral relativism. This ensemble encompasses a crude nominalism, pluralism, and skepticism, an extreme subjectivism, and even outright nihilism and antihumanism in various combinations.

In response to this confusion, mixed methodology approaches to action have flourished: if the facts are with you, but the majority are against you, you can be a positivist and argue that science dictates your action; if the facts are against you but popular opinion is with you, then you can be an interpretivist and argue that the people's truth must dictate action (for example, fascist popular government); if neither the facts nor the majority are on your side, then you can argue that the truth that guides your action is merely what you can get away with (for example, Rorty, 1980) and that truth is simply what the powerful say that it is (Hitler and Mussolini, in Sayer, 2000, p. 47).

To avoid this Machiavellian approach to epistemology, critical realism offers methodological unity in which the transitive aspect of truth avoids the ethical consequences of reifying facts (all facts have a social, historical component); the intransitive aspect of truth avoids the ethical consequences of relativism and fideistic absolutism (there is a real something apart from the knowers or text that refuses an anything-goes or fideistic attitude towards truthfulness); and the rejection of the idea that we can reduce causality to constant conjunctions in open systems (but rather to insist on a deep, transfactual causation) combined with the acknowledgement that there are deep ontological connections between things, such that the facts tell us something about our world, vindicates the move from facts to values and thus provides a non-relativist axiology (Price, 2014b). When these facts are non-reified it is possible to achieve greater transparency surrounding their usage in policy contexts, allowing for greater democracy in their application.

The Transformational Model of Social Activity

Another key critical realist concept is Bhaskar's Transformational Model of Social Activity (TMSA) (1998 [1979], p. 31) (Figure 2.2) which he developed further into his concept of four planar social-being (2008

[1993]). Since inaction towards a better world was a large part of the initial dilemma leading to critical realism, it makes sense that critical realism devotes significant attention to understanding agency. Bhaskar's colleague Margaret Archer (1995) has contributed a great deal to the critical realist debate in this area with her work on morphogenesis, and many of the researchers in this volume have been significantly influenced by Archer's work. By including time, and allowing for social structures and mechanisms to be real but not reducible to the individuals who nevertheless constantly reproduce those structures, the TMSA avoids reducing society to its individuals but also avoids determinist positions which suggest that individuals have no agency but are entirely determined by society. In the TMSA, structure pre-exists individual agents but is nevertheless emergent from agency (interactions) and is thus only relatively enduring. Agents are 'thrown' into pre-existing structure when they are born and are then either transformative or reproductive of that structure. Agents cannot be said to create structure but to reproduce it or transform it. According to Bhaskar (1998 [1979], p. 37): 'Society is both the ever-present condition (material cause) and the continually reproduced outcome of human agency.'

One might ask why we should care about the different models of structure and agency. A significant reason is that voluntarist and determinist positions offer questionable power[4] strategies to people and particularly governments. For example, determinist positions can play into the hands of governments as they deny the power of agency and can dissuade activists from acting; people nihilistically assume that there is no point in trying to change the status quo. Voluntarist positions can remove the responsibility of the state for the well-being of its citizens. For example, Margaret Thatcher removed the responsibility of her government to contribute to achieving well-being for its citizens, using a voluntarist perception of structure and agency, when she said 'There is no such thing as society.... Don't look back, or up; look inside yourself, where your own cunning, will and

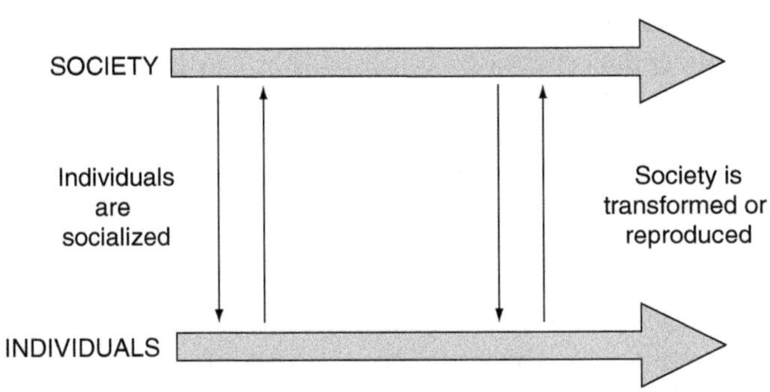

Figure 2.2 The Transformational Model of Social Activity (source: modified from Bhaskar, 1998 [1979], p. 40).

power – all the tools that life's improvement may require – are supposed to reside' (in Bauman, 2000, p. 30). When governments or other organizations try to implement changes in society through decisions cemented as policy, rules and regulations, without fully considering the resources, human and non-human, that will enable those decisions to be effected, they are taking a type of voluntarist stance (decisionism) which results in hidden stresses placed on the agents supposedly responsible to magically effect the changes. Not only is this disheartening, perhaps causing burnout amongst underresourced staff, but it also means that governments, organizations or individuals can be 'seen' to be doing the right thing, confident that their interventions are unlikely to make much difference to the status quo in practice. An environmental education example of this is when people are asked to reduce their environmental impact but are given no structural support or resources to achieve this. For example, a person may want to buy organic food, but if organic food is more expensive than non-organic food, due to the structure of agribusiness, this choice may be out of reach for low-income families. Organizations tasked with changing behaviour are unlikely to succeed when it is assumed that change will happen merely by changing hearts and minds (voluntarism). In environmental education, we often have good environmental legislation and excellent environmental awareness, but not much action towards improved environment because the deeper cultural and social structures do not support environmentally sound lifestyle choices. Activists who argue for the necessity to transform cultural assumptions of wealth and waste or who critique the global consumer society are suggesting an antidote to this problem. In terms of the TMSA, they are acknowledging that social structures are real and need to be considered in transformational politics.

The seven laminations of scale as a guide to interdisciplinarity

Given the layered, emergent nature of reality, and with the ecological crisis in mind, Bhaskar has suggested a model that can guide researchers and educators towards complex understandings of emergent issues, which he calls the seven laminations of scale (Price, 2014b; Bhaskar, 2010, pp. 9–10). The seven laminations of scale, where higher-order levels are emergent from lower-order ones, are:

1 the planetary (or cosmological) level concerned with the planet (or cosmos) as a whole, which is emergent from but also influences…
2 the mega-level of the analysis of whole traditions and civilizations, such as the South African trajectory from pre-colonialism, to colonialism, to apartheid, to independence, which is emergent from but also influences…

3 the macro-level orientated to the understanding way that whole societies or their regions function, such as the British economy, which is emergent from but also influences ...
4 the meso-level, studied by sociologists, which concerns the relations between functional roles such as manager and employee, husband and wife or MP and citizen, which is emergent from but also influences ...
5 the micro-level studied by ethnomethodologists and others, which is emergent from but also influences ...
6 the individual or biographical level, studied by existentialists and others, which is emergent from but also influences ...
7 the sub-individual psychological level, studied by psychologists and psychiatrists.

These laminations of scale are useful for researchers because they correspond to the major groupings of the disciplines. They are also more complete than the biophysical-socio-economic model guide to interdisciplinarity which has hitherto been the primary way to organize environmental education research.

The concrete singular and concrete universal

An irony of postmodernism is that its injunction against categorizing, designed to prevent oppression, in the end leads to the silencing of people from marginalized groups and thus enables oppression. Critical realism instead holds that it is not the existence of categories that is problematic. What is important is how we categorize and how we use those categories.[5] We can use categories to prevent oppression. For instance, by categorizing certain people as disabled, we can ensure that their special needs are met to allow their participation in public debate. For southern Africa, still reeling from the effects of colonialism, apartheid and a variety of social oppressions, the contribution of critical realism to better understanding how to deal with issues of oppression cannot be underestimated. Critical realist writing can sometimes be surprising for people who are used to the politically correct avoidance of categories. In theory, critical realists are comfortable to talk about, and even generalize about, different genders, races and sexual orientations provided the reason for such talk is to protect and care for those groups, rather than to oppress them. The real general of, for example 'women', is called a *concrete universal*, which will also have transitive and intransitive aspects. The concrete universal is an ontological category; it is not an abstract generalization. Thus the concrete universal is not the same as what the postmodernists called the 'abstract universal', which they were right to critique. However, although all women will have characteristics that allow them to be placed in this universal category, so each woman will be entirely individual, due to the unique mediations of her life (her birth-place, her race, religion, age, life history). Therefore every woman is both a *concrete singular*, and is part of the concrete universal. The concrete universal is emergent from the concrete singular. A critical realist would thus be likely to acknowledge that there are real differences, physical and psychological, between men and women: we cannot reduce these differences to social construction alone, whilst we must nevertheless acknowledge an important role for social influences. This is similar to ecofeminist Karen Warren's (1990) logic of domination compared to a logic of care. With the logic of domination, we can use the truism that women must take time off work to have children to justify seeing them as inferior and as second-rate workers. With the logic of care, we can use the same fact to make sure that women can continue to have the option to participate in work and social activities by providing adequate child-care facilities, assuming their equality with men. The denial of the differences between people as simply social constructions (abstract universals) was an understandable reaction to the logic of domination, but it resulted in the loss of language resources necessary for emancipation.

Dialectic critical realism

Marx's version of the dialect was about: knowledge production and reason (supposedly arriving at truth); and the changing of social structures (supposedly arriving at an emancipated society). However, this was not quite correct. The dialectic is not primarily about these things. It is primarily about one *thing*, which both knowledge production and changes in society happen to have in common. Indeed, all change has this one thing in common. This one thing is *absence* and absence is the key to dialectical critical realism. Whether we are moving from one version of inadequate knowledge to a more adequate version, or from one version of society to a more adequate version, we are absenting absences. Both Hegel and Marx were mistaken to have had such a purely positive, rather than a negative, account of the dialectic. The consequence of their positive view of the dialectic was that it was teleological and, as such, ultimately open to the potential to be overly directive and thus oppressive:

> Dialectic is the process of absenting constraints on absenting absences (ills, constraints, untruths, etc.). It is not in the business of telling people, in commandist (Stalinist) or elitist (Social Democratic) fashion what to do. Rather it is better conceived as an inner urge that flows universally from the logic of elemental absence (lack, need, want or desire).
>
> (Bhaskar, 2008 [1993], pp. 297–298)

Change as a response to lack is not only a vital aspect of the environmental educator's work in terms of changing the world into a better place (progressive social change). It is also integral to a theory of scientific epistemology since knowledge is advanced by absenting the absences within already-existing knowledge. Thus the critical realist version of the dialectic is applicable to both researchers and social change agents (Bhaskar, 2008 [1993]). We can summarize the methodology of both science and social progression as: incompleteness (absence) leads to inconsistency which leads to a movement to greater totality, that is a more inclusive or comprehensive theoretical or social situation (Bhaskar, 2008 [1993], p. 38).

Importantly for researchers, in dialectic critical realism, Bhaskar provides guidelines for both real and applied research which are equally applicable to natural and social science contexts, effectively removing the social science/natural science methodological divide whilst nevertheless acknowledging that the nature of the different subject matter in the natural and social sciences will require significantly different methods. Bhaskar's (2008 [1993], p. 101) outline of the scientific method when the phenomenon is being studied is relatively new (pure science) is known as the DREI(C) approach. It can be summarized as follows (bracketed inserts mine):

- *Description* (in describing it, you try find out everything hitherto known about the issue, perhaps through an initial literature review, but you may also need to spend time with the thing to be researched itself, gathering primary data – at this point you are likely to become aware of contradictions and other absences in current knowledge);
- *Retroduction* (creative aha! explanatory moment, whereby you surmise what would have had to have been for the thing/circumstances being studied to be as they are, thus absenting the absence of knowledge about it and/or absenting the contradictions in current understandings);
- *Elimination* (compare and eliminate competing theories – this will most likely require further engagement with the literature review, perhaps placed in the discussion part of a research document);
- *Identification* (if it is an invention, make it; if it is a theory, apply it to other contexts, i.e. use it, test it out – includes here deductive experimental tests if appropriate); and
- *Correction* (correct it based on evidence from using it – i.e. praxis).

However, Bhaskar also has an outline of the steps for applied science, which is where retroduction has already happened and one uses the resulting theory to explain other situations, although this process is also a way of testing the theory (for example, scientists continually apply, and thus test, Darwin's theories of evolution, resulting in corrections to his original theory). Bhaskar's (2008 [1993], p. 123) outline for applied science, known as the RRREI(C) approach, involves:

- Resolving the problem into its components;
- Redescribing these components in theoretically significant terms (using the transfactual, retroductively derived theory of efficacious

structures and mechanisms but further considering them in the light of the context in the situation at hand);
- Retrodicting back to possible antecedent causes (as suggested by the retroductive theory);
- Eliminating what must be a plurality of possible causes in an open system; eventually
- Identifying a full enough suite of causes to adequately achieve one's explanatory objectives;[6] and includes a
- Correction phase as the theories are adjusted in the light of further information.

The DREI(C) and the RRREI(C) are examples of the general model of dialectic provided by Bhaskar (2008 [1993], p. 187) that he calls the MELD (first Moment, second Edge, third Level and fourth Dimension). He explains (2008 [1993], p. 199) that they are a species of the genus of transcendental arguments (retroductive-explanatory arguments) familiar to science. His version of the dialectic of progressive social change is the Diagnosis Explanation Action (DEA) approach, which exploits the DREI(C) and the RRREI(C) (Bhaskar, 2008 [1993], p. 243).

Some misconceptions about critical realism by non-critical realists

Despite the wholehearted uptake of critical realist principles in some circles, critical realism remains misunderstood in much mainstream academia. For example, in *The Sage Handbook of Qualitative Research* by Denzin and Lincoln (2011) critical realism is described as follows:

> For some, there is a third stream, between naive positivism and poststructuralism. Critical realism is an antipositivist movement in the social sciences closely associated with the works of Roy Bhaskar and Rom Harre (Danermark, Ekstrom, Jakobsen and Karlsson).... Nevertheless, we do not think that critical realism will keep the social science ship afloat. The social sciences are normative disciplines, always already embedded in issues of value, ideology, power, desire, sexism, racism, domination, repression, and control. We want a social science that is committed up-front to issues of social justice, equity, nonviolence, peace, and universal human rights. We do not want a social science that says it can address these issues if it wants to. For us, that is no longer an option.
>
> (Denzin and Lincoln, 2011, p. 193)

Here critical realism seems to have been confused with Popperian post-positivism (based on Humean empiricism) which refuses to derive values from facts. The idea that critical realism is mainly a critique of post-modernism and a return to realism, where realism is assumed to be the same as postpositivism could not be more wrong. Furthermore Denzin and Lincoln (ibid.) seem to think that critical realism is uninterested in questions of emancipation and has no way of arriving at normative statements, which again, could not be further from the truth.

Another example of incorrect information about critical realism is to be found in *The New Blackwell Companion to Social Theory* by Turner (2009) who wrote:

> Although the critical realist view of science was a laudable attempt to escape the excesses of empiricist social research, it remained unclear how the notion of the openness of systems could be reconciled with their belief that the social scientists can use empirical research to test and validate their statements about the precise nature of the underlying mechanisms and their effects. Once the lack of synchrony between the actual and the deep is acknowledged (as the critical realists do), it seems no longer viable to argue that theories can be tested in a straightforward fashion with the help of empirical research.
>
> (Turner, 2009, p. 65)

This author seems to assume either that theory can be tested in a predictive, positivist way (his straightforward fashion) or that there can be no testing of theory. However, critical realists suggest another option. They test the power of their theories to explain empirical phenomenon through a logical but fallible process of elimination of competing theories (often this includes testing theories in practice by applying them, that is, through a process of praxis). Critical realism does not disallow the experimental method, provided that it is carried out in a closed system. It sees the experimental method as a special case of a more general scientific method, rather than as exhausting the scientific method. Closed systems are rather rare and are usually artificially produced, making them limited in terms of their practical application in society. Experimental, formally predictive tests are not technically possible in open systems. However, we do make informal predictions about our social interventions, that is, we assume that an action will led to an outcome, but unintended consequences are always a risk in our open system contexts. This is why fast feedback mechanisms are vital.

Some misconceptions about critical realism by critical realists

Just as the opponents of critical realism (above) found it hard to distinguish between critical realism and various forms of positivism, or found it hard to let go of positivism in the face of the critical realist critique of it, so too do many critical realists fail to move beyond positivism. I will discuss two common places where positivism remains entrenched even in critical realist thinking: in terms of the interpretation of the intransitive and transitive dimensions; and in terms of the interpretation of the Transformational Model of Social Activity (TMSA).

Misinterpreting the intransitive and transitive dimensions

Lilie Chouliaraki and Norman Fairclough (1999, p. 33) placed a positivist slant on the intransitive and transitive dimensions when they wrote: 'a problem, which may be either cognitive, for example, a misrepresentation or an unmet need (the former is transitive critique, the latter intransitive critique)'. However, it is impossible to carry out critiques of the transitive and intransitive separately. As soon as we, as humans, create knowledge about some aspect of our world, we must of necessity encounter the transitive dimension. Therefore, there are intransitive and transitive dimensions to both a misrepresentation and an unmet need. A misrepresentation does not only exist in a person's mind. It has its own ontological beingness. However, we can construe the misrepresentation differently and how it is construed will be different in different times and places. The unmet need similarly exists beyond what people say about it, but nevertheless, people can and will construe it differently. It appears that Chouliaraki and Fairclough (ibid.) are assuming that the transitive is epistemological (not really real – merely in our heads) and the intransitive is ontological (really real – out there and absolute).[7] Like the positivists, such authors can assume that there is some kind of absolutist trump card that ends the argument – tells it like it really is. So, positivists have access to the idea that 'you can't argue with the facts' and critical realists of this persuasion have access to the idea that 'you can't argue with the intransitive' (in the Fairclough and Chouliaraki example, they have access to the argument that 'you can't argue that an intransitive need has been unmet'). This is a kind of fetishism. Fetishism occurs when human-like characteristics are assigned to non-human objects. Facts are being fetishized when it is assumed that they can 'tell' us what is right; however, they are not human beings with the power of human communication. The 'telling' is always done by humans; this is the transitive aspect of facts. Activists who are tired of the powerlessness engendered by relativist positions may be attracted to critical realism, mistakenly thinking that it will return them to the (tyrannical) epistemological strength that they associate with positivism. This is simply not the case. However, critical realism does return our power to us. The difference between critical realist power and positivist power is more like the difference between street fighting and a martial art, such as judo. The street fighter's strategy is based on brute force; the judoka's strategy is based on using the opponent's aggression against them. In critical realism, the art of winning an argument depends on immanent critique or identifying theory/practice inconsistencies. Often it depends on an Achilles heel argument, whereby the defender attacks a position at precisely its strongest point, showing it in effect to be its weakest (Bhaskar, 2008 [1993], p. 66). Note how a critical realist, because of their concept of the dialectic as absenting absences can only act to transform a situation, like the judoka, in defence, specifically as a response to an absence or lack. The problem is that in the current mainstream world of socially acceptable adjudication, the critical realist may leave the opponent thoroughly vanquished, yet still lose on (questionably defined) points. This remains a challenge for critical realists, as we try to gain mainstream acknowledgement of the validity of our approach.

Misinterpreting the Transformational Model of Social Activity (TMSA)

Critical realists agree that the TMSA avoids both methodological individualism (resulting in voluntarism) and social holism (resulting in determinism), but there remains some confusion as to how it does this. Early on, Anthony Giddens' (1984) Structuration Theory was thought to be similar to the TMSA. A key assumption of Structuration Theory is that structures have a 'virtual existence' by which Giddens means that they cannot exist apart from their effects, that is, they exist only to the extent to which they are manifest as *rules and resources*, which can also be conceived as behaviour. Furthermore, in this approach structures supposedly do not exist apart from agents' conceptions. According to Porpora (2007, p. 424), the early Bhaskar also tended towards this position. Two critical realists who instead see structure as *systems of human relations among social positions* are Douglas Porpora (2007, p. 422) and Margaret Archer (1995). For these theorists, as well as the later Bhaskar, emergent structure has independent effects of its own and can exist without agents' conceptions of it. An example would be human relations such as sexism or racism that, even without the agents' knowing about them, are exploitative or otherwise place agents' interests in conflict (Porpora, 2007, p. 425).

In terms of Structuration Theory, then, to change society is simply a case of changing the rules. In practice, this amounts to little more than voluntarism or instrumentalist decisionism. It results in social activism that focuses on changing behaviour and policy or legislation, but does not challenge damaging social relations. It is a kind of actualism (positivism) because it assumes only the existence of the empirical behaviours and actual rules. A practical disadvantage, amongst many, of this approach is that rules about our behaviour cannot cover every eventuality, and we can quickly learn to tick the boxes of good behaviour yet find creative alternatives that continue to express the underlying structural relationships. So, for example, in the UK, instrumental interventions to reduce gender inequality have done little to solve the structural sexism and thus there remains a significant pay gap despite the Sex Discrimination Act of 1975 (Bovill, 2013).

However, whilst many critical realists in theory espouse a commitment to the TMSA or Margaret Archer's morphogenesis, in practice they only deal with rules and behaviours. Porpora (2007) describes this contradiction in the work of Dave Elder-Vass (2007). On the one hand, Elder-Vass explicitly defines social organization as a set of relations. However, on the other hand, he exclusively cites examples of rule-following to illustrate his arguments. I have found this contradiction to be remarkably common in research carried out by contemporary critical realists. I have tried to explore the issue with researchers whose works demonstrate this contradiction. One researcher, in her methodology section, mentioned sexism as an example of a structural issue. When she came to analysing her research context, even though sexism was clearly present, she failed to mention it. I asked her why. She admitted that in her workshops, she was afraid to bring up the issue with her participants because she felt that she would lose their respect if she began introducing feminist questions into their discussions. Other researchers exhibiting the contradiction, when asked about it, simply seemed to find the question confusing and personally challenging. They appeared to be unconvinced of the potential presence or relevance of structural issues, such as classism, racism or sexism and instead had defined structural issues as relating to, for example, organizational rules, or the behaviours of people in management positions. Like Elder-Vass, on the one hand they mentioned structure as relations between humans, but on the other hand, their examples suggested that what they truly meant by structure was rules and behaviour. Taking this evidence together, I suspect that we can at least in part explain this tendency to avoid the deep structural aspects of social issues as being both because of a fear of social sanction and because of a lack of reflexivity. Whilst the issue of social sanction seems fairly self-explanatory, I would like to explore the issue of reflexivity in more detail.

The absence of reflexivity in critical realist research

Elsewhere I have noted the strange absence of reflexivity including self-reflexivity in the work of many critical realists (Price, 2015). This is strange because reflexivity is a cornerstone of the dialectic. Bhaskar (2008 [1993], p. 273) describes reflexivity as follows: 'In its most basic form it specifies the capacity of an agent or an institution to monitor and account for its activity'. Bhaskar (2011 [1989], p. 114) also wrote that 'critique is part of the very process it is interested in and is therefore subject to the same possibility of unreflected determination that it describes'. Therefore, researchers, because they are part of the research context, are themselves limited in terms of their analysis to the extent that they themselves have been able to overcome the structural, ideological influences to which we are all subject. Another way of putting this is that, contrary to positivism which assumes a radical split between the researcher and the researched, critical realists assume that the researcher is part of their research context and therefore they must themselves be transformed by their research process (Price, 2015).

A reflexive agent will therefore try to avoid theory/practice inconsistency; that is, they will try to 'walk their talk'. However, this is not as easy as it may seem, since the pre-reflective reasonableness or understanding of everyday life – essentially based on a positivist world-view – readily tolerates contradictions without finding anything problematic in them: 'Positivism at once naturalises and normalises things and reflects in an endless hall of mirrors the self-image of Bourgeois Man' (Bhaskar, 2009 [1986], p. 308). The first aha! moment is simply noticing the contradictions (Bhaskar calls this the σ transform). Then there is another quantum leap required to explain those contradictions (Bhaskar calls this the τ transform). It is disingenuous to suggest that the σ and τ transforms are simple. Dialectic requires a great deal of stretching, of our concepts and our comfort zones.

The current unreflected tendencies in $subjects_2$[8] include personalism and decisionism.

> Personalism, perhaps the dominant moral ideology for $subjects_2$, is characterized by the attribution of responsibility to the isolated individual in an abstract, desocialized, deprocessualized, unmediated way, with blame, reinforced by punishment (rather than the failure to satisfy needs), as the sanction for default.
>
> [...]
>
> Decisionism is the ethics of the slaves of masters who are themselves masters of slaves.
>
> (Bhaskar, 2008 [1993], pp. 265, 366)

Decisionism and personalism entrench the status quo and often with the pre-reflective approval of the $subject_2$ themselves, who believe that they are fully to blame for their own or society's misfortunes, either through personal inadequacies or through bad decisions. Note, however, that personalism and decisionism are implied by the model of structure/agency that assumes that structure relates to rules and behaviour (Structuration Theory) rather than relations between people (TMSA). Porpora (2007) mentions how agents may not be aware of the structures to which they are subject. In terms of our work as environmental educators trying to improve society, each of us will, in one way or another, have to face the presence of ideology in our own belief structures. The only way to do this is to monitor our activity and check for theory/practice inconsistencies.

For example, if I am one of those $subjects_2$ who believes that the Capitalist machinery is a leveller of people – a meritocracy – and that it provides equal opportunities to all subjects to achieve the good life, then I will likely trust what I have been told: that if I follow the rules of society, I will be justly rewarded for my talent and hard work. In this case, I am simply not going to be able to see beyond rules and behaviours even if I am a critical realist and even if I have logically concluded that structure is about relations between people. Some examples of contradictions that might shake such a person from their unreflective state might be: they notice that there is no time to do the amount of recycling and gardening that they feel that they ought to do to prevent their household waste going to the tip; or it becomes clear to them that the environmental policy that they spent five years campaigning for will not lead to sound environmental practice because people can find ways to circumvent the legislation; or it dawns on them that despite having believed otherwise, the UK is not one of the most gender-equal places on earth and that their blindness to the inequality has led them to inadvertently uphold that inequality.

However, these contradictions are merely the beginning. Once one has realized the contradictions, there is another difficult journey to be made to work out how they occurred and then to ameliorative action. Most likely, the agent will only have a hunch that all is not right and that her trust has been betrayed. At this point, she could try to develop retroductive transfactual theories to explain the contradictions, and indeed she may need to. However, it would also be advisable to begin reading the vast literature that explores and tries to make sense of the contradictions. The critical theorists who have tried to explain both the oppression of people and the destruction of the Earth's ecosystem, include, to name a very few: Karl Marx, Martha Nussbaum, Ernesto Laclau, Luce Irigaray, Donna Haraway, Jürgen Habermas, Antonio Gramsci, Erich Fromm, Paulo Freire, Jacques Derrida, Murray Bookchin, Judith Butler, Simone de Beauvoir, Pierre Bourdieu, Zygmunt Bauman, Hannah Arendt, Louis Althusser and Theodor Adorno. This process of seeing the contradictions and then trying to explain them does not have to be linear; one could be oblivious to gender issues, read the work of a critical theorist such as Simone De Beauvoir, and thus be introduced to the contradictions in one's own life.[9] This lack, of both reflexivity and critical theory, in much critical realist research, is highly problematic. Indeed, there is a question mark as to whether such research warrants the label of *critical* realist, since it leaves out the critical side of the approach.

Conclusion

Critical realism does not return us to positivism. Instead, by maintaining a strong position against positivism, it reconnects us with our agency and gives us tools by which to approach social and environmental problems. It teaches fair play in the arena of knowledge production, denying access to the questionable, tyrannical strategies of knowledge production present in both absolutist and relativist approaches to epistemology. It does this by making the social aspect of knowledge production visible, but at the same time not denying the role of the intransitive (non-social) aspects. The reflexivity required of critical realist researchers is often challenging at a personal level as human researchers are not separate from what they are researching and therefore we are subject to the unreflected ideologies that we are seeking to unpack. In this short chapter it was not possible to give a complete explanation of critical realism. I nevertheless hope that environmental educators will find it to be a useful introduction to certain key concepts.

Notes

1 Sometimes people assume that the intransitive is the ontological and the transitive is the epistemological but this is technically misleading as an idea (knowledge) is also a being, so ontology is maximally inclusive and includes epistemology.

2 Transfactuality suggests that the laws of nature operate independently of the systems in which they occur. Induction results in inductive theories, deduction results in hypotheses and retroduction results in transfactual theories. The theory that climate change is caused by human activity is a transfactual theory based on, amongst other things, the principles of physics and chemistry. Even if we manage to stop global warming, the principle that global warming happens in a context of extra carbon dioxide in the atmosphere would remain a real law of nature. As a law it operates outside the system and does not need the system to be real.

3 However, recently there are some contenders for why this correlation exists, such as the presence of phytochemicals in certain vegetables, which affect gene function (Roysten and Tollefsbol, 2015).

4 Bhaskar (2008 [1993], p. 55) distinguishes between two types of power. There is $power_1$, which is the transformative capacity of agency, and there is a subset of $power_1$, namely $power_2$, which is the transformative capacity of agency expressed in structures of domination, exploitation, subjugation and control, which Bhaskar calls master-slave(-type) relations. In this case we are specifically referring to $power_2$.

5 Of particular relevance here is Bhaskar's (2008 [1993]) discussion of the concrete universal and the concrete singular.

6 To fully understand a social phenomenon in an open system we would need to address the complexity of the depth ontology of the situation, thus we might consider some version of the seven laminations of scale to obtain a full enough explanation of the issue at hand.

7 Another critical realist (Lipscomb, 2009, p. 45) put it this way: 'generative mechanisms refer to phenomena operative in the intransitive realm of the real.' Here Lipscomb mistakenly assumes that there is therefore no transitive aspect to generative mechanisms. He also mistakenly describes intransitive knowledge as 'true' knowledge (Lipscomb, 2013, p. 129).

8 $Subjects_2$ are oppressed people.

9 Note also that the more subjugated classes will see different contradictions from the other classes, and it will be easier for the subjugated to see contradictions since ultimately they are getting the sharper end of the stick. This point is also made by the stand-point feminist researcher Sandra Harding (2003), although she goes too far when she suggests that therefore the knowledge of the subjugated is the only valid knowledge (Haack, 2000 [1998], p. 97). Perhaps we might say that the subjugated class are better at the σ transform because their trust in the system is more likely to be seen to be betrayed (as suggested by the standpoint theorists) and the oppressor class (less subjugated?) are better at the τ transform because of their access to education and the ideas of the critical thinkers.

However, ultimately from whichever direction we come, we need to help each other. As Bhaskar (2008 [1993], p. 98) made clear: 'the free flourishing of each is the condition of the free flourishing of all.'

References

Archer, M. S. (1995). *Realist social theory: The morphogenetic approach.* Cambridge: Cambridge University Press.

Bauman, Z. (2000). *Liquid modernity.* Cambridge: Polity Press.

Bhaskar, R. (1998) [1979]. *The possibility of naturalism: A philosophical critique of the contemporary human sciences* (3rd edn). London: Routledge.

Bhaskar, R. (2008) [1975]. *A realist theory of science.* London: Routledge.

Bhaskar, R. (2008) [1993]. *Dialectic: The pulse of freedom.* London: Routledge.

Bhaskar, R. (2009) [1986]. *Plato etc: The problems of philosophy and their resolution.* London: Routledge.

Bhaskar, R. (2010). Contexts of interdisciplinarity: Interdisciplinarity and climate change. In R. Bhaskar, C. Frank, K. G. Høyer, P. Naess, and J. Parker (Eds), *Interdisciplinarity and climate change: Transforming knowledge and practice for our global future.* London: Routledge.

Bhaskar, R. (2011) [1989]. *Reclaiming reality: A critical introduction to contemporary philosophy.* London: Routledge.

Bookchin, M. (1999). History, civilization, and progress: Outline for a criticism of modern relativism. In J. Biehl and M. Bookchin (Eds), *The Murray Bookchin Reader.* Montreal: Black Rose Books.

Bovill, D. (2013). *Patterns of pay: Results from the Annual Survey of Hours and Earnings, 1997 to 2012.* London: Office for National Statistics. Retrieved 19 May 2015, from www.ons.gov.uk/ons/dcp171766_300035.pdf.

Bratby, P. (2010). *Memorandum submitted by Phillip Bratby: The disclosure of climate data from the Climatic Research Unit at the University of East Anglia – Science and Technology Committee* (Memorandum No. CRU 17). London: House of Commons. Retrieved 19 May 2015, from www.publications.parliament.uk/pa/cm200910/cmselect/cmsctech/387b/387we18.htm.

Denzin, N. K., and Lincoln, Y. S. (Eds). (2011). *The Sage handbook of qualitative research* (4th Edn). Thousand Oaks: Sage.

Elder-Vass, D. (2007). For emergence: Refining Archer's account of social structure. *Journal for the Theory of Social Behaviour, 37*(1), 25–44.

Chouliaraki, L., and Fairclough, N. (1999). *Discourse in late modernity: Rethinking critical discourse analysis.* Edinburgh: Edinburgh University Press.

Giddens, A. (1984). *The constitution of society: Outline of the theory of structuration.* Berkeley: University of California Press.

Graham, S., Dayal, H., Swanson, M., Mittelman, A., and Wilkinson, G. (1978). Diet in the epidemiology of cancer of the colon and rectum. *Journal of the National Cancer Institute, 61*(3), 709–714. doi:10.1093/jnci/61.3.709.

Haack, S. (2000) [1998]. *Manifesto of a passionate moderate: Unfashionable essays.* Chicago: University of Chicago Press.

Harding, S. (Ed.). (2003). *The feminist standpoint theory reader: Intellectual and political controversies.* New York: Routledge.

Layard, R. (2011). *Happiness: Lessons from a New Science* (2nd revised edn). London: Penguin.

Lipscomb, M. (2009). *The theory and application of critical realist philosophy and morphogenetic methodology: Emergent structural and agential relations at a hospice.* Unpublished PhD thesis, University of the West of England. Retrieved 19 May 2015, from http://eprints.uwe.ac.uk/18444/1/PhD%20Thesis%20-%20Martin%20Lipscomb.pdf.

Lipscomb, M. (2013). *A hospice in change: Applied social realist theory.* London: Routledge.

Locke, J. (1979) [1689]. *An essay concerning human understanding.* (P. H. Nidditch, Ed.). Oxford: Clarendon.

Lombardi, K. (2014). Benzene: One family's long fight to prove link to leukemia in children. *Guardian,* 8 December 2014. Retrieved 19 May 2015, from www.theguardian.com/environment/2014/dec/08/benzene-link-leukemia-children.

Meredith, C. (2014). Tory views on climate change are a 'huge embarrassment' for the PM. *Huffington Post,* 10 September 2014. Retrieved 3 November 2014, from www.huffingtonpost.co.uk/2014/09/10/climate-change-tory-mps-david-cameron_n_5796978.html.

Porpora, D. (2007). On Elder-Vass: Refining a refinement. *Journal for the Theory of Social Behaviour, 37*(2), 195–200.

Price, L. (2014a). Hume's two causalities and social policy. *Journal of Critical Realism, 13*(4), 385–398.

Price, L. (2014b). Critical realist versus mainstream interdisciplinarity. *Journal of Critical Realism, 13*(1), 52–76.

Price, L. (2015). Re-enchanting research: Transdisciplinarity as practical mysticism. In R. Bhaskar, S. Esbjörn-Hargens, and N. Hedlund-de Witt (Eds), *Metatheory for the 21st Century: Emancipatory praxis for planetary flourishing.* London: Routledge.

Rorty, R. (1980). *Philosophy and the mirror of nature.* Princeton: Princeton University Press.

Royston, K. J., and Tollefsbol, T. O. (2015). The epigenetic impact of cruciferous vegetables on cancer prevention. *Current Pharmacology Reports, 1*(1), 46–51. doi:10.1007/s40495-014-0003-9.

Sayer, A. (2000). *Realism and social science.* London: Sage.

Snyder, R. (2002). Benzene and Leukemia. *Critical Reviews in Toxicology, 32*(3), 155–210. doi:10.1080/20024091064219.

Turner, B. S. (2009). *The new Blackwell companion to social theory.* Chichester: John Wiley & Sons.

Warren, K. J., and Center for Environmental Philosophy, University of North Texas. (1990). The power and the promise of ecological feminism. *Environmental Ethics, 12*(2), 125–146.

Woods, J. (2014). Why does the Government want to teach mindfulness in schools? *Telegraph,* 13 March 2014. Retrieved 12 May 2015, from www.telegraph.co.uk/lifestyle/wellbeing/10694775/Why-does-the-Government-want-to-teach-mindfulness-in-schools.html.

3 Using critical realism to explain change in the context of participatory mapping and resilience

Million Belay Ali

Introduction

This chapter analyses the use of participatory mapping by a local organization in Ethiopia called 'MELCA-Ethiopia' (an NGO) for learning and change. It starts by briefly describing the context in which the mapping took place, which is mainly land degradation and the loss of biocultural diversity. Participatory mapping arose as a counter-hegemonic form of mapping in response to the dominant form of mapping that used trade for domination and subjugation of local people. It elucidates how counter-hegemonic mapping is useful to represent culturally and socially distinct understandings of landscape as it includes information that one cannot find in mainstream mapping. This form of mapping, which involves various tools, as used by local NGOs such as MELCA, involves mobilizing local people to articulate the change they want in order to develop plans for managing their biocultural resources. Such mapping processes can also be analysed from critical realist social learning perspectives (Wals, 2007; Archer, 1995). Archer's morphogenic cycle was particularly useful for providing insight into the agency and change emerging from the participatory mapping activities, showing structural elaboration (Archer, 1995) at the African Biodiversity Network level (which is where the author was exposed to mapping), at the MELCA-Ethiopia level (where the author is working), and at the community level as will be discussed in more detail below.

Context influencing land-use decision making and mapping at community level

With over 80 million people, Ethiopia is the most populated country in Africa next to Nigeria. It is endowed with significant reserves of natural resources and many endemic animals and birds. It is also ethnically diverse with over 80 ethnic groups. The country suffers from wide-ranging land degradation and loss of biodiversity. Gessese (2010) explains Ethiopia's land degradation in terms of socio-economic and institutional factors including: population pressure; poverty; the high costs of and limited access to agricultural inputs and credit; the low profitability of agricultural production; fragmented landholdings; short time planning horizons; and lack of information about appropriate alternative technologies for farmers. He elaborates the problems at the institutional level as including: little knowledge or capacity on land degradation among policy makers and professionals; lack of know-how on rehabilitation activities; top-down planning approaches which marginalize the participation of those who are at the forefront of the problem; weak linkages among government offices for the proper co-ordination of activities and approaches; limited capacity of government both in funding and in professional capital; lack of adequate information exchange among the stakeholders; and lack of implementation of policies even though the country has signed various national and international conventions. Frequent restructuring of offices directly responsible for land management leading to loss of experience and documents, and shortage of resources and incentives as the main problems is also a problem (Gessese, 2010, pp. 205–209).

Similarly, on threats to biodiversity, the Institute of Biodiversity Conservation (IBC, 2009, pp. 1–2) report states:

> The rich biodiversity of the country is under serious threat from deforestation and land degradation, overexploitation, overgrazing, habitat loss, invasive species and some water pollution. The underlying causes for these problems emanate from poverty, population growth, lack of alternative livelihoods, inadequate policy support, inappropriate investment and inadequacy of law enforcement.

Hoben (1995) has a different, more critical narrative describing land degradation in Ethiopia compared to the 'problem centred' narrative provided by researchers such as Gessese (cited above). He takes a more historical, political stand on the land degradation issue, and says that land expropriation was the first problem. Although farmers do have usufruct rights over their land, all land was declared public property and was repeatedly redistributed to landless claimants. This has decreased the incentive to invest in conservation activities.

Participatory mapping

In this chapter, I document participatory mapping exercises that were used to understand the complex problem of the causes of land degradation and loss of biodiversity from the perspectives of local communities in Ethiopia. The participatory mapping exercises were carried out in the Ormia Regional[1] state at Dinsho Wereda and at Wolmera Wereda. I facilitated the exercises in my capacity as director and researcher at MELCA-Ethiopia, which is an organization working on environmental education, agroecology and forest conservation. Traditional ecological knowledge cuts across the work of MELCA. The name MELCA stands for a crossing point on a river in some cultures of Ethiopia, mainly the Oromo. It is also a place of ritual for Oromo women. Finally, it is an abbreviation for Movement for Ecological Learning and Community Action.

The concept of participatory mapping is a combination of a much contested word 'participation' (Cooke and Kothari, 2001) and the word 'mapping'. Participatory mapping arose from the frustration of development thinkers and indigenous people who were critical of the appropriation and application of mapping technology for the purposes of domination and marginalization by those in power (Chambers, 1994). It followed what is called 'counter mapping' which arose from a common interest among geographers and indigenous people in the late 1960s 'for expressing tacit knowledge of natural resources and their cultural significance' (Crawhall, 2009, p. 8). In the late 1980s the practice of using a collection of methods called Rapid Rural Appraisal (RRA) and later Participatory Rural Appraisal (PRA) was started (Chambers, 1994) and participatory mapping featured as one of these collective methods. Corbett *et al.* (2006, p. 14) have adapted a number of participatory tools and methods that are used in practice (Rambaldi, 2005; Corbett and Keller, 2006). These participatory mapping tools include ephemeral or ground mapping, sketch mapping, transect mapping, scale mapping, participatory three-dimensional modelling (P3DM), GPS mapping, using aerial and remote sensing images, multimedia mapping, participatory GIS (PGIS) and internet mapping.

Participatory maps often represent a socially or culturally distinct understanding of landscape and include information that is excluded from mainstream maps, which usually represent the views and/or scientific traditions and methods of dominant sectors of society. Participatory maps can pose alternatives to the languages and images of the existing power structures and become a medium of empowerment by allowing local communities to represent themselves spatially. Participatory maps often differ considerably from mainstream maps in content, appearance and methodology (IFAD, 2009; Chambers, 2006; Chapin *et al.*, 2005). Critical realism allows researchers to negotiate the problem of relativism that can be associated with participatory approaches, wherein the reality of the participants is given epistemological privilege and participants are assumed always to be right. Critical realism provides an epistemology that allows for the voices of participants to be heard whilst nevertheless assuming their fallibility (as it assumes fallibility of all knowledge) claims.

Many organizations use participatory mapping to help communities articulate and communicate spatial knowledge to external agencies, to allow communities to record and archive local knowledge, to assist local communities in land-use planning and resource management, to enable communities to advocate for change, to increase capacity within communities and to address resource-related conflict (IFAD, 2009; Di Gessa *et al.*, 2008).

MELCA-Ethiopia (the organization I work with) started working with these maps in 2009. Since then, we have produced four computer-generated, three-dimensional models and scores of sketch maps. Participatory mapping has become one of the 'signature activities' for MELCA in Ethiopia. In 2011, we organized an international workshop on participatory mapping, attended by participants from Latin America, Africa and Europe (MELCA-Ethiopia, 2011). The stated purpose of the mapping exercises in MELCA-Ethiopia is to mobilize local communities that we work with so that they are more able to take action to change their livelihood and ecosystems, and through this, to develop social-ecological resilience (ibid.).

Alcorn (2000) comments positively on approaches such as those used by MELCA-Ethiopia, and notes that while researchers have long used mapping techniques and satellite imagery to analyse local situations for academic purposes and for making recommendations to donors and government, NGOs are now increasingly bringing this analytical power to the local level for improving local decisions and enabling local analyses to be shared with outsiders in order to improve national level policies. She states further: 'Maps communicate information immediately and convey a sense of authority. Mapping programmes can empower civil society efforts to bring accountability and transparency to local and national governments' (p. 2). A similar point has been made by authors of the participatory mapping tradition (Chambers, 2006; Rambaldi, 2005; McCall, 2004).

Participatory mapping used in the research

Sketch mapping

The participatory mapping activity used in the context of MELCA-Ethiopia was designed to allow participants to respond to the maps after they had made them. In the sketch maps, community members were asked to produce a map of the past and to reflect on what they remembered about the landscape and about their relation to it, with special reference to issues relating to biodiversity. The elders took the stage at this point, as they were the ones with longer memories of the past. After the past was visualized, they then constructed a map of the present. Following this, members of the community concerned were asked to develop a map of the future.

Participatory three-dimensional modelling (P3DM)

It was too expensive and time consuming to develop all three models (past, present and future) into the computer-generated three-dimensional model. Therefore, at the beginning of the process, the communities were asked which map they would like generated into the three-dimensional model. In all cases they chose to produce a map of the present. Three-dimensional (3D) modelling integrates spatial knowledge with elevation data in order to produce three-dimensional stand-alone, scaled and geo-referenced relief models. Geographic features relating to land use and cover are depicted on the model through the use of pushpins (points), yarn (lines) and paint (polygons). When the model is finished, a scaled and geo-referenced grid is applied to facilitate data extraction or importation. Data depicted on the model can be extracted, digitized and plotted.

The information that was used to obtain the participatory three-dimensional computer generated model came from the following mapping exercises: participatory sketch mapping and ecological calendars.

Participatory sketch mapping

We used a topographic map to produce the sketch map: 'Topographic maps make excellent base maps because they have sketch, north lines, and a co-ordinate system and they show natural features for local reference' (IFAD, 2009, p. 3). The study followed what Poole (2005) described as 'No Man's map'. This involves removing everything from the topo map except the rivers and contours. In the case of this research, we removed everything, except the rivers and roads. For Bale, one of the facilitators and the staff of MELCA, Tesfaye Tola, spent some time studying the rivers before the local community came to participate in the mapping processes. Therefore, with his facilitation, they could easily identify the rivers and it was easy for them to populate the map with their cognitive understanding of their biocultural landscape.

Ecological calendars

Maps usually depict the spatial relationship of people with nature. It is of course possible to represent this relationship as dynamic and changing on the maps through arrows connecting events. But it is much more explicit and visual if this dynamic, temporal relationship is represented in a circular way showing the changing patterns of practice according to the seasons. That is what is actually done during the mapping process. Eco-cultural calendars are developed in many ways. One way is to divide the circles vertically according to the number of seasons and to draw others circles in the big circle to show layers of expressions of nature and activities. The innermost circle can represent the season and then other circles expand to include people's livelihood, cultural expression, expressions of nature and climate, one after the other.

Participatory mapping learning interactions and agency

Even though theories of learning, including that of Vygotsky (1978) on acquisitive learning and Wenger (1998) on communities of practice, can be used to analyse the learning that has taken place during the mapping process, I will use mainly Wals' (2007) framework as it relates directly to social learning for social change. Wals (2007) sees dissonance as a prerequisite for the social learning process. He offers a framework suggesting how dissonance should be organized for maximum effect and suggests what should be done to facilitate the process and how to follow up. He presents a learning process depicted in Figure 3.1, which, in combination with his framework, highlights the major processes through which learning takes place.

Figure 3.1 is an adaptation of the Wals, Van der Hoeven and Blanken (2009) framework, drawing on the key learning processes associated with the participatory mapping in this study. It illustrates how routines (existing knowledge, fixed habits and norms, systems and training conditions) are put into a crisis situation (deconstruction, chaos, dissonance) through, in the case of participatory mapping, historical analysis and past-present comparative activities; how new learning takes place through meaning making processes that involve mobilizing and challenging existing knowledge, worldviews, institutions and norms; thus enabling emerging knowledge, new relationships, emergent norms and values, and experimenting to exist. This in turn becomes embedded in new knowledge, relationships,

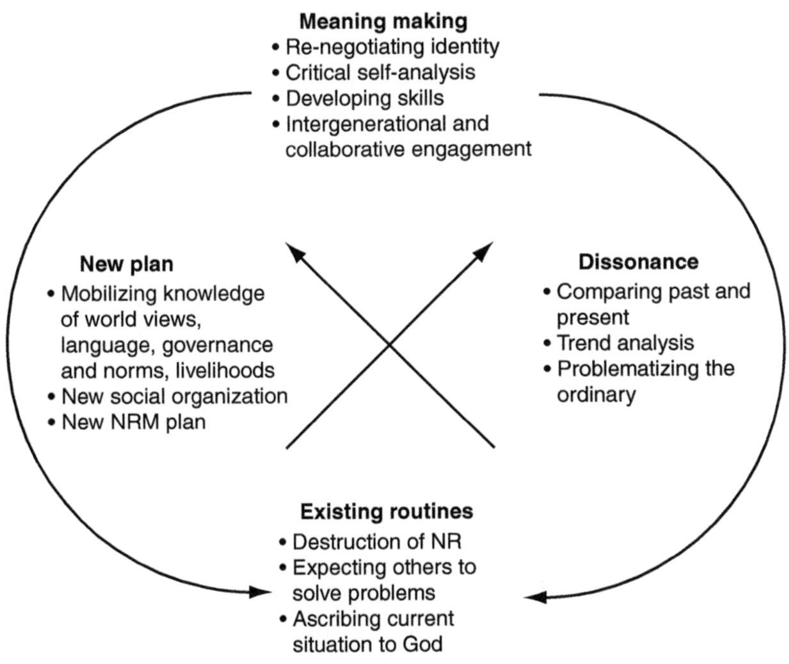

Figure 3.1 Learning system emerging from the mapping processes in the two case study contexts (source: adapted from Wals *et al.*, 2009; based on Hurst, 1995).

norms and values and routines. Put into a cycle, this involves ongoing processes of disruption and dissonance; demonstrations of how new knowledge becomes embedded and routine; and how, once subjected to dissonance, the result can be learning and a new form of embedding.

Of significance for this study is the manner in which the mapping processes appears to have created dissonance within such a learning cycle. Aman Mame of the Bale community said:

> This map has surfaced what we have forgotten. I did not think that we would finish our trees [deforestation] and be in a problem. Now we see everything. Forty and fifty years ago, we did not have that many people here. There were only three or four houses in a village. People used to use plants wisely. They used to listen to each other. It was forbidden to cut trees from the top. There was a law and people used to respect it. Now the government and the law say no but people do not listen. When we look at the maps, we see a naked place in the present and a forested one in the past. A forest is part of us and it even goes to our funeral with our body. During the cold windy season, people used to go to the forest. Nobody sits in exposed areas. Now there is no place to hide from cold wind. People just suffer from cold. People had no shoes before and they walked barefoot. They never got sick. They used to eat fat and ghee. The cattle used to graze on thyme. The ghee/butter had a particular smell. Now we eat ghee coming from another area. Now there is no ghee and fat from here and we are having a lot of illnesses. I think the people who have participated here should do something and be an example to others. This map has reminded us a lot of things. Whatever we do is for ourselves. Now because of the shortage of wood, hyenas eat people when they go to fetch wood. We are left with nothing. And we are exposed and we have no life. *The responsibility of a gourd is to keep the water that is left.* Now we have to start doing our share and we should not wait for others to act. We should start acting. Our fathers brought us to a plentiful world. We finished that. Now we should also leave something when we go. [metaphor emphasized]

This explains clearly that participation in the mapping process has created dissonance in the community. Comparing the map of the past with that of the present in the two-dimensional sketch mapping activity and in the eco-cultural calendar development processes seemed to be of most significance here, as the explicit visualization of past, present and future created dissonance. This same comparative process appeared, however, to take place in the dialogue and discussions during the P3DM processes: people discussed how things were then, and how they are now, and what has been lost or what has changed in the landscape.

In terms of the situated learning theories of Sfard (1998) and Wenger (1998), the learning became part of the community, as the existing norms and practices were challenged and responded to, for example in the case of Telecho where the farmers formed the Adere Environment Protection Association in which they wanted others to participate in the soil and water conservation practices; and in the case of Bale where the communities decided to stop cutting trees on their farm and rehabilitate their land, and new tree planting practices were initiated. These decisions were made after realization of the dissonance. Participation in the mapping activity created determination within the group and a sense of identity, expressed in these collective responses. Through this, it is also possible to see that the landscape is much more meaningful to them now and the processes of participation in which knowledge of the landscape, norms and practices were mobilized, helped to create this meaning. The mapping process also reflects Bhaskar's view of learning which sees identification of an absence as leading to a process of absenting absence in a desire for change (Bhaskar, 1997). The mapping process, through the dissonance created, revealed the incompleteness, the absences that have emerged, and the desires for change and further learning amongst the participants who were involved in the mapping process. As learning and agency can potentially also be related processes, I was interested to see, how, and if over time the mapping activities and learning had led to wider social change or not, and for this I used Archer's morphogenesis to interpret the structure-agency interactions and the emerging changes.

Morphogenesis

Archer's account of social structure was influenced by Buckley's (1967) concept of morphostasis and morphogenesis, Lockwood's (1964) concept of social and system integration and Bhaskar's (1978) critique of Anthony Giddens' Structuration Theory (Elder-Vass, 2007; King, 1999). Buckley (1967) used morphostasis to refer to those processes in complex system environment exchanges that tend to preserve or maintain a system's given form, organization or state; and morphogenesis to refer to those processes which tend to elaborate or change a system's given form, structure or state. The major theoretical claim for separating agency or individual action and structure rests, though, on Lockwood's temporal argument since it enables a visualization of the temporal interplay between agency and structure and their close interrelationships (King, 1999; Elder-Vass, 2007). Archer (1995, p. 76) says:

> Fundamentally the morphogenetic argument that structure and agency operate over different time periods is based on two simple propositions: that structure necessarily pre-dates the action(s) which transform it; and that structural elaboration necessarily post-dates those actions.

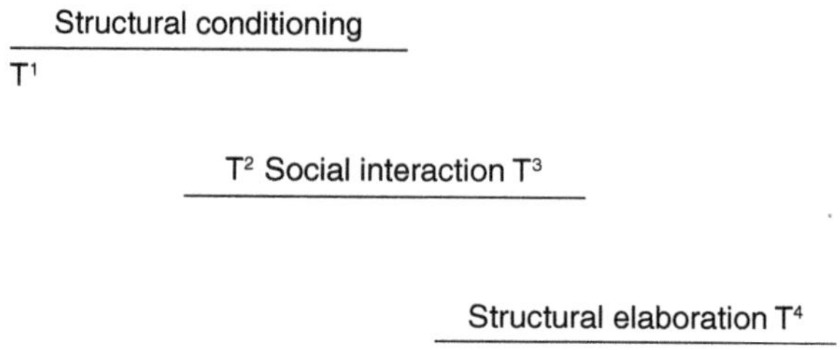

Figure 3.2 The morphogenetic cycle (source: adapted from Archer, 1995).

The morphogenetic cycle breaks down the structure and agency interactions into three temporal phases: structural conditioning (i.e. the context in which individuals find themselves), socio-cultural interaction (i.e. what individuals say, learn and do), and the resulting structural elaboration (morphogenesis or change) or structural reproduction (morphostasis or continuity).

According to Archer (1995, pp. 76–79, see Figure 3.2) T^1 represents the structure at the start of a process of change. These can be social, cultural and economic emergent properties. In the context of this chapter, some of these are elaborated at the beginning of this chapter. T^2 refers to the start of the action and refers to the agency exerted. Between T^2 and T^3 is social interaction such as the participatory mapping processes engaged in above, and between T^2 and T^3 agency exerts two independent influences, one temporal and the other directional. It can speed up, delay or prevent the elimination of poor structural influences. In our case it is the start of the mapping activity and the agency created and emergent from the mapping interactions, as will be discussed below. The speed of the change can depend on the willingness and the motivation of the actors to change and the political context in which they live. T^4 talks about transformation or the elaboration of a host of new possibilities if the action is successful. Simultaneously, however, structural elaboration restarts a new morphogenetic cycle, for it introduces a new set of conditional influences upon interaction, which are constraining as well as facilitating. T^4 is thus the new T^1, and the next cycle must be approached afresh analytically, conceptually and theoretically.

Agency, learning and change emerging from the mapping activities

Multi-leveled interaction and change potential

To shed light on processes of agency and change related to the participatory mapping activities I sought to analyse evidence of structural elaboration

(Archer, 1995) that took place as emergent from three substantive participatory mapping events. It explores the change that has come about due to the engagement of the actors in the participatory mapping activities and asks whether structural elaboration has come about due to the action of actors on structure (ibid.). It separates the agency of the actors through time and explores their interaction and finds out whether morphogenesis (structural elaboration) or morphostasis (structural reproduction) has resulted. Figure 3.3 shows three levels of interaction: at African Biodiversity Network (ABN), MELCA-Ethiopia, and at community level. The discussion that follows elaborates on Figure 3.3 and shows how the structural elaboration at ABN has acted as structural conditioning for MELCA-Ethiopia; and in turn how the structural elaboration at MELCA-Ethiopia has combined with the structural conditioning in the social-ecological contexts of the communities; and how this interacting process over time resulted in some morphogenesis, but also some morphostasis.

The discussion on morphogenesis starts at the African Biodiversity Network level (Level A in Figure 3.3). This is for three reasons. To begin with, it is ABN that created the opportunity for me to experience mapping. In this way, this network started the chain reaction for the actual mapping processes to happen. The second reason is that ABN organized a series of training opportunities and visits that sharpened my (as well as that of colleagues at MELCA-Ethiopia) understanding of mapping and broadened our view of what it means for local communities to engage with such activities. The last reason is that the mapping process also had an impact on ABN as an institution. MELCA-Ethiopia's practices influence ABN's

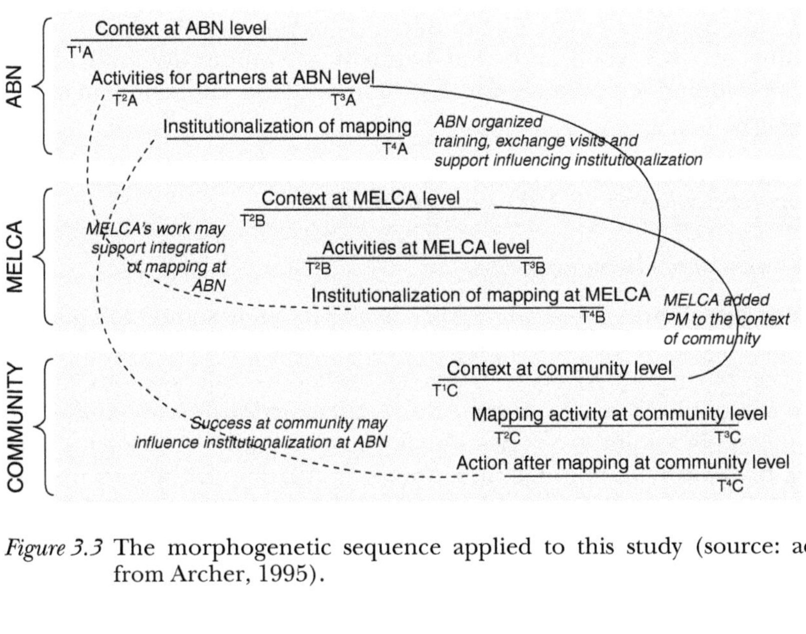

Figure 3.3 The morphogenetic sequence applied to this study (source: adapted from Archer, 1995).

practices: this is because MELCA-Ethiopia often hosts participatory mapping training for ABN members (to date four such training programmes have been hosted). I also serve on ABN's board of directors and am involved in designing its programmes. My engagement in the planning and execution of ABN's decisions means that the impact of MELCA on ABN could be direct. Therefore, I hoped that by starting with a morphogenetic analysis of the ABN's engagement with mapping and my role in it, I would be able to provide a more complete picture of the learning, agency and change relations than would be possible if I only analysed these at the level of MELCA-Ethiopia and the community alone. It also provides further methodological tools for causality descriptions as required by critical realism, which expects descriptions of why things happen the way they do, rather than simply what happens as a result of another action.

The second level of discussion on morphogenesis will be at the MELCA-Ethiopia level (Level B in Figure 3.3). This discussion will show how the structural elaboration of ABN forms part of the structural conditioning for MELCA-Ethiopia and what and how social interactions take place as a result. The morphogenesis that occurred will be explained, including the institutionalization of the method and the increased capacity of the staff. Morphostasis elements will also be explained where relevant.

The last level of discussion on morphogenesis will be at the community level (Level C in Figure 3.3). The social-ecological contexts of the communities involved in this study will be discussed, as this is part of the structural conditioning that influences community learning and agency development and resultant morphogenesis/morphostasis.

The process is not as linear as it appears, however. It is iterative. The process that happens in the community, for example, will affect what is happening at ABN level; and what happens at community level affects MELCA-Ethiopia etc. (see Figure 3.3). This will be explained in more detail below.

Morphogenetic cycle 1 (T^1A–T^4A) – the ABN level

T^1A – Structural conditioning at ABN

As Figure 3.3 shows T^1A is what ABN is and what it stands for and its reason for engagement in mapping. In short ABN is an African network working on biocultural diversity. It was established to address issues including: the loss of biodiversity and the culture associated with it; the influence of Western policies and strategies which impact negatively on Africa; and the lack of capacity to counter these forces in African civil society organizations. It works with 28 NGOs in nine African countries. The main activities are network development, catalysing wider action and co-ordination and capacity development. It is through ABN's capacity development exercise that I was introduced to mapping. ABN is engaged in mapping to increase the capacity of its members to use this methodology so that they can strengthen mobilization of local communities to recuperate their cultures (where relevant), support processes of intergenerational transfer of biocultural diversity knowledge and practices through learning, and rehabilitate degrading ecosystems in order to enhance social-ecological resilience. Their work is influenced by wider issues associated with biodiversity loss at a planetary level, and modern epistemological patterns that have separated culture and nature, treating them as dichotomous aspects of life, rather than as integrated.

T^2A–T^3A – Social interaction at ABN

Two critical processes occurred in 2006 with ABN, which led to my, and MELCA-Ethiopia's interest in mapping, and which ultimately led to this research. One of them was a trip that I undertook to Colombia. I could see in Colombia how local actors from communities, in collaboration with an NGO called Gaia Amazonas, used participatory mapping to mobilize their community for education, management of the environment, health and governance. ABN organized this meeting to introduce African NGOs to the thinking and philosophy of the Amazonians and to share their experience in using tools, including mapping, for social mobilization and environmental protection. In 2006, through ABN, I also participated in P3DM building in Kenya. This experience was foundational to MELCA-Ethiopia's efforts to try out P3DM in Ethiopia. I went back to Colombia in 2008 and this time I was introduced in a very direct way to how mapping was used for developing curricula. This deepened my understanding of the purpose and the process of mapping, and raised an interest in the role of mapping in facilitating learning and agency development. This ten-day experience gave me the practical knowledge of how to build a P3DM. This was the precursor for T^2B (activities at MELCA level) to be explained later. ABN also organized two mapping processes, one in Venda, South Africa and another in Tharaka, Kenya. Trainees of partners of ABN participated in these meetings, including MELCA-Ethiopia. A training manual is being prepared for partners to use when working with communities. ABN has also launched an eco-cultural calendar study in three countries (Kenya, South Africa and Ethiopia), and MELCA-Ethiopia is hosting one of these studies. These social interactions are building the capacity of the staff of MELCA-Ethiopia, which has translated into MELCA-Ethiopia's work and its engagement with communities, as described in detail in the two case studies.

T^4A – Structural elaboration at ABN

ABN has now made participatory mapping one of its methodologies. Therefore, we can see that over the period of time (2006–2011; T^1A–T^4A)

the mapping experience has changed ABN and now participatory mapping is becoming central to ABN's methodologies for working with communities. Mapping related activities are increasingly reflected in ABN strategy meetings and in its documents, including its plans. Quite recently, ABN organized a capacity building training programme based on the concept of resilience, where the relationship between mapping and resilience was discussed. This shows the internal change that has happened in ABN and the structural elaboration that resulted, as participatory mapping is now one of its key methodological tools.

Morphogenesis cycle at MELCA-Ethiopia

T^1B – Structural conditioning

MELCA-Ethiopia works in three places in Ethiopia: the Bale Mountains, the Sheka Zone and the Menagesha Suba area. It works on three of the seven thematic areas of the Convention on Biological Diversity, namely agricultural biodiversity, forest biodiversity and mountain biodiversity. It weaves traditional ecological knowledge into all three. When the mapping activities started, it had four programme areas, namely environmental education, advocacy and research, rehabilitation of degraded areas, and livelihood activities. MELCA-Ethiopia works to address: environmental degradation; loss and erosion of culture; poverty; and, as in case of this research, the failure of local communities to address their problems. Therefore, the mapping project arose in the context of MELCA-Ethiopia seeking a methodology to mobilize the local community both to rehabilitate their degrading land and to halt ongoing erosion of their culture. The idea was that the mapping would result in a management plan.

As part of ABN, and as Director of MELCA-Ethiopia, I have participated in the entire mapping processes organized to date. My participation in these mapping processes and the exchange visits reported on above, as well as the potential that I saw in these methodologies (when first observed via ABN social interactions) for addressing MELCA-Ethiopia's goals, influenced my decision to focus this research on the mapping processes. It is the added emphasis given to the mapping activities through this research process that has influenced MELCA-Ethiopia to take up mapping as one of its methodological tools to engage with communities. The morphogenesis at T^4A in ABN partly constitutes the structural conditioning at T^1B in MELCA-Ethiopia, although there are other important facts of structural conditioning in MELCA-Ethiopia that are not constituted directly through ABN's structural elaboration. For example, MELCA-Ethiopia is governed by Ethiopian government policies (which limits the range of involvement by local non-government organizations (NGOs)); and MELCA-Ethiopia is a funded NGO working at community level in Ethiopia (its practices are influenced by available funds); and their work is also influenced by the social-ecological histories in Ethiopia and responds to the contextual changes and challenges in the country (which is not as direct a feature of ABN's structural conditioning or elaborations).

T^2B–T^3B – Social interaction

The staff of MELCA-Ethiopia (other than myself) were trained in P3DM when it was first practised in Bale in 2009. That was their first encounter with this particular tool. As the model building went from one place to another, almost all of them had a chance to be part of it. They also participated in the construction of sketch maps. At first people participated out of curiosity and because I am the director, I was leading the process. They found the process engaging and educative, as reflected in these two quotes from an interview with MELCA-Ethiopia staff. Befekadu Refera, programme co-ordinator of MELCA, was somewhat sceptical in the beginning but later said that 'the process was active and real. I saw genuine participation. Result is unique and informative. I felt the power of maps for participation, visualizing biocultural diversity, planning, motivation of both the community and even donors.' Abebayehu Kassaye, environmental education co-ordinator at Bale, an active co-ordinator, facilitator and trainee of the Bale mapping processes, said that 'I was impressed with the knowledge of elders. Participation affirms MELCA's philosophy and can motivate stakeholders including government. I am now convinced of the power of mapping and its value for environmental education and conservation of biodiversity.'

We also had discussions on the effectiveness of mapping and in evaluating the impact of mapping at the community level. There were always questions about the established ways of working with communities and what the mapping process had brought. Because the process was intense and required a considerable amount of time, there was also reluctance and weariness on the part of some to track the results of the process. This internal struggle in the organization to accommodate and learn to use a new, demanding methodology, represented the social interaction that took place in the organization.

Furthermore, there was social interaction with a number of partners and funding organizations. These partners played an important role in helping to obtain resources for the mapping activities, and in helping document the processes. This extended the social interaction beyond ABN, to other partners and role players.

T^4B – Structural elaboration

Mapping has become one of the major activities of MELCA. It is shown in its website as one of the five programmes of MELCA-Ethiopia (see www.melca-ethiopia.org). In Bale, five schools used it as one of their environmental education activities. In Sheka and Bale, it is used for environmental education and every time a training exercise happens at the local level for schools and even for local militias, there is a visit to the model showing its usefulness in generating environmental awareness and MELCA's staff engagement with it. We have started generating funding for ongoing mapping activities, and associated community-based programmes linked to the mapping outcomes as part of the organization's ongoing programme.

Another significant aspect of morphogenesis in the organization is the increased capacity of MELCA-Ethiopia staff to support mapping processes within the broader goals of the organization, and in response to community demands and interests. The eco-cultural calendar mapping process was facilitated by the Bale and Menagesha branch offices and the Sheka branch office has also co-ordinated a sketch mapping process.

There was also morphostasis, as the following things have not changed as a result of the mapping activities in MELCA-Ethiopia:

- Some of the project's objectives and programmes remained the same as they were before the mapping activities, although a new programme on mapping was added (an example is the programme that we have for youth which we call SEGNI).
- Some aspects of the organization's ways of working remained the same (e.g. the way in which they viewed and worked within trusting relationships with communities).
- The funding and governance structures of MELCA-Ethiopia have not changed (as it remains an NGO working outside of government).

Morphogenesis at the local community level

T^1C^1 – Structural conditioning for community

The local communities in Horo Soba and Telecho have their challenges. In both environments, there is rampant felling of trees and bush for various purposes including selling to the market, for construction and fuel. Due to this and other factors, their environment is degrading rapidly. Rivers and wetlands are drying up, the size of the grazing area is decreasing, soil is washed away, the gullies are increasing and the productivity of the land is decreasing. They are losing their traditional belief system due to the increased radicalization of the major belief systems. Due to 'modernization' of their agricultural system, they are losing seed diversity and control over seeds and the knowledge associated with them. The gap between the youth and the elders is increasing due to many factors including formal education and cultural modernization, which presents children with different aspirations to those held by their parents. Children are increasingly respecting the knowledge gained at school in preference to that of their elders. Despite all this, there is little evidence that communities are organizing themselves to address these problems. Even though they come to meetings hosted by various government-organized forums, the relationship is top-down and not based on an understanding of the biocultural and the historical context of their landscape and the knowledge and practice related to it.

Natural resource management at the national level is beset with institutional weakness, confusion around policy generation and implementation, external influences and bad models of growth (Gessese, 2010; MELCA-Ethiopia, 2010).

It is in this kind of context that the social interactions in the mapping activities occurred.

T^1C^2 Structural conditioning – MELCA and community

As shown in Figure 3.3, structural elaboration at MELCA also forms part of the structural conditioning of the local community. This is because MELCA-Ethiopia has experience of biocultural diversity issues and rehabilitation, and methodologies for engaging communities in deliberation and change processes (such as mapping) that are situated in relation to the particular social-ecological contexts of the communities. Their interest in, and willingness to provide resources for, and engage with communities to address social-ecological degradation issues, through methodologies and approaches such as participatory mapping, formed part of the structural conditioning. The relationship of trust that existed with communities was also an important feature of the structural conditioning influencing the social interactions in the mapping processes. The contact between the two actors and the government participants during the mapping and discussions about the landscape were the initial starting points for social interactions around the mapping.

T^2C–T^3C – Social interaction

There were three mapping processes that took place in each of the two case study sites. Through these mapping processes, diverse and rich learning interactions and reflections occurred that influenced the identity formation and agency of members of the community involved in the processes (and others who later started to join them); and historical and cultural memory was mobilized. The story that emerged was deliberated upon and adapted and what was agreed upon was reified in the form of 3D models, 2D maps and eco-cultural calendars.

The mapping activities have initiated action, at the level of mapping practices, but also at the level of initiating wider changes in community practices.

There were those who supported the action that followed. Both individually and in groups, they started rehabilitating the land. There were, however, also those who wanted to keep the status quo as they benefit from cutting and selling trees and as they were not interested in starting soil and water conservation activities. This led to ongoing social interactions amongst the communities who were originally involved in the mapping activities. There were also further social interactions with MELCA-Ethiopia after the mapping activities, as further support was sought for implementation of the mapping activities; these processes are ongoing. This continuous social interaction (which is not without struggle) represents what occurred between T^2C and T^3C.

T^4C – Structural elaboration

There were changes after the mapping process. In the case of Bale, individuals started raising tree seedlings and planting the seedlings that they have grown. The community has also agreed as a group to stop cutting down trees or in their words ‘to rest their axe’. This represents a cultural elaboration in terms of morphogenesis.

In the case of Telecho, the community decided as a group to stop cutting down trees on their farm. They also formed an association called Adere Environment Protection Association. The Association had the chance to visit the northern part of the country after which they started a soil and water conservation activity. They now meet regularly, sometimes losing momentum, as they are busy in their own lives. MELCA-Ethiopia has contributed to the success of these structural elaborations, and has contributed seedlings of vetiver grass that were planted on their farms to stop soil erosion. They have undertaken soil and water conservation measures that are helping to stop the erosion of soil. There was also a youth group that was established, and they have engaged in various environmental activities. These all indicate that there was cultural, personal and structural elaboration occurring.

There was, however, also morphostasis (no change), primarily at the broader structural level. For example, it was not possible to change the fact that community members need income for livelihoods based on tree cutting and selling (due to the influence of the modern market economy). This led to some people continuing to cut down trees for the market economy. It was also not possible to change government policy on land distribution or biodiversity management (in the short time of this study). Such changes may gradually emerge from the interactions between community forums, government partners, MELCA-Ethiopia, ABN and others, but this remains to be seen over the longer term. Changes observed were primarily therefore short-term changes, visible at the local level. Other aspects such as the durability of the mapping influences on the youth’s cultural practices will need to be monitored over the longer term, as will the sustainability of the associations formed.

Iterative 'little' morphogenic cycles

Morphogenic processes, while broadly influenced by temporality, are not linear. They require a different, more cyclical concept of time, which influences how agency and learning are conceptualized in and through the mapping processes.

The process of morphogenesis at the three places described above is not as linear as it looks, although there is a clear pattern of emergence (social interaction emerges from and in relation to structural conditioning, but is not reduced to it; structural elaboration emerges from social interaction but is not reduced to it), both within the levels described, but also across them. There is also a pattern of iterations that occur across the levels. For example, the mapping processes at the community level have an impact on MELCA; mapping in the Telecho context was different to that of Bale. Based on the experience in Bale, for example, the approach to P3DM was modified. MELCA-Ethiopia is now more in favour of smaller-scale models as a result of experience gained over time. The experience gained at community level has been deliberated upon at the ABN level, thereby informing some of the principles of ABN practices, and the direction that ABN aims to take in future. For example, the mapping of MELCA-Ethiopia, Mupo (an NGO partner of MELCA from South Africa) and ICE (an NGO partner in Kenya) was discussed at the level of the ABN partners meeting. This has initiated discussion around the relationship between mapping and resilience. This shows how the morphogenetic process proceeds in 'little interacting cycles' and how this affects the institutionalization processes at ABN, at MELCA-Ethiopia and also the activity at the community level over time.

Archer's morphogenic theory has been critiqued for being 'linear', which can result in overdetermined or a structuralist bias on views of agency (Bates, 2006; Hay, 2002), but as shown above, this may be because the linearity of representation (as in Figure 3.2 and 3.3) and the broad description along these T^1–T^4 lines does not trace these 'little interacting cycles' in enough detail. In this regard, Bates (2006, p. 147) asserts that 'by employing morphogenesis and morphostasis, Archer is unable to provide [complex] concepts with which to explain change. Rather, they merely describe change by allowing the comparison of different historical periods.' (Note that this critique was produced before Archer's more refined work on agency and social interaction processes such as identity formation; learning; reflexive deliberation etc. that shape agency.)

To address this problem, Bates (2006) suggests adoption of Adam's (1990) notion of *circadian* time. This conceptualization views the passing of time as asymmetrical repetition with change being seen as rhythmical (not unlike the theorizing of change in resilience thinking[2]); which is partly represented by the concept of the 'little iterative morphogenic cycles' that I refer to above. Archer (2000) herself has also addressed some of these critiques by giving more attention to the processes of social interaction that influence agency.

Learning and agency development, as influenced by the mapping activities, therefore needs to take account of a more cyclical concept of time, which is nonetheless still influenced by temporality. The wider doctoral research study that informs this chapter provides substantively more detailed insights and descriptions of the social interactions that were occurring using thick descriptions at community level than is possible to include here (Ali, 2012).

Providing this level of depth of analysis at all levels is an enormous task and a longer-term analysis of the morphogenic process, especially in relation to broader processes of social-ecological resilience would need observations of more than the three- to four-year period that I was engaged with in this study, as resilience requires long-term monitoring. However, it was possible, through these smaller, iterative morphogenic analyses, to begin to conceptualize the relationship between the learning and the agency of people involved in the mapping processes at diverse organizational levels, and at a community level, showing some of the potential that learning approaches such as counter-mapping may hold for resilience building and more democratic forms of land use management that allow communities to reflexively engage with their concerns and the absenting of absences that they can identify via the mapping exercises.

Conclusion

The mapping exercise showed that people can easily understand how to do mapping and the process can be extremely shocking and painful. Wals' (2007) framework for learning was useful in explaining the state of people's condition before the mapping, the shock or dissonance that they experienced during mapping, and how this can be related to critical realist dialectical forms of learning. Researching the relationship between the learning activity of the mapping, and evidence of morphogenesis and/or morphostasis within a shorter term interactive cycle at multiple levels, also helped me to identify the agency and change potential of such learning approaches.

However, as discussed above, this approach was not sufficient for analysing or observing longer-term structural changes. It was, however, useful for providing a framework for explaining the change at various levels occurring in the shorter term including at the African Biodiversity Network, which facilitated my mapping experience, at the MELCA-Ethiopia level and at the community level. From this it was possible to see that there is potentially an open process and emergent relationship between learning and change processes. The annual soil and water conservation activity has resulted in the rehabilitation of degraded watersheds, the community has started implementing an agroecology project and they have organized themselves to store their traditional seed varieties in a seed bank that MELCA has constructed for them. This shows that the learning interactions around the mapping work have created some morphogenesis. However, morphostasis was also evident showing that learning does not always lead to change. This shows that the mapping activity by itself does not necessarily lead to change unless there is a collaborative engagement by all stakeholders to make the change happen, usually over much longer periods of time, especially if deeply rooted structural constraints are to be shifted or transformed. Learning processes, even the most participatory and innovative, must therefore be seen to be part of, and not the only condition for deep-seated structural transformation.

Acknowledgements

This paper is based on a PhD conducted by the author in the Environmental Learning Research Centre, Rhodes University (Ali, 2012). The study was partly supported by the Murray and Roberts Scholarship Trust Fund.

Notes

1 In Ethiopia the highest level of governmental organization is federal. The country is made up of nine regions and two satellite cities. The regions are divided into zones. The zones are divided into *weredas*. The *wereda* is an important administrative level structure for biocultural diversity as *weredas* are semi-autonomous in deciding on how the natural resources of their area should be managed. The next level is *kebele*. *Kebeles* are clusters of villages called *gots*. The Dinsho *wereda* was modelled in the three-dimensional mapping exercise focused on in this study; and there are nine *kebeles* included in the process. The modelling was done at the *wereda* level to make it useful for natural resource management planning.

2 Resilience thinking is a new body of theory being used in social-ecological studies as it helps researchers understand the implications of changes in the social-ecological condition, in the landscape, and in associated social systems that are of potential value to the future of communities in a context of rapid environmental change (Folke *et al.*, 2005; Gunderson *et al.*, 1995; Holling, 1973).

References

Alcorn, J. (2000). *Borders, rules and governance: Mapping to catalyse changes in policy and management*. Gatekeeper Series 91, London: IIED.

Ali, Million Belay. (2012). Participatory mapping, learning and change in the context of biocultural diversity and resilience. Unpublished doctoral thesis, Rhodes University, Grahamstown.

Archer, M. (1995). *Realist social theory: The morphogenetic approach*. Cambridge: Cambridge University Press.

Archer, M. (1998). Realism and morphogenesis. In M. Archer, R. Bhaskar, A. Collier, T. Lawson, and A. Norrie (Eds), *Critical realism essential reading* (pp. 356–381). London: Routledge.

Archer, M. S. (2000). *Being human: The problem of agency*. Cambridge: Cambridge University Press.

Bates, S. (2006). Making time for change: On temporal conceptualizations within (critical realist) approaches to the relationship between structure and agency. *Sociology, 40*(1), 143–161.

Bhaskar, R. (1978). *A realist theory of science*. Hassocks: Harvester.

Bhaskar, R. (1997). On the ontological status of ideas. *Journal for the Theory of Social Behaviour, 27*(2/3), 135–147.

Buckley, W. (1967). *Sociology and modern systems theory*. New Jersey: Prentice-Hall.

Chambers, R. (1994). Participatory Rural Appraisal (PRA): Analysis of experience. *World Development, 22*(9), 1253–1268.

Chambers, R. (2006). Participatory mapping and geographic information systems: Whose map? Who is empowered and who disempowered? Who gains and who loses? *Electronic Journal of Information Systems in Developing Countries, 25*(2), 1–11. Retrieved 13 December 2011, from www.ejisdc.org/ojs2/index.php/ejisdc/article/view/238.

Chapin, M., Lamb, Z., and Threlkeld, B. (2005). Mapping indigenous lands. *Annual Review of Anthropology, 34*, 619–638.

Cooke, B., and Kothari, U. (Eds). (2001). *Participation: The new tyranny?* London: Zed Books.

Corbett, J. M., and Keller, C. P. (2006). Using community information systems to express traditional knowledge embedded in the landscape. *Participatory Learning and Action, 54*, 21–27.

Corbett, J., Rambaldi, G., Kyem, P. A. K., Weiner, D., Olsen, R., Muchemi, J., and Chambers, R. (2006). Overview: Mapping for change – The emergence of a new practice. *Participatory Learning and Action, 54*, 3–6.

Crawhall, N. (2009). *The role of participatory cultural mapping in promoting intercultural dialogue: 'We are not hyenas'*. Paris: UNESCO.

Di Gessa, S., Poole, P., and Bending, T. (2008). *Participatory mapping as a tool for empowerment: Experiences and lessons learned from the ILC Network*. Rome: International Land Coalition.

Elder-Vass, D. (2007). For emergence: Refining Archer's account of social structure. *Journal for the Theory of Social Behaviour, 37*, 25–44.

Folke, C., Hahn, T., Olsson, P., and Norberg, J. (2005). Adaptive governance of social-ecological systems. *Annual Review of Environment and Resources, 30*, 441–473.

Gessese, B. (2010). Review of land degradation and land management in Ethiopia up to 2008/09. In S. Edwards (Ed.), *Ethiopian Environment Review, Vol. 1*, 187–214.

Gunderson, L. H., Holling, C. S., and Light, S. (1995). (Eds). *Barriers and bridges to the renewal of ecosystems and institutions*. New York: Columbia University Press.

Hay, C. (2002). *Political analysis*. Basingstoke: Palgrave.

Hoben, A. (1995). Paradigms and politics: The cultural construction of environmental policy in Ethiopia. *World Development, 23*(6), 1007–1021.

Holling, C. S. (1973). Resilience and stability of ecological systems. *Annual Review of Ecology and Systematics, 4*, 1–23.

IBC (Institute of Biodiversity Conservation). (2009). *Convention on Biological Diversity (CBD) Ethiopia's 4th Country Report*. Ethiopia: Addis Ababa.

IFAD (International Fund for Agricultural Development). (2009). *Good practices in participatory mapping*. Rome: IFAD.

King, A. (1999). Against structure: A critique of morphogenetic social theory. *Sociological Review, 47*, 199–227.

Lockwood, D. (1964). Social integration and systems integration. In G. K. Zollschan and H. W. Hirsch (Eds), *Explorations in social change.* Boston: Houghton Mifflin.

McCall, M. K. (2004). *Can participatory-GIS strengthen local-level spatial planning? Suggestions for better practice.* Chapter given at GISDECO 2004, Seventh International Seminar on GIS for Developing Countries, Skudai, Johor, May 2004.

MELCA-Ethiopia. (2010). Proceedings of National Workshop on Sacred Natural Sites and EIA. Addis Ababa, 1–2 July 2010.

MELCA-Ethiopia. (2011). *Report of the digitized P3DM models developed through community participation in Ethiopia.* Addis Ababa.

Poole, P. (2005). The Ye'kuana mapping project. In J. P. Brosius, A. Tsin, and C. Zerner (Eds), *Communities and conservation: Histories and politics of community-based natural resource management.* Walnut Creek: Alta Mira.

Rambaldi, G. (2005). Who owns the map legend? *URISA Journal, 17*, 5–13.

Sfard, A. (1998). On two metaphors for learning and the dangers of choosing just one. *Educational Researcher, 27*(2), 4–13.

Vygotsky, L. S. (1978). *Mind in society: The development of higher psychological processes.* Cambridge: Harvard University Press.

Wals, E. J. (2007). Learning in a changing world and changing in a learning world: Reflexively fumbling towards sustainability. *Southern African Journal of Environmental Education, 24*, 34–62.

Wals, A. E. J., Van der Hoeven, N., and Blanken, H. (2009). *The acoustics of social learning: Designing learning processes that contribute to a more sustainable world.* Wageningen: Wageningen Academic Publishers.

Wenger, E. (1998). *Communities of practice: Learning, meaning and identity.* Cambridge: Cambridge University Press.

4 Networking

Enabling or constraining institutionalization of environmental education courses in universities

Justin Lupele

Introduction

This chapter looks at how social networking can enable or constrain institutionalization of environmental education courses in universities. The chapter is based on a PhD study entitled *Networking: Enabling Professional Development and Institutionalisation of Environmental Education courses in Southern Africa*, carried out by the author, which involved 12 higher education institutions in eight Southern African Development Community (SADC) member states (Lupele, 2007). The 12 institutions were participating in a Course Development Network, a project of the Southern African Development Community Regional Environmental Education Programme (SADC REEP). This chapter provides an interdisciplinary explanation of the empirical findings, through a review of social, political, environmental and educational developments in southern Africa, with specific reference to post-colonial regionalization processes. A relational philosophy informed the research process and methodology. This relational philosophy included three distinct yet related theoretical perspectives, namely: critical realism, which provided the ontological perspectives, while aspects of actor network theory and community of practice provided epistemological perspectives. For the purpose of this chapter, however, only critical realism as it affected institutionalization of courses is discussed in detail.

Background

As southern African countries formed independent states after colonial occupation (from the 1950s onwards) they were also confronted by increased pressures of globalization. As a result, there was a need to form regional economic communities such as the Southern African Development Community (SADC). Regionalism was seen as a conduit for enhancing Africa's integration into the global political economy. Countries in regional groupings make commitments to enhance regional development and the integration of their economies. The end of the Cold War was a strong influence on political transformation in the region and this also led to the end of apartheid and ushered some of the SADC member states into a new of era of multiparty democracy. SADC is made up 14 countries: Angola, Namibia, South Africa, Lesotho, Swaziland, Botswana, Mozambique, Zimbabwe, Zambia, Malawi, Madagascar, Mauritius, the United Republic of Tanzania and the Democratic Republic of Congo. The region is home to approximately 300 million people and has been described as vulnerable to the impacts of climate change and other forms of environmental degradation, which are exacerbated by other concerns such as poverty and marginalization.

Post-colonial education

Southern Africa, like elsewhere in Africa, had very few universities established during the colonial era (Mamdani, 1994). Most of the universities in southern Africa, apart from some older universities in South Africa and the University of Zimbabwe (formerly University of Rhodesia), are a post-colonial phenomenon. At independence a number of countries in the region embarked on a rapid expansion of higher education institutions (HEIs). The state invested in human resource development to strengthen emerging government institutions and to implement development priorities (Africa Watch, 1991). As a result, in the last few decades there has been a mushrooming of universities. With neo-liberal policies introduced mainly after 1990, private universities have also been introduced into most SADC member states.

The post-independence nationalist agenda influenced debates on the curriculum and the institutional form and purpose of African universities. Education reform was primarily associated with nation building (Mamdani, 1994; Obanya, 2000). Obanya (2000) notes that this entailed a rethinking of national development goals, informed by the country's history and trends affecting society. He argues that genuine educational reform ought to be tied up with reforms related to the political, sociological and economic reorientation of the wider society. This was not the case in most post-colonial Africa. The focus was to train nationals to take over jobs left by the colonial government staff, i.e. universities had a primary function to provide government staffing.

Post-independence state control of universities reduced the possibilities for academic freedom. Under the nationalist leadership of post-colonial governments in southern Africa, heads of states became university chancellors and this, in the main, led to a somewhat 'utilitarian' model of higher education (Africa Watch, 1991). Protests against this model sometimes led to violence and human rights abuses. The utilitarian agenda of the development state, while supporting developmental objectives, effectively curtailed academic freedom and critical engagement with political, social, economic and environmental issues (Africa Watch, 1991; Mamdani, 1994).

The curricula and institutional forms were based on models that reproduced the intellectual and knowledge creation patterns characteristic of British, French, Portuguese, Belgian and other colonial universities and epistemologies (Mwanakatwe, 1968; Mamdani, 1994). As a result of this inheritance, HEI programmes and courses were developed by specialized people. These were often overseas consultants who were hired to develop courses for the newly established HEIs, or who came as experts through development aid (Van Staden, 2005). In some HEIs, course development is still a preserve of 'experts', often consultants or senior professors (Namafe, 2005; Van Staden, 2005; Nhamo, 2005).

Since the early 1990s when it became apparent that apartheid in South Africa was coming to an end, environmental educators started networking informally by sharing knowledge and course materials which were adapted from one context to the other across countries. These processes were strengthened by the establishment of the SADC Regional Environmental Education Programme (SADC REEP) in 1999. In 2001, Danida funded a Course Development Network (CDN) under the auspices of the SADC REEP. The research reported on here focused on the CDN by investigating professional development and institutionalization of environmental education in 12 higher education institutions in eight countries in southern African countries. Table 4.1 provides the composition of the CDN.

Table 4.1 Composition of the CDN

Partner institution	*Country*	*Type*
University of Botswana	Botswana	University
University of Malawi	Malawi	University
University of South Africa	South Africa	University
University of Swaziland	Swaziland	University
National University of Lesotho	Lesotho	University
Rhodes University	South Africa	University
Mufulira College of Education	Zambia	Teacher Training College
National Education Institute of Namibia	Namibia	Curriculum Institution
Polytechnic of Namibia	Namibia	Polytechnic
Swaziland National Trust Commission	Swaziland	Government agency
Mauritian Wildlife Conservation Clubs	Mauritius	Non-Governmental Organization
SEEN/DRFN[1]	Namibia	Non-Governmental Organization

Source: Developed by the author.

Note

1 SEEN – Support for Environmental Education in Namibia. DRFN – Desert Research Foundation of Namibia.

Institutionalization of environmental education

As already stated, the aim of the SADC REEP was to strengthen capacities of institutions and individuals in environmental education in southern Africa. Part of the rationale behind the establishment of the CDN was to contribute to the SADC REEP aim by supporting environmental education course development processes in the region. By implication, strengthening capacities and capabilities of institutions and individuals in environmental education processes entails making environmental education an integral part of the institutions and countries being supported by SADC REEP. The assumption is that this will enhance sustainability of the processes long after SADC REEP project life is over. It is assumed that institutionalization of environmental education processes takes place in the region as institutions and countries take over and sustain project benefits and activities. However, Schnack *et al.* (2004) warned that institutionalization is not an easy process and needs to be negotiated and implemented within specific institutions in nationally and institutionally specific ways. They argued that there must be a clear theoretical framework within a host institution that should be framed by the institution's policy (structure). They suggested too that where such policy does not exist or is unclear, the project needs to develop and articulate its theoretical framework. Further, they contended that a clear framework helps host institutions to understand, evaluate and possibly take on board project activities and/or innovations.

Schnack *et al.* (2004), writing in the context of four Danida funded projects in southern Africa,[1] suggested some broader points, which can be drawn on to enable institutionalization of environmental education in southern Africa. This chapter draws on these points to discuss institutionalization of environmental education courses in the context of the CDN. These are summarized below.

Political commitment at all levels of host institution

Projects are often negotiated, designed and agreed at a political level, but implemented at an administrative level. Schnack *et al.* (2004) note that it would be useful to talk to host institutions during the project design and implementation stage to ensure commitment at all levels, i.e. national and institutional levels. They observe that it is important that political commitment is secured at appropriate levels on an ongoing basis.

Institutionalization requires a post or a person

Sustainable project benefits often require some alterations in a host institution's structure, particularly if it is anticipated that selected project activities will be taken on board and implemented by the host institution after the project's life span. These alterations need to be agreed on and made before rather than during a project if people in the posts are to grow into their roles and responsibilities. Host institutions which allocate responsibilities for project activities to staff within existing posts need to factor in the costs in terms of time that this will incur, i.e. the person should be released from other duties in order to enable him/her to fulfill the expectations related to the host institution as well as the project.

Collaboration and synergies

Environment is a broad field, and there are often many projects working on related issues operating within the host organization and at a national level. Often valuable opportunities for collaboration are missed as each project works in isolation. Initial project design processes could explicitly work towards better integration with existing projects and programmes – bearing in mind reporting and other operational constraints required by funders.

Recognition of 'invisible work'

Relationships need to be built and trust established within host institutions before projects can operate at their full capacity. This is often 'invisible work', unrecognized in project documentation or reporting, but can be amongst the most important work a project engages in, in terms of sustainability. Some mechanism that recognizes and allows time for such work would be useful for institutionalization.

Capacity development – working within and for the system

The increased focus on building internal capacity of institutions rather than outsourcing functions needs to be explicitly recognized. The projects need to perform a valuable function as regards capacity building within their host organizations in both formal and non-formal ways. From the initial design, projects need to explicitly clarify internal capacity development processes in their project activities and outputs, what activities they will engage in and how they will engage in them.

Institutional structure and role conflict

One of the complicating factors in the process of institutionalization of ideas/activities is fitting into an institutional structure. This means that project staff have 'many bosses' both within and outside of their host organization. Clear responsibilities and lines of communication and 'command' should be negotiated prior to project implementation, and agreed to by all. This would help to avoid tensions and role conflicts that arise from projects being pulled in different directions to serve different needs.

Course development process

The course development process under the CDN involved networking institutions meeting at course development workshops twice a year over a period of four years. The focus of the collaborative work was around six generic issues associated with course development as identified by the CDN members at the network's inaugural meeting in July 2002, namely:

- curriculum deliberations;
- course materials development;
- course delivery strategies;
- assessment and accreditation of learning;
- monitoring and evaluation of courses; and
- electronic learning and web-based course design.

These generic issues formed the CDN Framework, which became the basis for networking and provided the structure for regional workshops. CDN members worked together as a community of practice to focus on one of each of these issues during regional workshops. They also worked in small groups around areas of interest.

The doctoral study (Lupele, 2007) that provides the background to this chapter resulted from a desire among environmental educators in southern Africa to exchange information, share good practice and enable collaboration in course development processes in the context of regionalism. While the doctoral study had a more elaborate purpose and aim, this chapter limits its discussion to factors enabling and restraining institutionalization of environmental education. Two key questions shaped discussions. These are:

1 How can networking enable or constrain institutionalization of environmental education courses in southern Africa in the context of the SADC REEP framework?
2 What structural and agential factors enable or constrain institutionalization of environmental education courses in universities?

Theoretical framing

Relational ontological view

The overarching philosophical orientation of the doctoral study (Lupele, 2007) was a relational philosophy. The philosophy underpinned three distinct theoretical perspectives including critical realism (which provided ontological perspective); Actor Network Theory (ANT) and Community of Practice (both provided epistemological perspectives). I looked at relational philosophy as it influenced the choice of the ontological and epistemological lenses for the study of the CDN.

According to Emirbayer (1997) a relational perspective emphasizes mutuality and reciprocity as the underlying principles of existence. This is a shift from a dominant rationality paradigm (which is inherent in the substantialist perspective), where dualism is an essential feature of the thought process. Substantialists view the world as a collection of objects to be analysed, compartmentalized, classified and controlled (Kumar, 2004). Kumar (2004, p. 76) notes that the dualistic world view gives the illusion that 'I exist independent of the Other. This attitude is founded on the belief that there is a substantial, separate, individual Self, which can act of its own accord, irrespective of the Other.'

Not only does the relational philosophy appeal to the relational nature of networking processes but it seems to be congruent with African worldview[2] as manifested in African philosophy and literature. Perhaps one of the classical examples from African literature can be drawn from Okri's (1991) post-colonial novel *The Famished Road*. Okri shares a vision of the world as one of infinite possibility woven in a string of relations. The novel's main character Azaro, the spirit-child, wanders in an idyllic world of his spirit friends and the world of the living. Azaro enables the reader to see the interconnectedness of life. The dead and the living, the humans and non-humans, all mingle together as exemplified in this quotation:

> I stared hard at the crabs clawing the edges of flower-patterned basins. After a while they left me alone. That was the first time I realised it wasn't just humans who came to the market-places of the world. Spirits and other beings come there too. They buy and sell, browse and investigate. They wander amongst the fruits of the earth and sea.
>
> (Okri, 1991, p. 19)

McLuhan (1994) explains a number of relational perspectives from the African point of view by drawing on African philosophers and literature. These include the interconnectedness of humans and the land, which traces 'roots of existence as well as binding them mystically to their departed' (Mbiti, cited in McLuhan, 1994, p. 278). Many African traditional societies such as the Kikuyu of Kenya see the earth as the supplier of material needs of life, through which spiritual and mental contentment is achieved. Communion with the ancestral spirits is perpetuated through contact with the soil in which the ancestors of the tribe are buried (McLuhan, 1994; see also Price, Chapter 17, this volume).

Another relational perspective between the earth and humans is manifested in the customs surrounding the disposal of the placenta and the umbilical cord of a newborn. McLuhan writes that 'in a large number of the traditional cultures both the umbilical cord and placenta are immediately reconnected to the Earth' (McLuhan, 1994, p. 279), in reference to the practice of burying the placenta and the umbilical cord.

In sub-Saharan African narratives, relational philosophy is best described by the word *Ubuntu* which refers to people's allegiances and relations with each other. *Ubuntu* is translated differently in different parts of southern Africa but refers to life as a web of relationships (Kumar, 2004). Thus it does not simply refer to humanity; it means the whole of existence is together, interlinked, totally connected (ibid.). The word *Ubuntu* has its origins in the Bantu languages of southern Africa.[3] However, it has gained currency in post-apartheid South Africa as it is been used as a unifying concept among people of every colour, creed and culture. It was used as a guiding principle by the South African Truth and Reconciliation Commission, the South African Constitution and other legislation. *Ubuntu* has been used by the South African government as an inspirational motivation for solidarity, co-operation, the state of being and becoming, and as a process of self-realization through others (Louw, 1998).

Louw (1998, p. 3) notes that '*Ubuntu* dictates that if we were to be human, we need to recognize the genuine otherness of our fellow citizens ... we need to acknowledge the diversity of languages, histories, values and customs all of which constitutes South Africa.' Maphisa (cited in Louw, 1998) argues that transformation of an apartheid South Africa into a democracy is a rediscovery of *Ubuntu*.

I found traditional African practices such as *Ubuntu* to be an appropriate lens for understanding networking in the context of the CDN. This followed the insight that in traditional African societies *Ubuntu* fostered kinship networks (which moved beyond family relationships to include all human beings) and associated values. *Ubuntu* promoted human solidarities which help to solve social issues such as poverty and hunger as people shared problems and worked together to resolve them. Used in this sense, *Ubuntu* inspired and motivated this research to contribute to the notion of research as a process of social transformation, an agenda of southern African environmental education articulated by other scholars in the region (see Janse van Rensburg, 1995; Lotz, 1996; Taylor, 1997).

Although the idea of *Ubuntu* was not entirely used as an ontological and epistemological approach per se, it inspired me and to some extent provided a lens for my thought processes as I worked on constructing the theoretical framework, research questions and data interpretation. The relational nature of knowledge as described in this section has implications for understanding of institutionalization of environmental education. Further, accepting the notion of the construction of knowledge as a social process helped me to identify some of the enablements and constraints associated with knowledge construction in a network such as the CDN that may affect such a process.

Critical realism as an 'underlabourer'

Having discussed the relational view of knowledge, I now turn to discuss critical realism as 'underlabourer' to further elaborate the relational ontology perspective. Archer (2000) notes that the emergence of our human properties and powers are relational in nature: stemming from the way our species is constituted, the way the world is and the necessity of their mutual interaction. The roots of critical realism can be traced to the work of Roy Bhaskar (Sayer, 1984; Yeung, 1997; Pawson and Tilley, 1997; Schostak, 2002). There are a number of varieties and versions of critical realism which are not entirely consistent with each other (Yeung, 1997). It is not the intention of this chapter to engage in a full-fledged review of the realist philosophies or to provide space for evaluation of the different views. Realist perspectives provided insights into the research questions based mainly on what Yeung (1997) calls the Bhaskarian version of scientific realism in social science. That is, critical realism that argues for the existence of reality independent of human consciousness (realist ontology) and that which reorientates the social sciences towards its emancipatory role (realist epistemology). The study also drew on realist social theory, which seems to further clarify earlier Bhaskarian critical realism in sociological context, as articulated by Archer (1982).

Stratified ontology

Critical realists argue for the existence of reality outside human thoughts about it (Sayer, 1984, 2000; Archer, 2003). Danermark *et al.* (2002) note that critical realism provides an answer to the nature of reality. They argue that there exists both an external world independent of human consciousness and at the same time a dimension which includes our socially determined knowledge about reality. This is congruent with the *Ubuntu*/African philosophy that sees the world as one of infinite possibility woven in a string of relations.

Viewing reality in this manner has implications (in research) in terms of data generation, inferences and analysis. Sayer (2000) warns that critical realism should not be confused with empiricism – which identifies the real with the empirical, i.e. with what we can experience. This assumes that the world happens to correspond to the range of our senses and to be identical with what we experience. Critical realism distinguishes not only between the world and human experience of it, but between the real, the actual and the empirical (ibid.).

According to Sayer (2000) a crucial implication of this ontology is the recognition of the possibility that powers may exist unexercised, meaning that what is known to have happened does not exhaust what could have happened. He observes that the nature of the real objects present at a given time constrains or enables what can happen but does not predetermine what will happen. This realist ontology makes it possible to understand how we can become many things which we are not. For example, how novice course developers under CDN can become expert course developers or how a weak network of course developers becomes a strong network, or alternatively does not fulfil its potential.

By distinguishing the real, the actual and empirical, critical realism proposes a 'stratified ontology' in contrast to 'flat' ontologies populated either by the actual or the empirical, or conflation of the two as noted by Archer (2003) and Sayer (2000). Sayer explains that empirical realism assumes that what we can observe is all that exists, while 'actualism' assumes that what actually happens at the level of events exhausts the world of powers which can be either activated or remain dormant. Bhaskar accepts the existence of the thing in itself and says that it is knowable, but nonetheless asserts that there are 'structures' and 'mechanisms' which exist beyond empirical reality and which are not knowable (Bhaskar, 1998).

Central to realist social theory is the argument that 'the causal power of social forms is mediated through social agency' (Bhaskar, cited in Archer, 2003, p. 2). Archer (2003) observes that the above statement, while specifying that the causal power of social forms is mediated through social agency, does not tell us anything about the mediatory process involved. She argues that the word 'through' in the statement denotes conditioning, and that since conditioning is not determinism, the process necessarily involves the interplay between two different kinds of causal powers – those pertaining to structures and those belonging to the agents. She explains that an adequate conceptualization of conditioning must deal with the interplay between these two powers which involves a specification of how:

- structural and cultural powers impinge upon agents, and
- agents use their own personal powers to act so rather than otherwise, in such situations (Archer, 2003).

Applied to the study of the CDN, Archer's notion helped to tease out how structure (network, institutions) impinges upon the agents to enable or constrain institutionalization of environmental education courses. It also provided space for understanding how agents in the course development network used their own personal powers to participate in the network and make decisions that guided their actions related to course development and related institutionalization processes.

Since structure and agency are said to be irreducible to their individual elements, Archer (2003, p. 203) argues that their irreducibility to one another entails examining the interplay between them. Explanations of actions of individuals therefore requires not a 'micro (reductionist) regress' to the inner constitution (though this may be relevant too) but a 'macro regress' to the social structures in which they are located.

Archer (2003) further explains that the two different causal powers of structural objectivity and agential subjectivity are both entailed by the concept of 'constraints' and 'enablements'. She notes that constraints and enablements do not possess an intrinsic capacity for constraining or enabling in abstraction. She notes that 'for anything to exert the power of a constraint or an enablement, it has to stand in a relationship such that it obstructs or aids the achievement of some specific agential enterprise'[4] (Archer, 2003, p. 5). She postulates that the account of how structures influence agents is entirely dependent upon the proposition that our human powers of reflexivity have causal efficacy – towards ourselves, our society and the relations between them. She observes that reflexivity, which is held to be one of the most important of personal emergent properties, is often denied to exert causal powers. As a result of this, reflexivity becomes considerably less interesting or of no importance at all in accounting for any outcome. Archer (ibid.) advocates for personal reflexivity as a means of mediating between structure and agency. She argues that there should be acceptance of the need for identifying a mediatory process that links 'structure' and 'agency'. This, she calls the internal or reflexive deliberation.

Internal deliberation

In order to examine the role of reflexive deliberation (internal deliberation) in mediating between structure and agency, Archer (2003) argues that one has to first understand the philosophy of the mind. She draws on Descartes' work of how we obtain knowledge of our own. Descartes (cited in Archer, 2003) refers to ideas of 'seeing with the mind's eye' or 'looking inwards' to obtain transparent self-knowledge from a privileged standpoint. He calls this 'introspection'. But Archer (2003, p. 21) argues that 'the idea of "looking inwards" implies that we have a special sense, or even a sense organ, enabling us to inspect our conscious states, in a way which is modeled upon visual observation.' She notes that behaviourism is impotent to deal with covert mental acts as they have no behavioural manifestation. She, however, explains that there are many mental states and activities that are inaccessible to 'extrospection' – about which a person could normally claim knowledge about thinking, imagining, believing and feeling – but where introspection can still be invoked. Of importance is the fact that although the CDN was set up with tangible outputs (i.e. course materials and institutionalized programmes running) in mind, it would seem unwise to restrict the measure of success of the network to behavioural manifestations such as change in behaviour, as some members may have acquired other forms of knowledge and skills which cannot manifest in behaviour or product, as was found to be the case.

Archer further explains that the main activity in the domain of mental privacy is reflexive deliberation. Charles (cited in Archer, 2003) postulates that through reflexive deliberation, people use their powers as strong evaluators. Perhaps a good example of reflexive deliberation is the processes we go through when say we want to buy a pair of shoes. Before the action of buying takes place, we reflexively deliberate in our mind. We ask ourselves such questions as 'Will this pair of shoes last?', 'Is it worth the price?' We ask ourselves whether or not the particular shoe is in fashion. 'Who else has it?' And so on. It is only when we satisfy ourselves with the answers to these questions that the action of buying takes place.

This example demonstrates how we start to define our positions from internal deliberation to external reality (Charles, as cited in Archer, 2003). Reflexive deliberation in the context of mental privacy is not easily accessible to outside inspection, because internal deliberations have no necessary behavioural manifestations. Therefore the agent is a crucial source of self-knowledge. Archer further postulates that the private life of the mind is not a passive matter of the 'looking inwards' but an active process in which we continuously *converse* with ourselves in order to define what we believe, desire and intend to do. It is important to acknowledge the fact that the end of internal deliberations may well lead to our decisions and actions being manifested in society. I have no doubt that CDN members, apart from collaboratively working together, engaged in individual reflexive deliberations as they examined what they were learning against their own reality, in the institutions they came from and their countries' reality.

Research process

As already mentioned, the research process was informed by a stratified ontology, drawn from critical realism, which argues for the existence of reality independent of human consciousness (Archer, 2003; Sayer, 1984, 2000). In most studies the empirical (what we experience) is associated with reality (Sayer, 2000). This is based on the assumption that the world happens to correspond with the range of our senses and to be identical to what we experience. The stratified ontology of critical realism views the world not only in terms of empirical and actual (as it relates to human experience) but includes the real. Bhaskar (1997) argues that mechanisms, events and experiences constitute three overlapping domains of reality, i.e. the domain of the *real*, the *actual* and the *empirical*. This relationship is presented in Table 4.2.

Table 4.2 Mechanisms, events and experiences

	Domain of real	*Domain of actual*	*Domain of empirical*
Mechanisms	✓		
Events	✓	✓	
Experiences	✓	✓	✓

Source: Adapted from Bhaskar, 1997.

Table 4.3 Summary of data analysis process using Nvivo

Step	*Detailed analysis*
1	Edited transcripts were numerically marked to identity of the interviewees, emails and field notes
2	Coding of key words/phrases from each script to get general ideas from the data
3	Broad brush coding of each of the scripts
4	Coding by means of main ideas from each of the questions under the broad categories
5	Revising and merging of codes where appropriate for consolidation
6	Emerging theory

Source: Developed by the author.

The major research methods used to examine the CDN were mainly document review and analysis, interviews and focus group discussions. Data was analysed by means of qualitative software, Nvivo, as summarized in Table 4.3.

Findings

Although institutionalization of environmental education courses at an institutional level was viewed differently among CDN members, general understanding was expressed in aspects such as owning the courses, recognizing the courses, taking responsibility for the courses and enabling continuation.

Recognition and ownership of courses

Recognition and ownership of a course developed under the CDN was seen as key to institutionalization. Participants understood a course to be institutionalized when a university or institution of learning took it on as an integral part of its own programmes. The participants seemed to suggest that when a course is said to be institutionalized, it ought to be incorporated in the institutional course programme and lead to some qualification. It was suggested that recognition and ownership is exhibited by the institutions' willingness to approve and register, accredit and provide certification for the course, like any other courses run within the institution. While recognition and ownership of courses seem to have been less problematic with universities and tertiary education institutions, they proved very challenging to non-governmental organizations and non-traditional institutions of learning such as government departments.

Approval and registration

Three of the thirteen courses developed through the CDN were approved and registered as components of full-fledged courses or as stand-alone courses within three different CDN member institutions (National University of Lesotho, University of South Africa and the National Polytechnic of Namibia). Three other institutions (Rhodes University, University of Botswana and University of Swaziland) approved and registered the courses developed and revised the courses (in case of Rhodes University) as short courses under the institutional short course policy. Of the five non university-based courses, three were accepted into the mainstream education systems in Namibia (two courses) and Zambia. It also seemed that approval of courses developed under the CDN were easier and faster in institutions which were already running similar courses or were contemplating developing courses in environmental education than in institutions where these courses were seen as completely new.

Curriculum structure

Institutions where the curriculum was rigidly structured or had several cross-cutting themes had difficulties accepting the new course. Government departments and NGOs were said to have several activities/programmes that were competing for space in the teacher training curriculum and introducing new innovations called for long and often frustrating negotiations. Sometimes they simply did not bother to respond to new innovations, especially if these came from outside their management framework. This called for course developer's power and agency to convince the authority of the importance of the 'new' course.

Accreditation and certification

As was the case in the process of approval and recognition of the courses developed under the CDN, there were a number of problems (underlying causal mechanisms) associated with accreditation and certification of courses. For example, local universities seemed not to have entertained the notion of accrediting and certification courses developed by non-traditional institutions of learning – a legacy inherited from the post-colonial curriculum development processes. This was mainly due to the fact that the tradition and culture of universities did not provide for such processes. Furthermore, there were no institutional mechanisms for accrediting and certificating courses developed outside the university.

Institutional support

Institutional support emerged as one of the key factors towards institutionalization of environmental education courses developed under the CDN. The research data indicates that institutional support ranged from institutional culture and politics of who is mandated to develop courses; availability of a short course support structure; support from colleagues and superiors; to institutional commitment.

Institutional culture and politics

It emerged that within university culture and politics (inherited from traditional scholarly thinking of what is knowledge) courses are usually developed by 'experts'. In many instances, course development was controlled by rigid administrative structures, which spelt out the qualifications and procedures of how to introduce a new course.

Due to the short time that the CDN project was to be implemented, approval of courses (within the project life) was not always possible. The bureaucracy of having new courses approved by various bodies and committees in the institutions was a big challenge. Thus some members did not even invoke their symbolic agential power to submit their courses through the approval committees within their institutions, fearing the length of the course approval process. They therefore opted to run the courses as short-term courses, which did not seem to be a problem with most of the universities. Most university CDN partners had support structures such as short course policies that enabled the development of short-term courses for professional development.

Short course support structure

Although some courses developed under the CDN could not be taken on at the same level as degree courses in some universities, they were accepted and became institutional courses under short course policy and other short course support structures aimed at supporting professional development. This was the case in the University of Botswana, University of Swaziland and Rhodes University (although the Rhodes course had long been institutionally situated, and the CDN helped to strengthen the course). Short course support structures such as policies and establishment of in-service centres that run short courses provided an opportunity for speedy approval and acceptability of courses developed under the CDN. Of course much depended on the commitment and agency of the course developer. In cases where universities lacked a policy or structure that supported short courses, ownership and acceptability of CDN-supported courses became difficult as it was almost impossible to transform the existing structures.

Lack of knowledge in environmental education

It was also argued that approval of courses became difficult because some members of the approval committees might not have had knowledge in the field of environmental education. The lack of knowledge in the field by the approval committees might have led to non-approval (or delays in approval) of good courses in environmental education in some institutions. In some instances, the philosophical orientation of the courses (which was participatory in nature and built on relational ontology) went against the set standards and traditions of what constitutes a quality course. Courses were defined with the colonial lenses of what constitutes a standard course. Some institutions faced challenges of how courses developed with outside support would be administered within a more structured tradition. These challenges were further enabled or constrained by the existing internal relationships among colleagues and superiors.

Support from colleagues

Built on the spirit of *Ubuntu*, the level of involvement of colleagues was central to the institutionalization of the courses. Data reveals that in situations where the CDN representative involved colleagues, the course was easily accepted and owned by the institution. Institutionalization of the courses was achieved when the developer(s) solicited support from colleagues by orienting and involving them in the course development processes at an early stage. Developers who constantly engaged their colleagues seemed to have received support and the courses they were working on became institutionally located. However, in some cases CDN members struggled to motivate interest and mobilize support among colleagues. Apathy towards participation in the CDN courses was also attributed to what one participant called 'academic short sightedness or academic jealousy' (interview data).

Support from superiors

Nearly all the CDN members whose courses were accepted and institutionalized seemed to have enjoyed the support of their superiors at the head of department and dean level (in the case of the universities). Non-university course developers also seemed to have been supported by their immediate supervisors, directors or principals of institutions. In cases where the superiors understood and were willing to support the process, bureaucracy of having new courses approved by the various bodies and committees became the hindrance to institutionalization. This was attributed to the short project span, as normal degree courses took much longer to be approved. Due to the urgency and the short time available for course development in the context of the CDN, some heads of department or deans (in university situations) went against laid down procedures to have the courses approved.

In some cases course developers were not able to have their courses institutionally located due to lack of support from their superiors who seemed not to understand fully what the course was all about. The authority and power of the course developers were constantly challenged. In such cases course developers summoned (or at least they attempted to) SADC as an intergovernmental organization to intervene. The mere mention that the programme was a SADC initiative helped institutionalization in some institutions.

Discussion

To aid in the discussion of the findings on the institutionalization of environmental education courses, I draw on Schnack *et al.*'s (2004) institutionalization framework (introduced above). The main focus of the discussion is to reflect on the structural and agential factors that enabled/ constrained institutionalization of courses in the CDN.

Political commitment at all levels of the host institutions

Political commitment was captured under the theme 'institutional support'. Two sub-themes emerged: (1) institutional culture and politics, and (2) support from colleagues and superiors. The institutional culture and politics enabled and in some cases constrained the process of institutionalization of courses. Some institutions had strict policies as to who was qualified to develop courses at a given level. Therefore courses developed by those perceived as 'unqualified' members of staff could not easily be recognized and accepted into the mainstream institutional course development framework.

It was also clear that lack of support from superiors in some institutions was sometimes more political than professional. Some course developers managed to overcome the hurdle of not being supported by their immediate supervisors by coercing support from a higher authority, through invoking the name and power of SADC.

Post or a person in charge of institutionalization

At the initiation stage of the CDN, the assumption was that the selected course developer in each participating institution would lead the process of course institutionalizing. A number of course developers had little or no authority to influence institutionalization of courses. Although Schnack *et al.* (2004) suggest making alterations to the structure of host institutions in terms of job descriptions and procedures in order to enable institutionalization, institutional structural changes were not possible in the short and intermediate terms. The CDN did not have enough power and authority to influence structural changes in the course approval systems. The process of institutionalization was left to the course developers' mobilization of their own agency. Although all the participating institutions appointed representatives or endorsed those initially approached by the SADC REEP officials, time as a cost on the part of course developers was not factored in. The course developers continued with their normal workloads in addition to the CDN activities. This seemed to have had caused role conflicts with the individuals who had to attend to many responsibilities. Conflicting roles of individual course developers made the process of institutionalization more complex.

Collaboration and synergies

The ability to identify and mobilize stakeholders from within and outside the institution seemed to have helped a number of course developers to achieve institutionalization of courses. Course developers who collaborated and created synergies with other similar projects within their institutions or at national level had the greatest chance of having their courses accepted and institutionalized. Collaboration with other stakeholders seemed to have given credibility to the courses developed under the CDN and subsequent recognition and registration. Most course developers worked with an array of experts in the environmental field.

Recognition of 'invisible work'

Schnack *et al.* (2004) postulate that relationships need to be built and trust established within host institutions before projects can operate at their full capacity. They call this 'invisible work', which is often not recorded in the project documentation or reporting, but can be amongst the most important work a project engages in for the purpose of sustainability. However, the CDN did not allow much time for such work, except for a recruitment process by a consultant and the few visits made by the SADC REEP senior staff to some institutions. These visits focused on marketing the network and selection of members. These visits seemed to have helped a number of CDN members in the process of institutionalization of environmental education courses they were working on. Following such visits, CDN members said they had received recognition and support from their colleagues.

It would have been desirable to visit and discuss the CDN and its progress with heads of institutions more regularly. It also appears most CDN members did very little 'invisible work' to negotiate the process of institutionalization.

Institutional structure and role play

Schnack *et al.* (2004) note that one of the complicating factors in the process of institutionalization of ideas/activities is fitting into an institutional structure. This observation is true of the CDN. The CDN was a funded project with time and budget limitations: most course developers could not even attempt to submit their courses through the institutional approval committees, as they tended to operate within a much longer time cycle than that allowed by the CDN. Course approval processes in most of the higher education institutions was very laborious and could take many years. It was not possible for the CDN to imagine that the approval of institutional structures could be changed. The CDN worked with the existing structures and depended on the course developers' agency to navigate and negotiate the institutional structures for institutionalization of the courses. The short course policy or structure became the alternative to the long procedures of approving courses in higher education institutions.

Given the above research findings which are mainly based on the empirical data and participants' own experiences, I am tempted to advance the following retroductive theories about underlying causal mechanisms that affected institutionalization of environmental education courses in southern Africa.[5]

- It would seem that in countries and institutions undergoing curriculum transformation, new courses and innovations in course development are more likely to be institutionalized. This was the case in the National University of Lesotho and the University of South Africa.
- It would also seem that countries with greater commitment to democratic processes would be more likely to accept a participatory course than countries with a weak democratic governance record. Although results from the empirical data showed support from colleagues, existence of short course policies and negotiations with superiors enabled institutionalization of the courses; no amount of these aspects was going to be helpful if the country was undemocratic. Democratic governance at the level of government is a precondition for participatory courses, or processes in general in a country.
- Institutions in countries at the top of the power hierarchy in the SADC region and which are perceived to 'set the standards' (such as those in South Africa), have a higher probability of accepting unconventional courses than those in the lower power hierarchy. The latter countries are likely to be more reticent when it comes to novel courses, due to a desire to be seen to be maintaining standards and due to a lack of confidence.
- The family symbol is one area that would further enhance the institutionalization of courses, especially in the SADC region which was marred by years of suspicion and conflicts along racial lines under

colonial and apartheid segregation. Every member state and institution would want to show family unity and solidarity by working together on a regional programme. In many ways the family symbol is informed by the foundation of *Ubuntu*.

- In some SADC countries which have conservative educational systems, novel courses such as those developed under the CDN are unlikely to be accepted. This is because curricula and institutional forms in such countries are based on reproductive models, reproducing intellectual and knowledge creation patterns characteristic of former colonial universities and epistemologies.

Conclusion

Institutionalization of environmental education courses was influenced and shaped by several underlying internal and external relations. It is not possible to explain the structural and agential factors that influence institutionalization by solely depending on empirical and actual data (as it relates to human experience); there is a need to include the real (the structure and power of objects that exists outside human experience). Critical realist ontology presented a useful theoretical and analytical framework, which probed institutionalization of environmental education beyond the empirical observations, experience and actions of those involved in the CDN.

The critical realist ontology domains of the empirical, actual and real provided an analytical framework for probing ontological depth. The usefulness of probing such causal mechanisms is that they insist on a different level of action to make things better, and they give us greater understanding of our constraints and enablements. We can only do what we can do, with what we have. Therefore, for some course developers operating in undemocratic environments, no amount of institutional rearranging, or negotiation with superiors, or discussion with colleagues, was helpful in the end. Their efforts went to waste and their courses never became part of the institutional course framework. It might also be useful to note that the regional power balance between countries (even under the guise of regionalism) is significant; perhaps this would suggest a strategy in which top administrators in some countries are shown what the trendsetters are doing to reassure them that these courses are not out of line with good regional practice.

Acknowledgements

This chapter is based on a PhD study undertaken by the author in the Rhodes University Environmental Learning Research Centre (Lupele, 2007). Aspects of the study were supported by the SADC Regional Environmental Education Programme. The contribution of participating universities in the CDN is also acknowledged. An earlier version of this paper was published in the *Southern African Journal of Environmental Education.* It has been reworked for this book, but permission to use parts of it has been obtained from the Journal Editors.

Notes

1 The four Danida funded projects were:

- Regional Environmental Education Support Project (SADC REES)
- Supporting Environmental Education in Namibia (SEEN)
- Lesotho Environmental Education Support Project (LEESP)
- National Environmental Education Project – General Education and Training phase (NEEP-GET).

2 'African worldview' as used here does not refer to a common agreed philosophy. Rather, my use of the term is meant to explain that the way of life in most African traditional societies thrives on a relational orientation that is shaped by ongoing contact and interactions among individuals as well as with nature, as exhibited in the extended family value system, for example. I am also aware that the beliefs and practices that embrace what I have called African worldview are not unique to Africa, as these can also be traced in various Eurasian philosophies (Louw, 1998).

3 There have been some appropriations of the word *Ubuntu* by associating the term with either Zulu or Xhosa speaking people among South African scholars due to the currency the concept has gained in post-apartheid South Africa (Louw, 1998). It is common to read statements such as 'the word *Ubuntu* comes from the Zulu and Xhosa languages' (Christians, 2004). It is my humble view that this appropriation may lead to the loss of the principles and value of the concept as non-Zulu or Xhosa feel segregated. However, the word and its related interpretations is embedded in Bantu languages (as well as cultures) of southern Africa. *Ubuntu* is differently referred to as *Unhu* or *Hunhu* in Shona (Zimbabwe); *Umunthu* in Chewa (Malawi); *Ubuntu* in Bemba (Zambia), among others.

4 The generic name given to agential enterprises is 'project' (Archer, 2003).

5 Critical realism allows a researcher to advance such causal mechanisms, even if there are exceptions, since potentials may exist even if they are not yet actualized. The validity of these assertions comes from their ability to explain certain empirical or actual events.

References

Africa Watch. (1991). *Academic freedom and human rights in Africa. An Africa Watch Report.* USA: Human Rights Watch.

Archer, S. M. (1982). Morphogenesis versus structuration: On combining structure and action. *British Journal of Sociology, 33*, 455–483.

Archer, S. M. (2000). *Being human: The problem of agency.* New York: Cambridge University Press.

Archer, S. M. (2003). *Structure, agency and the internal conversation.* Cambridge, UK: Cambridge University Press.

Bhaskar, R. (1997). *A realist theory of science.* London: Verso.

Bhaskar, R. (1998). *Critical realism essential reading.* (M. Archer, Ed). London: Routledge.

Christians, C. G. (2004). Ubuntu and communitarianism in media ethics. *Ecquid Novi, 2,* 235–256.

Danermark, B., Ekström, M., Jakobsen, L., and Karlsson, J. (2002). *Explaining society: Critical realism in the social sciences.* London: Routledge.

Emirbayer, M. (1997). Manifesto for a relational sociology. *American Journal of Sociology, 103,* 281–317.

Janse van Rensburg, E. (1995). *Environmental education and research in southern Africa: A landscape of shifting priorities.* Unpublished doctoral thesis, Rhodes University, Grahamstown.

Kumar, S. (2004). *You are therefore I am: A declaration of dependence.* Foxhole, UK: Green Books.

Lotz, H. B. (1996). *The development of environmental education resource materials for junior primary education through teacher participation: The case of the We Care Primary Project.* Unpublished doctoral thesis, University of Stellenbosch, Stellenbosch.

Louw, D. J. (1998). Ubuntu: An African assessment of the religious other. Retrieved 16 November 2010, from www.bu.educ/wcp/Papers/AfriLouw.htm.

Lupele, J. (2007). *Networking: Enabling professional development and institutionalisation of environmental education courses in Southern Africa.* Unpublished doctoral thesis, Rhodes University, Grahamstown.

Mamdani, M. (1994). Introduction. The quest for academic freedom. In M. Diouf and M. Mamdani (Eds), *Academic Freedom in Africa* (pp. vii–xix). Dakar: CODESRIA.

McLuhan, T. C. (1994). *The way of the earth: Encounters with nature in ancient and contemporary thought.* New York: Simon & Schuster.

Mwanakatwe, J. M. (1968). *The growth of education in Zambia since independence.* Lusaka: Oxford University Press.

Namafe, C. M. (2005). Postgraduate course development processes in geography and environmental education in the University of Zambia. In J. Lupele (Ed.), *Monograph: Cases of course development in environmental and sustainability education in southern Africa* (pp. 13–25). Howick: SADC Regional Environmental Education Programme.

Nhamo, G. (2005). Principles of environmental education course, NUST, Zimbabwe. In J. Lupele (Ed.), *Monograph: Cases of course development in environmental and sustainability education in southern Africa* (pp. 26–34). Howick: SADC Regional Environmental Education Programme.

Obanya, P. (2000). Patterns of educational reform in Africa. In N. Alexander (Ed.), *Educational innovation in post-colonial Africa* (pp. 1–6). Cape Town: PRAESA.

Okri, B. (1991). *The famished road.* London: Vintage.

Pawson, R., and Tilley, N. (1997). *Realistic evaluation.* London: Sage.

Sayer, A. (1984). *Method in social science.* London: Hutchinson.

Sayer, A. (2000). *Realism and social science.* London: Sage.

Schnack, K., Parker, B., and Squazzin, T. (2004). *Reflecting back ... and looking forward: Learnings from four Danish-funded environmental education projects in SADC region.* Howick: SADC Regional Environmental Education Project.

Schostak, J. (2002). *Understanding, designing and conducting qualitative research in education.* Buckingham: Open University Press.

Taylor, J. (1997). *Share-Net: A case study of environmental education resource materials development in a risk society.* Unpublished doctoral thesis, Rhodes University, Grahamstown.

Van Staden, S. (2005). Industrial safety, health and environmental management: Courses for Namibian vocational instructor trainees. In J. Lupele (Ed.), *Monograph: Cases of course development in environmental and sustainability education in southern Africa* (pp. 60–67). Howick: SADC Regional Environmental Education Programme.

Yeung, H. W. (1997). Critical realism and realist research in human geography: A method or a philosophy in search of a method. *Progress in Human Geography, 21*, 51–74.

5 Underlabouring systems thinking with critical realism in understanding Rhodes University's response to the sustainability imperative

Muchaiteyi Togo

Introduction

This research was based on a case study of Rhodes University (RU) and was carried out within the context of an international sustainability mainstreaming initiative, called the Mainstreaming Environment and Sustainability in African (MESA) Universities Partnership. The partnership was developed as a flagship project of the United Nations Environment Programme (UNEP). It was a response to the objectives of the Decade of Education for Sustainable Development. At start of the study, universities participating in the programme had limited themselves to introducing sustainability issues into individual work contexts (UNEP, 2008). There was therefore a need for a more inclusive '*systems-wide approach to mainstreaming* in participating universities' (UNEP, 2007, p. 4, emphasis in original). In this chapter, I describe the systems theory led implementation of sustainability 'mainstreaming' at Rhodes University. I explain how my methodological approach to this task benefited from a critical realist perspective.

The study context

Rhodes University was established in 1904 in Grahamstown, a small rural town located in the Makana District in the Eastern Cape Province of South Africa. The founding and early history of the university reflects unfolding dominant political ideologies in South Africa (UNEP, 2006). For instance, the university was established as an English University as part of a project to suppress the Dutch/Afrikaner culture in the Western Cape which was viewed to be threatening British supremacy (Maylam, 2005). At the same time, the university was obliged to observe segregation policies after the ruling apartheid[1] government introduced racial segregation in all public institutions including areas of living, marriage, education, transport, and even toilets and cafés, to mention a few (Valin, 2003). Rhodes University therefore initially operated as a segregated white university and was described as 'acquiescent' and 'accommodating' towards apartheid (Maylam, 2005, p. 18). Blacks had their own separate institutions which were under-resourced where they received inferior education called Bantu education to enable them sufficient skills to work only as labourers (ILO, 2010).

In addition to, and often exacerbated by, the legacy of apartheid, there were other issues facing Grahamstown. They included issues such as poverty, declining water quality, unsustainable exploitation of resources, increasing pollution, declining air quality with negative impacts on heath, and land degradation (Makana Local Environmental Action Plan, 2005; Lotz-Sisitka and Wilmot, 2007). There was therefore growing pressure from local and international donors, government and from within Rhodes itself for the university to be involved in community development initiatives. This led to the development of a variety of community engagement-related initiatives including auditing and reporting, and policy development among others.

At the international level, education was given a key role in addressing sustainable development. Some discourses emphasized the role of the universities, for example the Tbilisi Declaration (1977), the Talloires Declaration (1990) and the Lüneburg declaration (2001). A summary of the themes emerging from the declarations show that together they were encouraging mainstreaming of sustainability across all university functions and operations which include teaching, research, community engagement, physical operations, student initiatives as well as in different kinds of collaborations (Wright, 2002, 2004). This chapter explores Rhodes University's response to this call to address issues of both social justice and sustainability. Specifically, I investigated the extent to which the initiatives at Rhodes University were in line with the needs of its immediate community and the sustainability themes defining the roles of universities (Wright, 2002, 2004). Equipped with critical realist tools for analysing causality, I then went on to develop an explanation of the possible causal mechanisms that influenced the implementation and/or mainstreaming of identified sustainability initiatives.

Methodology

Initially I drew on systems thinking to collect my empirical data. I looked at the Rhodes University system as being composed of many parts/components responsible for different functions. I also looked at how Rhodes University is framed as a sub-system of the broad environment in which it is situated. I viewed education as being interrelated with socio-economic and ecological environments. Two of the systems models that I employed are the systems-environment model and the functions/structure model (Banathy, 1992). According to Banathy (1992, p. 22) the systems-environment model is 'a lens that projects a bird's-eye view of the landscape in which the system is embedded' and thus enables the researcher to describe the system within the context of its environment. Figure 5.1 models the main functions at Rhodes University which were investigated in the study.

The main shortfall with systems thinking is that it assumes that only empirical knowledge is allowable. Thus, it does not provide the epistemological framework to arrive at transfactual theories of causality in complex and messy open social systems where a variety of causes can lead to similar outcomes. It also does not provide a way of understanding the historical emergence of phenomena.

Due to the deficits of systems thinking and the multi-dimensionality of sustainability issues, I found it necessary to underpin this methodology with critical realism. Critical realism makes a distinction between the objects of knowledge that science seeks to discover, and the theories that

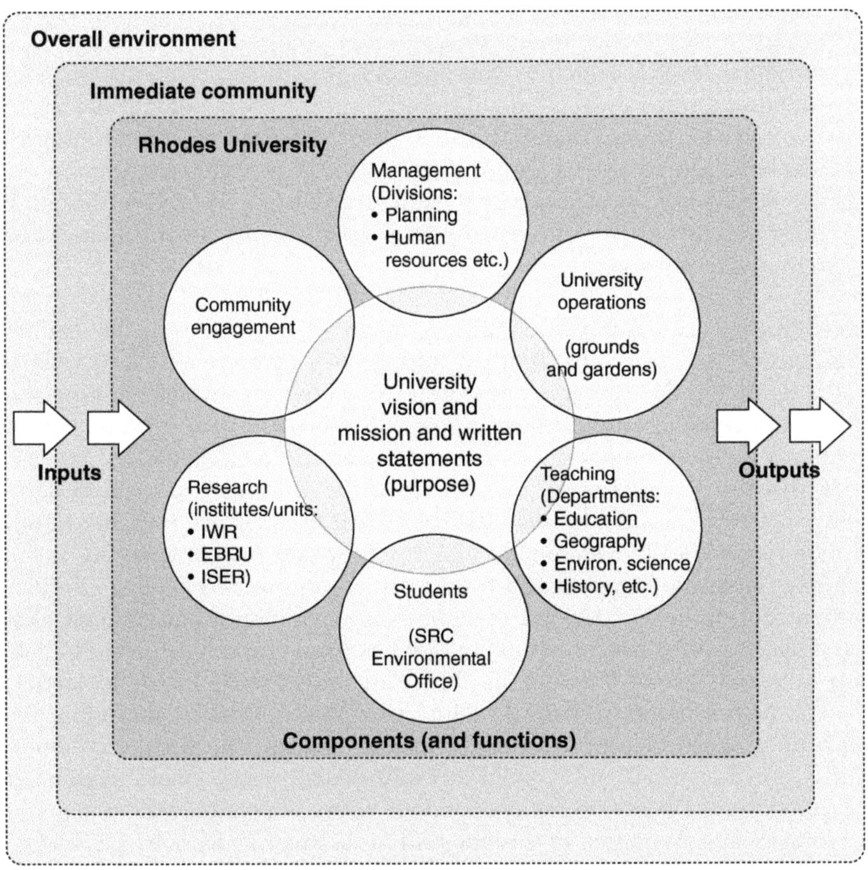

Figure 5.1 Systems view of Rhodes University capturing its functional components (source: developed by the author).

we use to explain them. Reality is therefore stratified into levels, that is, the empirical level of our experiences; the actual level of flows and consequences; and the real level of objects, their structures and powers. Critical realism seeks explanations beyond empirical evidence to develop explanations of causal mechanisms that shape human experience (Danermark *et al.*, 2002). It provides the epistemological framework to arrive at transfactual theories of causal mechanisms and structures to explain empirical evidence. The critical realist concept of a 'laminated system' was particularly relevant to this research. Bhaskar (2010) explains the laminated system as the organization of mechanisms in a hierarchy with one mechanism being the precondition for the other. Bhaskar (2010, pp. 9–10) defines the seven layers of a laminated system as:

- the sub-individual psychological level;
- the individual or biographical level;
- the micro-level studied, for example, by ethnomethodologists and others;
- the meso-level at which we are concerned with the relations between functional roles such as capitalist and worker or MP and citizen;
- the macro-level orientated to the understanding of the functioning of whole societies or their regions, such as the Norwegian economy;
- the mega-level of the analysis of whole traditions and civilizations; and
- the planetary (or cosmological) level concerned with the planet (or cosmos) as a whole.

According to critical realism, society consists of two theoretically distinct elements of structure (institutions) and agency (the creative individual), where agency is the capability of doing things or of actively changing the course of events by causal intervention (Giddens, as cited in Weik, 2006). Agency is however limited by structure (Carter and New, 2004).

In this study, I therefore first described the Rhodes University sustainability initiatives at the empirical level and then used critical realism's depth ontology, specifically its laminated system, to guide my explanation of the possible causal mechanisms that resulted in the empirical evidence. Besides recognizing the importance of context in understanding phenomena, critical realism, unlike systems thinking, goes beyond context to seek explanation in the history of phenomena. The morphogenetic approach (Archer, 1995), which is based on critical realist principles, offers a methodology for explaining social change over time through tracing the history of emergence of events which are created through structure-agency interactions in morphogenetic cycles. The morphogenetic approach enabled me to explain the historical emergence of sustainability initiatives at Rhodes University, providing a social change vantage point for systems thinking.

I have already mentioned that the use of a systems thinking approach alone was not adequate due to its empiricism. At the same time, while critical realism, like social systems thinking, is based on relational thinking and a concept of open systems, and thus argues against reductionism, it is a philosophy and therefore does not provide everything that a researcher needs. I chose systems theory as my research tool to help in the description of the initial situation at Rhodes University. As demanded by critical realism, I also used other theories to explain the empirical data, such as theories of psychology, the marketization of education, and theories that critique capitalism as being exploitative of human and environmental resource, and the theory of climate change. It was therefore clear from the beginning that systems theory and critical realism complemented each other.

The data for the study was collected using a Unit-based Sustainability Assessment Tool (USAT) (Togo and Lotz-Sisitka, 2009) developed as part of the study. I complemented USAT data by interviews and content analyses of documents and observations. In analysing the collected data, I employed morphogenetic analysis (Archer, 1995), and inductive, abductive and retroductive modes of inference (Danermark *et al.*, 2002). The morphogenetic analysis of social transformation/reproduction enabled me to trace the historical emergence of sustainability initiatives at Rhodes University. Induction facilitated reorganization of the data into themes which particularly represent the main sustainability activities at Rhodes University. Abduction, through recontextualizing data in a systems thinking framework, enabled further insights into the phenomena. Retroduction enabled identification of causal mechanisms which influenced the emergence of sustainability initiatives at the university.

Results and discussion

Rhodes University's responses to the call for sustainability

In the functions/structure systems model, functions are used to explain what the university has to do in order to fulfill its goals (Banathy, 1992). I therefore reviewed what the university was doing and the nature of its activities in relation to each of the defined roles of universities. Figure 5.2 is a composite model building in the components of a systems-environmental model but also illustrating the functions of the university in which various sustainability initiatives were being integrated. It shows that Rhodes University was integrating sustainability challenges from its environment into its functions (inputs) based on policies and other requirements also defined in the environment. It also shows the result of sustainability mainstreaming activities, which are the outputs that directly benefit the environment (e.g. having sustainable physical operations and community engagement initiatives that solve some of the sustainability challenges in the community), and others that indirectly benefit the environment (e.g. interdisciplinary curricular, inter-university collaboration in sustainable development, etc.).

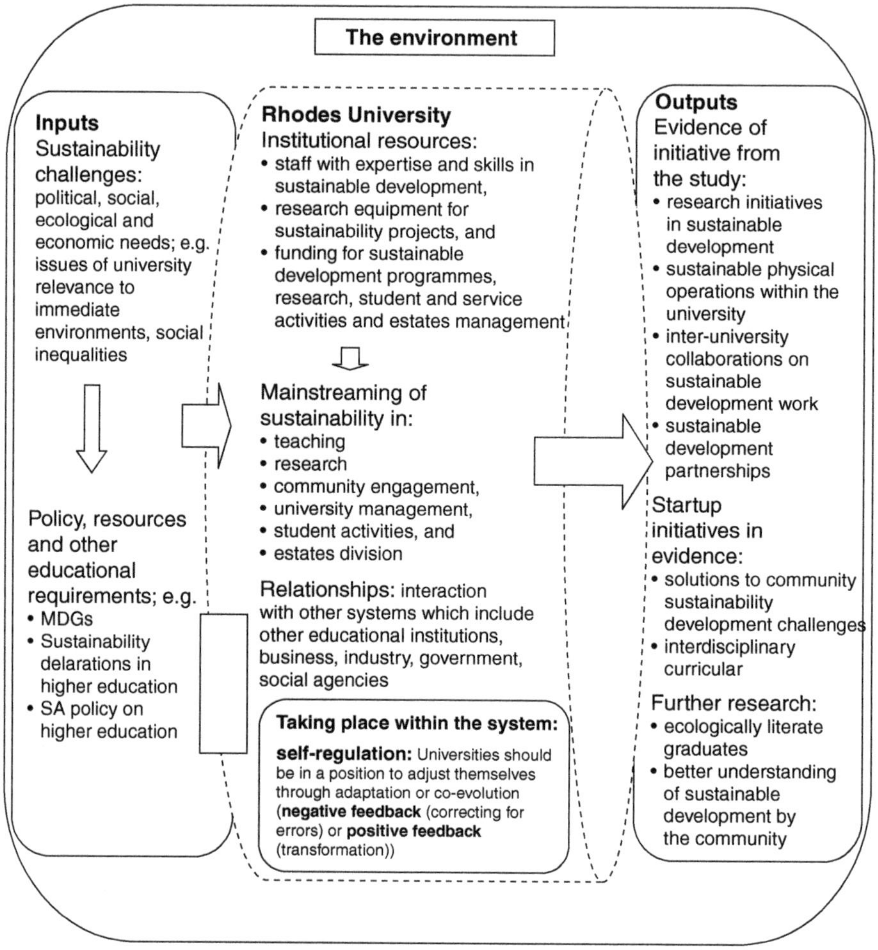

Figure 5.2 Functions of the university in which various sustainability initiatives have been integrated (source: developed by the author).

Explaining the emergence of sustainable development challenges

Morphogenesis provided the understanding that, with regard to sustainability, the university was an emerging system and was likely to continue realigning itself with emerging sustainability challenges as it has been doing since the 1990s when it embarked on a number of initiatives to address sustainable development (examples are the establishment of the Environmental Sciences programme, signing of the Talloires Declaration, establishment of an Environmental Policy and setting up of a Chair in Environmental Education).

Identification of causal mechanisms that influenced mainstreaming of sustainability at Rhodes University

Figure 5.3 illustrates, as a hierarchy, causal mechanisms which influenced the sustainability initiatives at Rhodes University. The mechanisms are presented in a spiral but hierarchical relationship according to the seven layers of a laminated system (Bhaskar, 2010). The mechanisms influence one another and were emergent from each other in the given order of scale. For example, while the ultimate decision to mainstream sustainability may depend upon an individual's dispositions and choices, this can only happen where the norms and values as defined by an institute are permissive. The institution in turn operates within an environment governed by national policy which in turn is shaped by global trends.

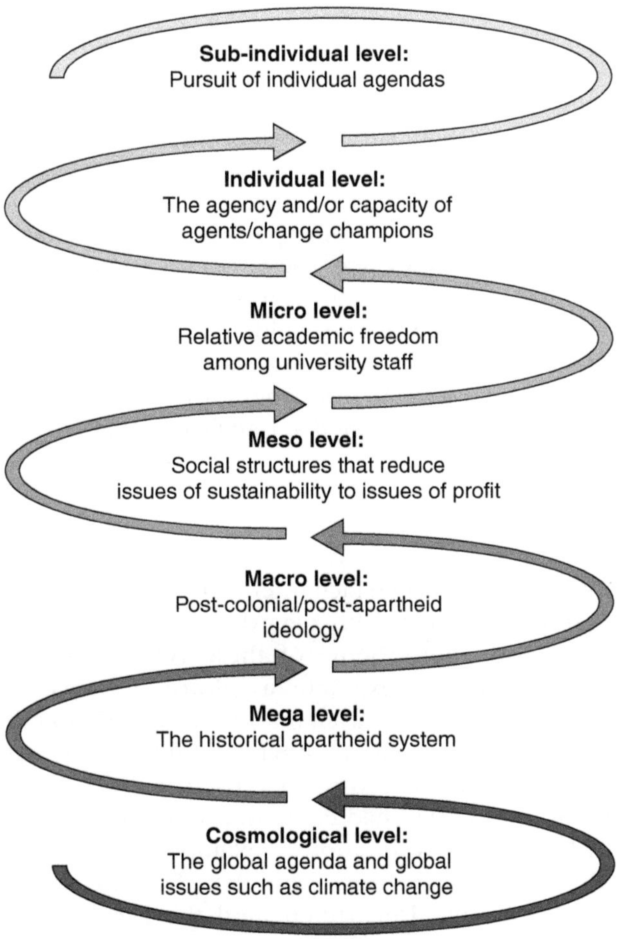

Figure 5.3 Hierarchy of causal mechanisms (source: developed by the author).

The seven laminations of scale

Sub-individual level

I theorized that, at the sub-individual level, there were (conscious or unconscious) psychological motivations amongst the staff to explain both their interest, and at times their lack of interest, in engaging in sustainable development related research. These may possibly include personal or altruistic agendas. Such motivations most likely included: improving the quality and relevance of one's teaching (e.g. through developing an interdisciplinary curriculum), achieving a qualification, achieving promotion, expanding their CVs, or creating opportunities to write research papers for publication.

Individual level

Without capacity, the pursuit of the individual's agenda would not be possible. The question here is whether the individual has the skills or resources to carry out the (sub-individual psychological) desire to contribute to the sustainability initiatives of the University. For example, in the Estates Division, despite the existence of inhibitive factors (for instance the escalated costs of harnessing solar energy), alternative means were sought to green university operations (e.g. use of energy saving bulbs, use of biodiesel in lawn mowers and some of the campus vehicles). Inadequate human resources were mentioned as the major reason the Accounting Department was failing to focus on and integrate environmental and sustainability issues in their activities. In the Geography Department, shortage of human resources was also mentioned as one factor negatively affecting sustainability initiatives.

Micro level

Agency at this level depends on face-to-face interactions and relationships between actors. Organizational rules with regard to behaviour and work responsibilities were relevant here. Some of these rules enabled, and others constrained, agency. In one example, a culture of respect for at least a certain amount of freedom for academics to decide what and how to teach and research made it possible for them to initiate sustainability practices. Freedom of choice applied to other university divisions as well (e.g. the Estates Division) where staff had the power to propose and implement environmental best practices. In another example, in some departments, especially those where lecturers were less free to determine curriculum (e.g. the Accounting Department and the Faculty of Pharmacy) there was reluctance among staff members to mainstream sustainability as they considered sustainability issues not to be part of their role as lecturers. Thus, environmental initiatives were constrained by a culture of keeping to job descriptions. Alternatively, the problem could be seen to lie in an institutional culture in which job descriptions were not modified to take account of higher-level organizational support for sustainability issues. Furthermore, it seemed that face-to-face interactions between individuals (one party indicating to the other party that there was a need to include sustainability) did not always lead to action. However, there was evidence that strong leaders were able to influence their staff to go beyond their job descriptions to carry out sustainability activities without necessarily changing institutional rules. The university's environmental policy was an enabling factor in that it supported the implementation of sustainability practices. However, some research funding was provided for 'sustainable development' which limited the researchers to 'development' research.

Meso level

At this level it was relevant that there existed ideologies that shaped the educational processes at Rhodes University. According to SASCO (2009, unpaginated), education is shaped by dominant ideologies and is mostly accommodative of 'the needs and dictates of the ruling class'. Specifically, the South African rulers are largely committed to a neo-liberal model of economic practice (Nattrass, 2014). In this model, profit is of primary importance and resources, whether human or environmental, tend to be exploited to increase capital accumulation. In the context of the university, this has resulted in an increasing marketization of tertiary institutions in South Africa (Sehoole, 2013). Universities are no longer perceived as public goods, but rather as profitable enterprises. Within this framework, the allocation of resources (human or financial) to sustainability initiatives becomes constrained if it interferes with profitability. The lack of resources to support sustainability initiatives, frequently mentioned by the staff, can be explained to some extent as being emergent from this move towards marketization.

Macro level

Unemployment is very high in the former black area in Grahamstown and many households rely on government grants. The former homelands were ravaged by the struggle for land and mismanagement of resources as a result of apartheid policies and remain wasted. Amid all these challenges, the 'imposed' predominantly Eurocentric university education in South African institutions, according to Nwanosike and Onyije (2011), did not respond to the African contexts and alienated graduates from their culture. The post-apartheid polices (macro level) are therefore intended to correct these past mistakes. Many of the challenges in the Makana District are a result of apartheid policies. Educational quality problems among black communities, for example, are remnants of the system of segregation which saw black people exposed to inferior Bantu education in under-resourced institutions to enable them to work only as labourers (ILO, 2010). Unsustainable exploitation of resources can also be explained by referring back to the policy of segregation which saw the blacks moved to and concentrated in homelands. The consequence was a struggle for resources, especially land, in the homelands which resulted in over-utilization of natural resources, hence current sustainability challenges. It is in response to this socio-political historical context that some of the sustainability practices emerged at Rhodes University and in particular, unlike in some Western countries, linked issues of social justice to issues of sustainability.

Mega level

Rhodes University exists in a post-colonial/post-apartheid context. Sehoole (2013) explained how the marketization discourse has been interpreted in South Africa in a way that reflects the influence of apartheid and colonialism. Furthermore, the legacy of poverty and inequality was discussed in this chapter's introduction (Makana LEAP, 2005; Lotz-Sisitka and Wilmot, 2007). The apartheid policies led to, among other challenges, the mismanagement of natural resources in homelands, poorly resourced schools and inferior education among blacks. Rhodes University was established as a whites-only institution to promote British supremacy (Maylam, 2005) and still has a strong Western presence; it has recently been critiqued again by its students for colonialism and ongoing forms of colonial education by the Black Students Movement under the #RhodesSoWhite and the associated University of Cape Town 'RhodesMustFall' campaigns, showing ongoing continuities with these histories. For example, there remain aspects of the Geography curriculum which reflect a Western agenda, with the main focus being environment and development, whilst social justice is ignored. In comparison, programmes that were established in a different historical context, post-apartheid, show a keen interest in social justice. A good example is the Environmental Science programme established in 1998 which teaches learners to view the environment as a dynamic, integrated social-ecological system.

Global and cosmological level

The response to global warming and other environmental issues led to international events such as the Rio Earth Summit in Rio De Janeiro in 1992. These gave momentum to sustainability mainstreaming in the early 1990s. This contributed to the establishment of the Rhodes University Chair of Environmental Education in 1990, and the Environmental Science Department in 1997, which paralleled increased national and international institutional development in relation to environment and sustainable development challenges. Sustainable development as an agenda for education was defined at a global level (Clugston and Calder, 2002; Wright, 2004). The expectation at a global level, for university graduates to be competitive in a global market (Chinnammai, 2005) makes it necessary for education to be responsive to global agendas like sustainable development. Additionally, international donors make research funding dependent on the involvement of the community in sustainability projects. Because of the system of apartheid, South Africa was subjected to international boycotts and isolated from participating in international events (DEAT, 2004). This therefore adds impetus for the country to respond to global agendas.

Conclusion

Through showing that Rhodes University is in a co-evolutionary nested relationship with its environment, this chapter demonstrates the importance of systems theories in studies that seek to understand relations and interconnections among phenomena. Systems models also help in establishing the roles that are played by systems or their functions in addressing the challenges in their environments. However, considering that education systems are open, different causal mechanisms can lead to the same outcome while different outcomes can result from similar causal mechanisms, systems theories do not provide tools to identify possible causal factors that influenced sustainability mainstreaming at the university. It was therefore important to employ ontological tools provided by critical realism to probe causal factors behind sustainability practices. It turned out that these are more complex and deeper than would have been reflected by information established through systems theories only, which was limited to the empirical level of reality. Systems theories can therefore benefit from ontological theories of emergence that take account of a 'laminated totality' as described by Bhaskar.

Acknowledgement

This chapter draws on a PhD study undertaken by the author in the Rhodes University Environmental Learning Research Centre (Togo, 2009). The study was supported by the Environmental Learning Research Centre and the Murray and Roberts Scholarship Trust Fund.

Note

1 Valin (2003) defined apartheid as a policy of segregation based on the colour of skin which resulted in territorial segregation and ensured the monopoly for Europeans of economic, political and social power.

References

Archer, M. S. (1995). *Realist social theory: The morphogenetic approach.* Cambridge: Cambridge University Press.

Banathy, B. (1992). *A systems view of education: Concepts and principles for effective practice.* New Jersey: Englewood Cliffs.

Bhaskar, R. (2010). Contexts of interdisciplinarity and climate change. In R. Bhaskar, C. Frank, K. G. Hoyer, P. Naess, and J. Parker (Eds), *Interdisciplinarity and climate change: Transforming knowledge and practice for our global future.* London and New York: Routledge.

Carter, B., and New, C. (Eds). (2004). *Making realism work: Realist social theory and empirical research.* London: Routledge.

Chinnammai, S. (2005). Effects of globalisation on education and culture. ICDE International Conference. New Delhi. University of Madras, Chennai, Tamil Nadu, India, 19–23 November 2005.

Clugston, R. M., and Calder, W. (2002). The World Summit on Sustainable Development and Higher Education for Sustainable Development. *The Declaration,* 6(1).

Danermark, B., Ekström, M., Jakobsen, L., and Karlsson, J. (2002). *Explaining society: Critical realism in the social sciences.* London: Routledge.

DEAT (Department of Environmental Affairs and Tourism). (2004). A National Climate Change Response Strategy for South Africa. Pretoria. Retrieved 9 October 2012, from http://unfccc.int/files/meetings/seminar/application/pdf/sem_sup3_south_africa.pdf.

ILO (International Labour Organisation). (2010). *The implementation and impact of National Qualifications Frameworks: Report of a study in 16 countries.* Geneva: ILO.

Lotz-Sisitka, H., and Wilmot, D. (2007). *The Makana Regional Centre of Expertise: An emerging community of practice?* Makana RCE Occasional Paper no. 1. Grahamstown.

Makana Local Environmental Action Plan. (2005). Executive summary. Grahamstown: Makana.

Maylam, P. (2005). Rhodes University: Colonialism, segregation and apartheid, 1904–1970. *African Sociological Review, 9*(1), 14–22.

Nattrass, N. (2014). A South African variety of capitalism? *New Political Economy, 19*(1), 56–78. doi.10.1080/13563467.2013.768610.

Nwanosike, O. F., and Onyije, L. E. (2011). Colonialism and education. *Mediterranean Journal of Social Sciences, 2*(4), 41–47.

SASCO (South African Students Congress). (2009). *16th National Congress discussion document: Education transformation.* Retrieved 8 August 2014, from www.sasco.org.za/docs/congress/16_congress/inst_autonomy.pdf.

Sehoole, M. T. (2013). *Democratizing higher education policy: Constraints of reform in post-apartheid South Africa.* London: Routledge.

Togo, M. (2009). *A systems approach to mainstreaming environment and sustainability in universities: The case of Rhodes University, South Africa.* Unpublished doctoral thesis, Rhodes University.

Togo, M., and Lotz-Sisitka, H. (2009). *Unit-based sustainability assessment tool. A resource book to complement the UNEP Mainstreaming Environment and Sustainability in African Universities Partnership.* Howick: Share-Net.

UNEP (United Nations Environment Programme). (2006). *Education for sustainable development innovations: Programmes for universities in Africa.* Share-Net: Howick.

UNEP (United Nations Environment Programme). (2007). *Mainstreaming Environment and Sustainability in African (MESA) Universities Partnership project, Phase 2.* Project document draft, August 2007.

UNEP (United Nations Environment Programme). (2008). *Mainstreaming Environment and Sustainability in African Universities Partnership: Supporting universities to respond to environment, sustainable development and climate change challenges. 2004–2008 Report.*

Valin, L. (2003). What's in a word? The semantics of 'Apartheid'. *The McGill Tribune*, online edition, 25 November 2003. Retrieved 11 December 2008, from http://media.www.mcgilltribune.com/media/paper234/sections/20031125features.html.

Weik, E. (2006). Working relationships: A meta-view on structure and agency. Retrieved 17 September 2007, from http://theoryandscience.icaap.org/content/vol. 7.1/weik.html.

Worldwatch Institute. (1990). Apartheid devastating South African environment. Retrieved 25 June 2008, from www.worldwatch.org/node/1587.

Wright, T. (2002). Definitions and frameworks for environmental sustainability in higher education. *International Journal of Sustainability in Higher Education, 3*(3), 203–220.

Wright, T. (2004). The evolution of sustainability declarations in higher education. In P. B. Corcoran and A. E. J. Wals (Eds), *Higher education and the challenge of sustainability* (pp. 7–19). Dordrecht: Kluwer Academic Publishers.

6 Bhaskar and collective action

Using laminations to structure a literature review of collective action and water management

Jane Caroline Burt

Water as a social object

Water is a fascinating, life-giving being. It flows. It cuts into the earth to create its passage by finding the places where the earth will give. As it meanders it passes through many different contexts, whether at a fast-rush-full-blown flood or the trickle of an underground desert stream. It does not demand that the space it moves through is homogenous. It adapts to whatever terrain it finds itself journeying through.

Human creatures, dependent as they are on water for their own beings and for life, flock to water like flies to food. We build our cities, our factories and homes on the banks and corners of rivers and draw and drink and use (and abuse) water. We throw in and we take out and the river keeps going until sometimes it doesn't. Then we worry. Sometimes it is not the lack of flow that worries us but the fact that when we drink it babies become sick, some die and this causes a stir.

In South Africa this security, or lack of it, is embedded in the political-historical landscape of this land. Water is classified as scarce. Our rainfall is low in comparison to other countries. In South Africa, during apartheid this scarce resource was not available in equal measure to all. The rulers of apartheid South Africa were farmers and miners. They were of British decent, sent to farm the new colony. They were also descended from first colonial settlers arriving from the Netherlands and then Germany, known colloquially as the Boers. South Africa was divided and redivided until in 1948 the Boers gained independence from British rule and finished the job that the British had started by legalizing the separation of races. Farming was still core to the Boer way of life and this is reflected in the laws that dictated resource use and management, such as the 1956 Water Act No. 54 (RSA, 1956) which gave riparian rights to those that owned land. Water was also subsidized to boost the economy, particularly for the large mines, and parastatals such as Eskom (Byrnes, 1996). Water was not only used to ensure that the economy of white South Africa flourished, it also was the landmark of division. The British used the Great Fish River as its point of division from the Xhosa (Mostert, 1992) and this continued with black homelands being crouched between rivers and mountains in what came to be known as 'homeland states' under the separatist land policies of the apartheid state.

In 1994 the government of South Africa changed. For the first time in its history, South Africa had a democratic government, representing the majority of its people. In today's South Africa, water is an actual and symbolic representation of betterment. It is a government promise that has come to symbolize more than water itself. It has come to symbolize that we are living in a democratic country and that we matter, all of us matter, as can be seen in the Department of Water Affairs slogan 'Some for all, forever' (Palmer *et al.*, 2002). The new National Water Act (RSA, 1998) embraces the principles of equity, sustainability and efficiency and the old riparian rights have fallen away. In their place water is seen as a right with the environment having an equal claim on this right. The environment's share is known as the ecological reserve. In addition to this reserve there is also a human right of some 25 litres per person per day. Only after these two amounts are accounted for is water allowed to be licensed for economic use and this licensing process is to be done in a way that shares the water equally and fairly (Palmer *et al.*, 2002).

The new Water Act also embraces the global trend towards Integrated Water Resource Management (IWRM). In the Act it is recognized that water management includes the integrated management of all aspects of water resources with the aim of striking a balance between supporting livelihoods and ensuring the protection of river systems for future generations (Pollard *et al.*, 2011). South Africa is also regarded as one of the forerunners in terms of embedding IWRM into statutory reform as well as developing sophisticated methodologies for determining environmental water requirements (EWR) (Pollard *et al.*, 2011). This mechanism for ensuring the projection of water resources is known as the 'ecological reserve', which can be defined as a dynamic quantity and quality of flow for a water resource (Pollard *et al.*, 2011, p. 1). The ecological reserve is not only a mechanism for protecting the rivers, it also protects and ensures ecosystem services which many South Africans rely on for their day-to-day survival. Ensuring that the ecological reserve is met is dependent on the implementation of a collection of practices envisaged by the National Water Act and the National Services Act. This means that ensuring the ecological reserve is not only a matter of assessing the river and setting EWR, it requires the collective action and participation of many different stakeholders to ensure that integrated water resource management is indeed achieved (Pollard *et al.*, 2011).

The reality of South Africa's water situation is, however, far from this dream. Compliance with the ecological reserve is rare and the tension between keeping the old economy of South Africa going and issues of redress are continual, with the needs of the old economy still emerging as a priority. Service delivery riots have become common in South Africa and recently in the town of Grahamstown (where I received my education), the Makana municipality has been put under administration (Velaphi, 2014). Before this happened the city experienced debilitating water cuts which forced the Vice-Chancellor of the University to lead a protest through the main streets of Grahamstown and to publish an open letter to the municipality in the local newspaper warning of the closure of the entire University if the water crisis was not dealt with (Grocotts Mail, 2013). Riots in the small town of Addo in 2014 led to the burning down of both the municipal and African National Congress (ANC) offices (SABC, 2013). These are just two examples of many protests happening in South Africa today.

Collective action as a core competency for IWRM

The success of IWRM seems to rest not only on the shoulders of good biophysical scientists (of which South Africa has plenty) but on our ability to understand the social mechanisms which would inhibit or enhance the collective action and management of water in South Africa. It is this understanding which spearheaded the work of the non-governmental organization, Association of Water and Rural Development (AWARD) to initiate an action research project to unearth what inhibits and enhances the sustainable management of water (Pollard *et al.*, 2011). One of the mechanisms identified as necessary for the implementation of the sustainable management of the ecological reserve was the need for collective action at a catchment scale. This led to a research project to explore the possible manifestations of collective action in the lowveld catchments and what constrained or enabled collective action (du Toit *et al.*, 2013).

The Shared Rivers Initiative study of AWARD explores what enables and constrains the ability to implement the processes needed to protect our river resources. It focuses on the process known as the ecological reserve (mentioned above). One of the findings of the Shared Rivers Initiative was that people view the ecological reserve as the water that is taken from them for the environment. This perception is far from the truth. The ecological reserve, if correctly implemented, ensures the sustainability of water resources and thus also ensures the myriad array of ecosystem services that the river provides to people every day. Without it people would inevitably suffer (Pollard *et al.*, 2011).

The South African National Water Act has set up measures and controls for ensuring the fair and sustainable use of water. The two measures to ensure the integrated management of water are resource directed measures and source directed controls. Resource directed measures are for ensuring the protection of the river by setting objectives for the desired state of the river. Source directed controls are measures to control use and to limit impacts to an acceptable level. Resource directed measures require the participation of multiple stakeholders in deciding the kind of river they want through a process known as classification. Depending on the classification people get a different complement of ecosystem services from the river. There is also a recommended ecological category and reserve (quality and quantity of water) for each classification. This forms the basis for planning and implementing IWRM at a catchment scale and ensuring the recommended environmental flows (Pollard *et al.*, 2011, pp. 4–5).

The Shared Rivers Study investigated what enables and constrains a series of catchments in the lowveld in South Africa. A key finding was that:

> Transformation towards a collective, catchment based vision can only be achieved through a collective understanding and approach. Whilst localized platforms do exist, these often reflect single-sector interests or focus on specific sections of the river. In general collective action towards IWRM is weak and requires attention.
>
> (Pollard *et al.*, 2011, p. 7)

AWARD applied for a new research grant to investigate the idea of collective action as key for ensuring the sustainability of rivers and the implementation of the ecological reserve and IWRM. For this new project I was tasked with doing a literature review of collective action for the study. It was for this literature review that I turned to Bhaskar's laminated framework (Bhaskar *et al.*, 2010, pp. 9–10) for understanding the social phenomenon of collective action.

When I took the consultancy job at AWARD, I had only a superficial knowledge of Bhaskar's theories. I had been tentatively reading Bhaskar, or rather reading Bhaskar via Leigh Price's work (Price, 2005, 2006, 2007, 2013, 2014). It was through this reading, particularly of Price (2014), that I became familiar with Bhaskar's laminated framework.

As I began my task of working through the literature on collective action I noticed that there were contradictions in what different scholars argued enabled and constrained collective action. Some spoke about collective action in terms of how individual psychology would determine how an individual would collaborate; others spoke of how a particular institutional context enabled or constrained collective action. It became confusing to work out what theories had relevance for the work AWARD was doing in the lowveld in South Africa. It then occurred to me that what I was seeing as conflicting theories may simply be theories that spoke to a different layer of reality described by Bhaskar and Danermark in their paper on disability (Bhaskar and Danermark, 2006).

The laminated framework is consistent with some versions of systems thinking (Morin, 2002) which forms the foundational theoretical frame for AWARD's work. System thinking takes cognizance of drivers at all scales and how these impact on each other and on socio-ecological systems (Morin, 2002). The laminated framework helps us understand what these drivers (or mechanisms) are and how they influence integration and collective action.

It was these insights that led me to experiment with the laminated framework for writing a literature review on collective action where, at different layers, collective action could be enabled or constrained in different ways. There is obviously interplay between the layers as reality is not actually split up into little bits. What was useful about using a layered approach is that it helped me understand what we know about collective action at different scales. This helped diffuse the debate about whether individual choice and agency has more of an effect than policy changes and institutional structure. By doing this I hopefully avoided one explanation of collective action being reduced away by another (Price, 2014). An example from research into the nature of disability explains this: in the 1950s and 1960s the neurophysiological explanation dominated (disability was seen as a problem of the physical body). This was seen in opposition to socio-economic explanations (which highlight problems of the person who is disabled as societal). Finally, cultural, constructivist explanations saw the use of language and discourse as directly impacting on the experience of being disabled and how this is dealt with in society (Bhaskar and Danermark, 2006). Bhaskar and Danermark (2006) argue that for the disabled person, the experience of being disabled is the interplay of all these factors. One explanation cannot be reduced away by another nor does one explanation fully capture what it means to be disabled and the challenges a disabled person may face.

The literature review on collective action

As mentioned above, there are many studies on what enhances or inhibits collective action in a variety of contexts and in terms of the management of different resources. Studies range from case study research to far more quantitative studies such as models that draw on game theory and analyse human propensity to engage in collective action. For the literature review I drew very loosely on Bhaskar's laminated system and worked with the scales that make the most sense in terms of collective action. The scales of Bhaskar's framework, as interpreted by Price (2014), are as follows. In brackets is my interpretation of this scale in terms of collective action:

- *Psychology of individuals* (Collective action in response to a reading of an individual identity, this can be conscious but equally unconscious);
- *Individual material circumstances* (Collective action in response to an individual context);
- *Human face-to-face interaction* (Collective action as interactions between individuals. This can be one-on-one interactions, interactions between neighbours, between groups);

- *Structure-culture of society* (Collective action and the influences of social structures such as institutions, e.g. local government or cultural beliefs);
- *Society as a whole* (Collective action and the influence of ideas that are held by a particular society. An example of a society would be southern Africa. The influence of this society on collective action would be that it is an emerging democracy, adopts a human rights approach and, in terms of water, embraces the principles of IWRM);
- *Geo-historic trajectories* (Collective action as it relates to broad geological/geographical and historical influences such as a degrading environment, a closed catchment, rainfall patterns, apartheid and colonialism, land use practices);
- *Global trends* (Collective action as it refers to global movements both socially and ecologically that affect all of us such as climate change, democracy, patriarchy, the rise of the internet, social media, economic rationalism).

I reviewed the literature as it referred to each level and the kinds of explanations that were given for what inhibits and enhances collective action.

How people's perceptions of themselves will influence collective action (psychology of individuals)

I did not find many studies that spoke directly to how an individual's psychology will enhance or inhibit collective action. I drew on my psychology background and my interest in neuropsychology to explore what behaviours are a prerequisite for collective action and related these to studies on collective action that made reference to the way people's perceptions of themselves influenced the way they engaged collectively. A possible reason for the lack of research into people's perceptions of themselves in the field of collective action is because individuals' view of themselves is tied to other levels of understanding reality such as social and cultural beliefs and interactions, which are in turn influenced by global movements such as patriarchy, democracy and economic rationalism. Collective action studies tend to focus on these levels. What is worth noting though is that much current research in neuroscience, neuropsychology and evolutionary psychology points to two factors within human psychology that would influence collective action.

How we act is often based on how we want other people to see us (Gilbert, 2010). For example, in the context of big business we do not want to be seen as weak. We want to be viewed as competitive. Working together is encouraged within the in-group of fellow employees but even this is not assured. Employees can easily be fired and/or replaced depending on the dictates of the balance sheet or institutional politics. Being generous and working collaboratively is not valued as much as being seen as ambitious and achieving set goals. However, in the context of our domestic lives being seen as ambitious and self-promoting is not necessarily valued.

In this context we want to be seen as kind, generous, strong, maybe funny. This desire to be seen in a particular light with a particular identity will impact on how we act collectively. A study done by Padmanabhan (2006) in India on the sharing of rice seeds within a collective showed that villagers identified themselves as generous and this impacted on the rules they developed around sharing seeds. This trait, based on a group practice, is being threatened by the need to grow cash crops to survive in a changing economic environment (Padmanabhan, 2006). Binenbaum and Pardey (2005), in a study on collective action and plant breeding, used a quantitative model from game theory to understand the likelihood of individuals to engage in collective action. One of the variables of the model is how our propensity to participate within a particular collective increases as we are influenced by those close to us. Berejikian (1992) also explores this from the viewpoint of revolutionary collective action.

Ostrom (2004) adds that collective action is influenced by how people view each other and the future. This in turn is influenced by their sense of place within the environment and their connection to a vision of themselves as part of a larger system. Ostrom explains that high discount rates usually mean people are only interested in short-term goals. Things like secure property rights, which would influence people's sense of place within the environment, can influence a long-term vision (Ostrom, 2004). Most people can imagine a future for themselves but seldom reflect on this future in relation to the collective. It takes effort to expand one's view of oneself to include seeing one's individual self as interconnected with others and to consider how the future of a catchment is deeply connected to one's own future. Values and attitudes can also be affected by one's view of oneself in relation to others and the environment, for example, a catchment under stress can affect people at a psychological level. They can see themselves as under threat of losing something or, because the catchment is no longer cared for and in a state of degradation, it can be seen as something to be further abused and used at will.

Another important factor is that a human brain has three main emotional systems that are influenced by different hormones. These emotional systems are:

- *Threat system*: focused on averting threat or danger. In the water sector this is apparent in the focus on risk management approaches.
- *Drive system*: focused on gaining resources, this is realized through the drive to compete. This drive is apparent in the water sector where stakeholders try to maximize yield and continually compete for higher water allocations.
- *Affiliation/contentment system*: focused on feeling safe, content and connected. This drive is apparent in the water sector amongst traditional healers and spiritual groups who view water as embodied with voice that heals and protects them (Gilbert, 2010).

Collective action research tends to focus on how the first two emotional systems influence collective action (see Swallow *et al.*'s three motivations for collective action in the next section). Studies tend to ignore the third, which is alluded to in Padmanabhan's study (2006) where a group of people adopt an identity of generosity. This gives them a sense of belonging and connectedness to each other and those outside their defined group.

Cultural spirituality also feeds the affiliation system through connecting people to the environment. This is often expressed in terms of sacred spaces, for example, the Kerala villagers view the rice paddies not only as a source of food but as sacred (Padmanabhan, 2006).

As mentioned above, strong game theory (a voluntarist theory), which has had a significant influence on understanding collective action, tends to focus on the human tendency to want to avert danger and compete for personal or collective advantage. It does not acknowledge the equally strong evolutionary human need to feel connected and safe.

What is interesting to contemplate is how a catchment that has become toxic or polluted would affect these drive systems. If the way in which people behave in the catchment is predominantly within the competitive drive system, how will their relationship towards the river change if it is polluted? Would they want a high stake in a polluted catchment? Would they still want to compete for a resource, which is no longer seen as valuable? If people are operating from the contentment drive system and their sense of belonging and safety are linked to the health of the catchment, how would this influence their behaviour and ability to function as a collective?

What is important to note is that all these drives are working in the individual simultaneously and, in different contexts, different drives will come to the fore. The point that is being made here is that the literature on collective action tends to focus on how we harness the first two drives without considering how the third could play a role in collective action both in terms of enhancing people's willingness to work together as well as inhibiting it (if people no longer feel that they know their catchment or belong to it).

How and why human interaction affects collective action

The world of human action is not only influenced by an internal world but also by the material world into which we are born. People find themselves in a material world in a particular context which includes a particular status in society. People have very little say over this context and their roles are in response to and limited by this material reality.

Game theory has had an impact on the study of collective action and our attempts to understand what makes an individual want to act within a collective. It does this by suggesting that individuals take on different metaphorical roles, for example in certain situations they may respond as a prisoner. Benvenisti (1996) uses game theory (which claims that when certain conditions are met, co-operation between certain actors will happen) in his paper on riparian states and shared freshwater to explain the kinds of problems that usually trigger a need for collective action. He uses three games as examples to identify the problems that need collective action.

First, he looks at the Prisoner's Dilemma that says that in order to ensure sustainability of the riparian states, the actors/stakeholders should co-operate through communication and be loyal to their commitment. Second, the Chicken Game proposes that the actors/stakeholders involved want the other to solve a problem or sustain a resource, but since neither party wants to do that, to co-operate would be the best solution in managing the resource. Third, the Stag Hunt game suggests that several actors/stakeholders would have to sacrifice some of their control to meet the needs of the collective.

Swallow *et al.* (2002) also use game theory to argue that self-motivation to engage in collective action is stronger when participants can see how they will directly benefit from participating. They identified three key motivations: (1) reduced risk; (2) increase in cash crop production; (3) avoiding punishment. This is important for water resource management where the direct benefit of participating in sustainable action or a task of water resource management is not always clear. For example, deciding the classification of the reserve will directly impact on the availability of ecosystem services to residents of a catchment at different time and space scales. If residents do not understand these implications they will be less likely to participate in collective decision making.

Ostrom (2001) argues a similar point but from a different standpoint. Instead of it being important for individuals to understand benefit, she claims that there should be an agreement that there is a problem and that it is important. Bahuguna *et al.* (1994) mention that it is a crisis or collective problem that clearly leads to collective action. Understanding that a river system is stressed and the effects this would have on livelihoods may enhance collective action.

The added complexity of IWRM in relation to people's interactions with each other is that a crisis in a water resource, although originating in one point in the catchment, transports itself to another point in the catchment and changes the material context in which people live. This unique quality in water resource management means that people from many different contexts need to be able to act together in order to address the arising concerns of the catchment as a whole. These are all considerations that are brought to the fore in water resources management where there is an assumption that the collective will be able to make collaborative decisions, develop a common vision and take collective action without some form of co-ordination. The kind of choices that humans make about how they interact could be linked to the context they find themselves in. Understanding this context may explain why actors/stakeholders behave in a particular way as is highlighted by game theory and whether they are motivated by benefit (game theory) or a wish to solve a collective problem (Ostrom, 2004).

Structure and culture of society affects collective action

There is the misguided assumption that all of society functions according to the same rules and obligations. Governments and global institutions often accept Western rules and ideologies as the norm. Pollard and Cousins (2014) challenge this dominant perspective and argue that many assumptions within this dominant ideology can inhibit collective action.

The way that a group of people structure themselves and their understanding of themselves in relation to their environment will impact on the way in which collective action practices emerge. This feeds directly into issues of governance.

An example is work by Cousins *et al.* (2010) on wetland management and governance. This study highlights two arguments that would affect collective action. First, the way property and the ownership of property are understood impacts the governance of a resource. In Western economic terms, property is viewed as the fixed asset of an individual, company or government. Meinzen-Dick and Nkonya (2007) argue that property rights are actually a bundle of entitlements that define the owners' rights and responsibilities for using the resource. When ownership consists of a group of people sharing a common resource then the need to establish rights and responsibilities becomes a collective activity. Clarifying how a resource is understood and the language used to describe the resource will affect the kind of structures that are put in place for the collective governance of a resource. Second, Pollard and Cousins (2014) argue that there is a flawed perception of how commonly held natural resources are perceived as belonging to no one, are subsequently not managed (according to Western notions of ownership/tenure) and are thus open to abuse. They argue that in many cases these resources are held under traditional systems of tenure that are as securely governed as property under title deed.

Different cultures also have different ways of viewing or understanding their relationship to a resource, whether shared or otherwise. Berejikian (1992) argues that at the heart of understanding collective action is the role of structure and agency. He argues against individual and structural reductionism; instead he proposes that individuals evoke alternative rules in different structural contexts, which leads to an exploration of the role of ideology in explanations of collective action. These rules are related to individuals' understanding of current and past structural relationships (Berejikian, 1992). This relates to Pollard and Cousins' (2014) research on property ownership and the management of a common resource. They argue that if governance structures are developed that function according to rules that are based on a misunderstanding of property and property ownership, this will have direct implications for the kind of collective activities that develop. They recommend that it is important to understand the kind of structures that are already in place and use these as a basis to define rules for collective action together with residents.

Ostrom (2004) also argues that, contrary to classical game theoretical situations, people in the natural world do not make decisions independently. They engage by sharing and withholding information within a broader context of local institutions and larger governmental structures. It is these structures which could be in place to improve the likeliness of people to behave in a predicted way (Ostrom, 2004).

Effects of society as a whole on collective action

The literature reveals various challenges that confront collective action and community co-operation. Amongst these are ethnic and cultural division, class differences, corruption and gender inequalities.

Waring (2011) demonstrated his theory on ethnic diversity and the lack of co-operation in Tamil Nadu, India, by showing that there was little voluntary contribution of labour in irrigation systems by a great number of castes in the area. Trust issues among ethnic groups, for example, and a lack of mutual understanding among people with different socio-cultural backgrounds make the management of a natural resource difficult (Varughese and Ostrom, 2001).

A case study on women's collective action in sustainable water management in Gujarat, India, suggests that gender also plays an important role in collective action and that in most cases the impact of some of the gender-related issues that would eventually enhance the management of natural resources is ignored. For example, Panda (2006) found that: the claim of success is insignificant when gender inequality exists; that the lack of promotion of women's efforts might be significantly different (and more successful) than that of men; and that there is a lack of understanding about the different ways men and women organize for natural resource management.

Geo-historic trajectories affect collective action

One of the complications with activities (tasks) that need collective action is understanding where to place boundaries. Boundaries can exist at two scales: the scale of structure-culture of society in terms of socio-political boundaries (e.g. local villages, towns, municipalities, provincial governments or countries) or at the scale of the geo-historical (e.g. geographical boundaries, hydrological boundaries, ecosystem, biome or biosphere, and historical boundaries and colonial influences such as the splitting up of southern Africa using colonially defined national borders). The boundaries for collective action could be defined by stakeholders themselves, particularly when this comes to collective action around a particular task and or practice that is already in place (for example, farmers involved in a particular crop). In this case the boundary is set around an object (the crop), the user group (farmers) and in terms of scale (this settlement or this group of farms).

Swallow *et al.* (2002) argue that mobilizing political support for collective action across political boundaries is difficult. Although it seems logical in the water sector to set boundaries around catchments, research shows that the most successful collective action happens at the village level and scaling up from this level is extremely rare without co-ordination of some kind. Collective action at a larger scale is more successful when areas are small and fall into a single government area. It is for this reason that he suggests that practices that need collective action are situated within social and administrative boundaries rather than hydrological boundaries (Pollard *et al.*, 2011).

This has implications for the water sector in South Africa where collective action is needed at all scales if we are going to implement our current policies and strategies. One of the issues with current practice is the mismatch between the boundaries of water resource management structures, which are based loosely on large catchment scales, and water service delivery structures, which are based on political boundaries. As Swallow *et al.* (2002) highlight, collective action in these circumstances will not be possible without some co-ordination. Knox *et al.* (1998) also argue for the need for co-ordination, particularly when collective action is needed across private and communal property boundaries. His case refers to farming technologies and, in particular, controlling the use of fertilizer, but it could equally apply to the activities of IWRM. The nature of water as a resource means that collective action needs to be facilitated. Unique aspects of water resource management include the size of geographic area, diversity of stakeholders involved, combination of both private and common property (Johnson and Baltodano, 2004) and a disruption in the integrity of the resource happening at one point in the catchment and moving to another, thus affecting a completely different context.

Bahuguna *et al.* (1994) argue that as long as a resource is plentiful and treated as an open access system it is only when it is under stress that people want to control and manage it and thus the potential for conflict or collective action arises. This gives rise to the fact that the resource unit itself can enhance or inhibit collective action. Ostrom (2004) identifies four attributes of a resource unit that enhance collective action:

- Focus of the resource unit is predictable. In the case of water in South Africa, availability is variable and volatile particularly with the growing threat of climate change, which exacerbates dryland conditions (IPCC, 2014).

- Resources are scarce but not destroyed. Collective action responses depend on the level of degradation.
- Reliable and valid indicators of the condition of the resource are available at the local level at affordable cost.
- Resource system is moderately sized.

This has implications for collective action in South Africa. As a system becomes more stressed it will change the dynamics of collective action and the ability of the collective to reach resolution on an action.

Global trends that influence collective action

Among the social theorists who use complexity theory are Urry (2005), Byrne (1998) and Harvey and Reed (1996). The strength of a systems approach is that it acknowledges drivers at multiple levels. Margaret Archer warns that some systems approaches are merely a new name for voluntarism and central conflation or irrealism in social theory (Archer, 2012). In the environmental context, they can also present a new form of determinism, if not accompanied by adequate social theory to explain the social in the social-ecological system.

When dealing with collective action at a local scale, the effect of global trends are often forgotten or ignored, but these powerful trends have huge impacts on the way in which water resource management is played out at a local scale. The very fact that South Africa has adopted an Integrated Water Resource Management approach is due to the global trend towards this approach, which in turn is influenced by global trends in democracy, participation and a rights-based approach to governance.

Other global trends that impact at a local level are global environmental challenges such as climate change. As mentioned above, climate change will increase the variability of rainfall leading to unpredictability and possibly more stress on already stressed catchments, especially in southern Africa, which is classified as a dryland country with multiple stressors that will exacerbate the impact of climate change (IPCC, 2014).

Padmanabhan (2006) explores the effects of gender on collective action. She asks the question: since action resources are very much determined by the existing construction of gender, how does collective action enhance or inhibit the choices of men and women? She tracks the influence of global economic trends on local agricultural practices and how these change the activity of labour leading to women being excluded from the production of food and thus the rules that govern collective action. Here two global trends are being played out at a local level, that of patriarchy and the current neo-liberal economic regime. This is a case isolated to India. Chikunda (2011) reports that in Zimbabwe the government efforts to empower people and alleviate poverty by giving subsidies for cash crops such as cotton farming has had detrimental gender consequences, with women becoming a labour force for a large industry rather than controlling the production of their own food. As the cotton industry boomed, incidences of women committing suicide has increased and systematically children have become more and more malnourished. Padmanabhan (2006) concludes that neglecting gender (a globally constituted force) as an analytical dimension leads to a flawed assessment of community institutions in terms of participation, distributional equity and efficiency.

What does a laminated analysis of literature tell us about collective action?

In summary, when we are dealing with collective action we need to appreciate that all the levels of reality are acting simultaneously on a given context and cannot be resolved in isolation. For example, global trends will have an impact on people's response at a socio-cultural level. An example of this is the concept of poverty traps (Holling, 2001) where feelings of disempowerment further inhibit collective action. Poverty is not only a local concern but the mechanisms for its continual re-enactment are located at all levels of society and within dominant global structures and ideologies. Expecting poverty to be resolved through local collective action is naive. This is not to say that collective action at a local level cannot impact all levels of society, but it is less likely to do so if there is no understanding of how global movements influence a local context – for example, ignoring globally constituted influences on gender or, on a positive note, drawing on strongly constituted principles of democracy that also respect principles of social and environmental justice.

Collective action is inhibited or constrained at different levels and scales by different things. A group of individuals who have an undeveloped contentment drive system, such as communities that have been through trauma or are influenced by a strong consumer culture, could potentially engage in collective action out of fear or for self-benefit. Patterns of interaction will also affect collective action and research into these patterns, such as Ostrom's research and research by game theorists, can potentially help us understand the kinds of interaction patterns people are most likely to choose in a given situation (even if these approaches draw on vastly different orientations). The rules of governments and institutions also have an affect on the way in which collective action will play out. For example, if government institutions see water as a valuable economic commodity then support, and the encouragement of collective action, will be guided by this value system. Ostrom (2004) also shows that the state of the resource itself can have implications for how people interact collectively. Finally, global trends also influence the kinds of rules and policies that governments adopt.

Global ideologies can directly impact on the way in which people perceive themselves, their resource and their interactions with each other, such as the exclusion or inclusion of women in decision making.

Another key observation is that collective action is not suspended in a fixed context. Although there is a growing understanding as to what enhances or inhibits collective action, there is no formulaic way in which collective action can happen. With a continually changing context, learning to adapt becomes a core principle of collective action.

What the laminated analysis of the literature shows is that these collective relationships are influenced by a variety of factors at different levels and scales. Different researchers have tried to understand what constrains and enhances this collective interaction at different levels. Some have focused on understanding individual behaviour and choices, for example game theory. Ostrom and others focused on understanding the role of individuals, local custom and law, institutional frameworks and the resource itself on relationships. Some researchers, such as Panda (2006), try and understand how global ideologies of gender inequality play out at a local level when it comes to collective action and how women can be excluded from the rules that govern collective action. Drawing on this research and experience of how collective action can be supported or inhibited gives insight into our current limitations in supporting collective action and in understanding the kinds of collective action encounters that are occurring in catchments in South Africa. These understandings have implications for how we consider learning, and the potential contributions of learning-led change.

A reflection on undertaking a laminated system inspired literature review

The laminated system framework gave me an intelligent way of sifting and weaving through the literature on collective action so as to create a rich and layered picture of a concept that is seen as a core competency in achieving resource management and water resource management. This layered picture reminds us to take into consideration all aspects of society when learning to work together, including the context of the resource itself. Although my literature review was not intended to be a critical review of the literature, the laminated framework also provided me with tools to take this review further by developing an explanatory critique of the concept of collective action based on multiple levels and scales and potential emergent processes that intersect these laminations.

Acknowledgements

The author acknowledges the work of AWARD's Shared Rivers Initiative Phase 1 and 2 that has spearheaded research in the area of collective action in the South African water sector. The research reported in this chapter forms a small part of this bigger body of work. The author also acknowledges the Water Research Commission who funded the Shared Rivers Initiative.

References

Archer, M. (2012). *The rise of complexity theory in sociology*. Accessed 12 February 2015, from http://international-criticalrealism.com/2013/06/10/audio-margaret-archer-complexity-theory.

Bahuguna, V. K., Luthra, V., and Rathor, B. (1994). *Collective forest management in India*. United Kingdom: Ambio.

Benvenisti, E. (1996). Collective action in the utilization of shared freshwater: The challenges of international water resources law. *American Journal of International Law, 90*(3): 384–415.

Berejikian, J. (1992). Revolutionary collective action and the agent-structure problem. *American Political Science Review, 86*(3): 647–657.

Bhaskar, R., and Danermark, B. (2006). Metatheory, interdisciplinarity and disability research: A critical realist perspective. *Scandinavian Journal of Disability Research, 8*(4): 278–297.

Bhaskar, R., Frank, C., Hoyer, K., Naess, P., and Parker, J. (2010). *Interdisciplinarity and climate change*. London: Routledge.

Binenbaum, E. and Pardey, P. G. (2005). *Collective action in plant breeding*. Paper prepared for the American Agricultural Economics Association Annual Meeting, Providence, Rhode Island, 16 May 2005.

Byrne, D. S. (1998). *Complexity theory and the social sciences: An introduction*. London: Psychology Press.

Byrnes, R. M. (1996). *South Africa: A country study*. Washington: Federal Research Division.

Chikunda, C. (2011). A gender agenda in sustainable education. In *Dialogue*. Grahamstown: Environmental Learning Research Centre.

Cousins, T., Pollard, S., du Toit, D., Maaboyi, J., Mkhabela, B. B., Mweli, M., Williams, C., Maluadzi, T., and de Wolf, J. (2010). *Community based governance of wetlands in Craigieburn village: Learning about facilitating local adaptive management of a complex system*. Final technical report. Msinga: LEAP-Mdukatshani.

du Toit, D., Pollard, S., Burt, J., and von Balkom, M. (2013). *The Shared Rivers Initiative Phase ii: Part 1 – Collective action for improved water resources management*. Technical report. Pretoria: Water Research Commission.

Gilbert, P. (2010). *The compassionate mind: A new approach to life's challenges*. Oakland: New Harbinger.

Grocotts Mail. (2013). Water crisis: Rhodes University's open letter to Makana Municipality. Accessed 3 February 2015, from www.grocotts.co.za/content/water-crisis-rhodes-universitys-open-letter-makana-municipality-14-08-2013.

Harvey, D. L., and Reed, M. (1996). Social science as the study of complex systems. In D. L. Kiel and E. Elliot (Eds), *Chaos theory in the social sciences* (pp. 295–324). Ann Arbor: University of Michigan Press.

Holling, C. S. (2001). Understanding the complexity of economic, ecological, and social systems. *Ecosystems, 4*(5), 390–405.

IPCC (Intergovernmental Panel on Climate Change). (2014). *Climate Change 2014: Impacts, adaptation, vulnerability*. Cambridge: Cambridge University Press.

Johnson, N. L., and Baltodano, M. E. (2004). The economics of community watershed management: Some evidence from Nicaragua. *Ecological Economics, 49*(1): 57–71.

Knox, A., Meinzen-Dick, R., and Hazell, P. (1998). *Property rights, collective action and technologies for natural resource management.* SP-PRCA Working Paper (1). Washington: CGIAR.

Meinzen-Dick, R. S., and Nkonya, L. K. (2007). *Understanding legal pluralism in water and land rights: Lessons from Africa and Asia.* London: CABI.

Morin, E. (2002). *Seven complex lessons in education for the future.* Paris: UNESCO.

Mostert, N. (1992). *Frontiers: The epic of South Africa's creation and the tragedy of the Xhosa people.* London: Jonathan Cape.

Ostrom, E. (2001). Vulnerability and polycentric governance systems. *IHDP Update, 3*(1): 1–4.

Ostrom, E. (2004). *Understanding collective action.* Vision Focus Brief 11(2). Washington: International Food and Policy Research Institute.

Padmanabhan, M. A. (2006). *Collective action in plant genetic resources management: Gendered rules of reputation, trust and reciprocity in Kerala, India.* Technical report. Washington: International Food Policy Research Institute.

Palmer, C. G., Berold, R., Muller, N., and Scherman, P. A. (2002). *Some, for all, forever: Water ecosystems and people.* Pretoria: Water Research Commission.

Panda, S. M. (2006). *Women's collective action and sustainable water management: Case of SEWA's water campaign in Gujarat, India.* Washington: International Food Policy Research Institute.

Pollard, S., and Cousins, T. (2014). Legal pluralism and the governance of freshwater resources in Southern Africa. In M. Snowman and R. Wynberg (Eds), *Governance for justice and environmental sustainability: Lessons across natural resource sectors in Sub-Saharan Africa* (pp. 220–241). London: Routledge.

Pollard, S., Du Toit, D., and Gouws, S. (2011). *Towards the sustainability of freshwater systems in South Africa: An exploration of factors that enable or constrain meeting the ecological reserve within the context of integrated resources management in the catchments of the lowveld.* Report to the Water Research Commission. Pretoria: Water Research Commission.

Price, L. (2005). Social epistemology and its politically correct words: Avoiding absolutism, relativism, consensualism, and vulgar pragmatism. *Canadian Journal of Environmental Education, 10*(1): 94–107.

Price, L. (2006). Joking around in Zimbabwe, undoing and redoing participation. *Southern African Journal of Environmental Education, 23*: 156–161.

Price, L. (2007). *A transdisciplinary explanatory critique of environmental education.* Unpublished PhD thesis, Rhodes University, Grahamstown.

Price, L. (2013). Naked science: Avoiding methodolatry in an environmental education context. *Southern African Journal of Environmental Education, 29*, 10–24.

Price, L. (2014). Critical realist versus mainstream interdisciplinarity. *Journal of Critical Realism, 13*(1): 52–76.

RSA (Republic of South Africa) (1956). *National Water Act No. 54 of 1956.* Pretoria: Government Gazette.

RSA (Republic of South Africa) (1998). *National Water Act No. 36 of 1998.* Pretoria: Government Gazette.

SABC (2013). Protesting Addo residents burn buildings. Accessed 15 January 2015, from www.sabc.co.za/news/a/0b9d7f0041fb9c65b832bd1c2eddf908/Protesting Addo-residents-burn-buildings-20132711.

Swallow, B. M., Garrity, D. P., and Van Noordwijk, M. (2002). The effects of scales, flows and filters on property rights and collective action in watershed management. *Water Policy, 3*(6): 457–474.

Urry, J. (2005). The complexity turn. *Theory, Culture and Society, 22*(5): 1–14.

Varughese, G., and Ostrom, E. (2001). The contested role of heterogeneity in collective action: Some evidence from community forestry in Nepal. *World Development, 29*(5): 747–765.

Velaphi, S. (2014). *Makana under administration for three months.* Grahamstown: Grocotts Mail. Accessed 15 January 2015, from http://thenewage.co.za/136231-1016-53-Makana under administration for 3 months.

Waring, T. M. (2011). Ethnic forces in collective action: Diversity, dominance, and irrigation in Tamil Nadu. *Ecology and Society, 16*(4): 1.

7 Absenting the absence of parallel learning pathways for intermediate skills

The 'missing middle' in the environmental sector in South Africa

Presha Ramsarup

Introduction

Environment and sustainability are continually recognized for significance to the future of planetary well-being. But the complex, cross-cutting transversal nature and associated 'newness'[1] of environment and sustainability concerns within education and training systems raise a number of challenges for education and training systems. In this chapter I explore how critical realist dialectics can help to more fully explain the absence of intermediate pathways in the environment and sustainable development 'sector' in South Africa and through this analysis raise opportunities for creating more seamless environmental learning pathways into green jobs, enhancing social justice potential and public good concerns.

The chapter situates the discussion within the South African policy discourse of meaningful learning pathways (DHET, 2010) through a study of two priority scarce skills occupations in the environmental sector (environmental scientist and environmental technician). This is used as an example to illustrate systemic disjunctures that demonstrate how environmental learning pathways in and for sustainable development emerge. Using a critical realist lens to understand the absences that denote a relationship away from being allows the chapter to conceptualize absence as central to the real and hence to being (Lotz-Sisitka and Ramsarup, 2012; Bhaskar, 1993). Privileging absence allows me to develop a vantage point that connects being to becoming and hence underlines the intent for change inherent within this research. Norrie (2010, p. 28) states that 'understanding change as a process of absenting of absences as well as the absenting of those structural constraints that keep absences in place ... lies at the core of change'.

Introducing learning pathways

Within South Africa, where 3,200,000 young people in 2010 did not gain access to post-school education, Cloete and Butler-Adam (2012) highlight the landscape that paints the picture for access. They argue that there are 2.5 times the number of people out of education as there are in education within the 18–25 year old cohort. Against this background, the notion of seamless learning pathways has been integral to the emergence of the South African National Qualifications Framework (NQF), with its foundational commitment to objectives of facilitating access, mobility and progression within education, training and career path development, as reflected in the NQF Act No. 67 of 2008 (RSA, 2009) and the underpinning NQF objectives (Allais, 2009; RSA, 2009). The Cloete and Butler-Adam (2012) insight into the scope and scale of lack of access, progression and mobility, especially amongst youth in South Africa, provides some foundation for thinking through and understanding a multiple pathway framework that incorporates broad options, non-traditional pathways as well as second-chance pathways. All these will enable broader access to post-school education, which Hoppers (2009) argues is essential for transformation in Africa.

A rapidly emerging Green Economy discourse in South Africa and elsewhere has elevated the profile of environment as an economic driver (DEA, 2010a; Maia *et al.*, 2011), pointing out that learning pathways into environmental jobs need more careful consideration. Envisaged environmental practices cross elementary, intermediate and high skills levels and will therefore require varied and multiple learning pathways. This will necessitate transversal engagement across the NQF as sustainability practices are located across schooling, higher education and occupationally directed training (Lotz-Sisitka, 2011). This emerging scenario means that tomorrow's jobs will be different from those we know today, which has implications for in-service up-skilling and re-skilling, as well as for new learning pathways development. As interest in the opportunities presented by the green growth path escalates, so too is the recognition that investment in green skills planning is critical for long-term performance of a greener and more sustainable economy and society. This requires a reconceptualizing of environmental skills development needs; skills development planning systems need to prepare themselves for an emergence of non-traditional, new and multidisciplinary environmental learning pathways.

More broadly, however, and of relevance to the critical and emancipatory interest in this research, is the point made by Death (2014) about the emergence of the 'green economy' in South Africa. He suggests that there is a need to consolidate a particular kind of commitment to the green economy that should ideally be pursued in a manner that does not produce 'new power relations of inequality and injustice' (p. 1). In this chapter, I argue that central to this would be the manner in which the emergence of the green economy enables capacity building and skills development in South Africa in a sustained, socially just manner, within a concept of sustainability that incorporates economic possibilities for the poor and marginalized, and wider public good interests and concerns such as conservation of biodiversity. In proposing this I suggest a concept of sustainability that encompasses the flourishing of all, now and in future.

A critical realist dialectical lens on environmental learning pathways research

Critical realism proposes a three-part 'depth ontology of the real (structures, underlying generative mechanisms), the actual (phenomena and events, generated by underlying structures/causal mechanisms) and the empirical (phenomena and events experienced by the knowing subject)' (Bhaskar, 1993; Norrie, 2010). This underlabouring enabled me to explore a perspective that goes beyond the colloquial currency that the concept of learning pathways has gained in South African education policy, which has focused mainly on 'pathway engineering', representing fairly 'flat, actualist accounts', or a 'purely positive account of reality' (Bhaskar, 1993, p. 400).

This chapter draws on absence as the primary category of dialectical critical realism as the conceptual tool this chapter; it sets out to investigate the deep-seated absences and the dialectical process of change that a focus on absence can enable. This prioritizes absence as a central concept in being and implies that presence is inherently part of absence, hence giving primacy to absence. The Bhaskarian dialectic is concerned with the 'absenting of constraints on absenting absences or ills' (Bhaskar, 1993, p. xxxiii). Bhaskar's (1993) use of the term 'absence' refers to 'absence of an entity (including structures and mechanisms, events, and experiences) at any level of any region (including being)' (Shipway, 2011, p. 94). This helps one to conceive of being as a process of becoming.

Bhaskar (1993) argues that change must be conceptualized in terms of absence as absence is ontologically prior to presence, which means that any world capable of change contains absences, and agency can thus be viewed as an act of absenting (Shipway, 2011, p. 96; Bhaskar, 1993, p. 240). Bhaskar (1993) recommended identifying *absences*, as real absences (things that are simply not there, e.g. the economic structures that enable green learning pathways), through real possibility (e.g. exploring how it might be possible to support the emergence of learning pathways through green jobs analysis and assessments or other possibilities) and then realizing these in practical terms (i.e. actually implementing green training, once conceptualized). Absence therefore becomes a way of beginning to conceptualize what Bhaskar (1993) terms 'eudemonia', or a better society as is anticipated via environment and sustainability thinking and praxis. In adopting a position of absenting absences we take the perspective that surfacing and removing the constraints will allow what already exists to be.

To help me identify the absences, I drew on Bhaskar's idea that existence of an absence of an entity (including structures, mechanisms, events, experiences) is possible at any level or any region – the entity could be absent from the region as Shipway (2011 p. 94) suggests. It may be:

- never anywhere – does not exist;
- sometimes somewhere else – finite existence;
- spatio-temporally distant – meaning it was away for now (drawing on Bhaskar, 1993, pp. 38–39).

Drawing on this framework, to understand what is absent in the South African skills system context, I was able to look at what is present somewhere else, e.g. studying how occupations are differentiated in agriculture gave me opportunities to understand what is absent in the environmental sector, or to study qualification routes within engineering helped me to understand the nature of what is absent within the environmental sector. Exploring how absences may be spatio-temporally distant, or present elsewhere, helped me to formulate possibilities, re-imagine practices and conceptualize a more coherent learning pathways framework for environmental skills in South Africa.

The dialectical critical realist perspective necessitates in-depth research and deliberation that requires one to look beneath the empirical (the way a learning pathway is shaped or not shaped). Thus I have utilized a more complex notion of learning pathways as neither completely individualistic nor wholly structurally determined. The concept of learning pathways used in this study is informed by the social realist recognition that structures precede human actions *and* that agents may influence structures and exercise reflexivity as they 'make their way through the world' (Archer, 2000; Bhaskar, 1975; Archer *et al.*, 1998). Based on this emergent view of human development, the perspective on learning pathways used in this chapter recognizes that systemic and structural factors shape learning pathways in certain ways, but agents (the individual learner or other corporate agents) may shape the learning pathways in certain ways too.

Understanding the presence of an absence

This chapter takes a critical look at provisioning for intermediate skill needs in the environmental sector. 'Intermediate skills' are defined by the National Qualifications Framework (NQF) as people whose qualifications are sandwiched below university-acquired 'high skill' degrees but above the entry-level skills and qualifications (Kraak, 2008). The Organising Framework for Occupations (OFO) classification classes this group as 'workers who perform a variety of skilled tasks applying broad or in-depth technical, trade, or industry specific knowledge, often in support of scientific, engineering, building, manufacturing and processing activities' (DHET, 2012).

The Environmental Sector Skills Plan for South Africa (DEA, 2010a), as well as the South African Green Accord (DED, 2011), have highlighted a shortage of over 1,500 environmental technicians in South Africa at present indicating a severe shortage of technical skills in the public sector, where such skills are critical to service delivery. Thus was further corroborated by the Local Government Sector Education and Training Authority (LGSETA, 2012) Sector Skills Plan (SSP) which highlights that shortages of artisans in certain trades are widely reported and municipal planning functions (urban planning, local economic development and environmental planning) are all currently underserved. It reports further that municipal technical services and engineering services are coming under increased pressure as the demand for service delivery increases (ibid., p. 37). The Energy and Water Sector Skills Plan (EWSETA, 2012) and the Culture, Arts, Tourism, Hospitality and Sport Sector Education and Training Authority Sector Skills Plan (CATHSSETA, 2012) both report that the technical intermediate cohort is by far the largest section of its employee base (Ramsarup, 2015).

The research reported on here sought to explore how the technical, intermediate skilled cohort in the environmental sector is being catered for within the systems of provisioning. It sought to probe what types of pathways exist and how occupational progression for this cohort is envisaged. To understand this, I explored some of the platforms of provisioning and how these are articulated within the education and training system. Although multiple generative mechanisms and system factors impact on pathways, the chapter opens up only a selected few of these.

Occupational differentiation (or the absence thereof)

The Organisational Framework Occupations (OFO) is a coded occupational classification system. The OFO groups jobs into occupations based on similarity of tasks, skills and knowledge. This grouping provides a common language across the system when talking about occupations (DHET, 2012). Figure 7.1 shows how some of the occupations in one environmental sector are grouped according to different codes and categories that show different levels of occupation. Related to this is the National Occupations Pathways Framework (NOPF), which reflects progression pathways across the different major groups of the OFO. The NOPF clusters into occupational families and occupational clusters to show occupational progression and articulation. These clusters inform employers, human resource professionals and education and training professionals of potential progression pathways and career pathways which assist occupational qualification developers to lay the foundation for

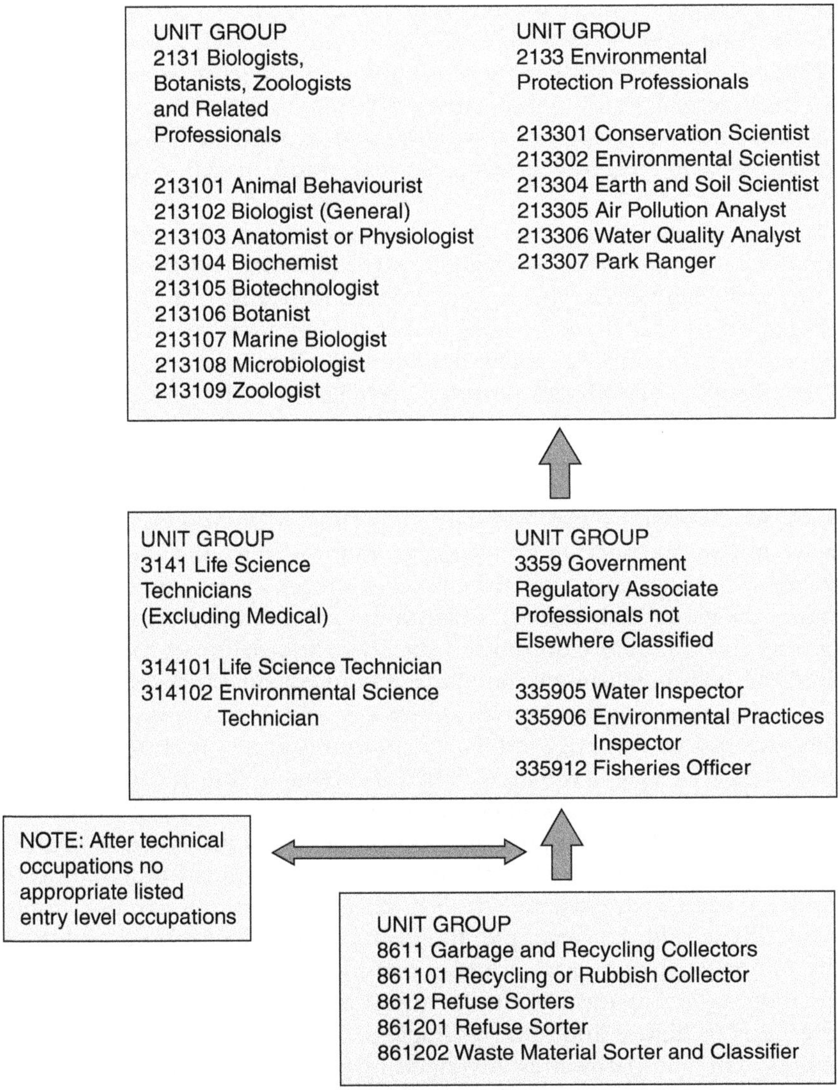

Figure 7.1 Depiction of occupational families from OFO showing some possibilities for horizontal and very limited opportunities for vertical progression (source: developed by the author).

vertical progression when developing occupational qualifications. The example shown in Figure 7.1 depicts how NOPF works and how well developed occupational pathways can be represented showing opportunities and linkages for occupational progression.

A careful study of the most recent OFO (DHET, 2012) reflects a growing emergence of new environmental occupations as compared to earlier versions, e.g. Environmental Impact and Restoration Analyst and Environmental Engineer, but these occupations reflect mainly at the high skills/professional levels. What emerges is a picture of a highly skilled sector, and there is fairly extensive coverage of high skilled professional occupations reflecting the growing specialization in the sector (DHET, 2012).

Of concern, however, is the fact that there is *very little representation of environmental occupations at intermediate and elementary skill levels.* Considering this elite, high skills phenomenon more critically, retroductively one might hypothesize that (at the level of the real) modern social institutional arrangements and cultural beliefs assume that all natural resources are or can be owned. Therefore, resource management is, and has clearly become, the preserve of the elite (the owners of the resources) (Martinez-Alier, 2002), just as managing the economy is only for the owners of the businesses. With this worldview predominating in modern capitalist systems, why would a refuse collector need to understand environmental issues? They simply need to know how to follow orders and the elite management will make sure that they protect the environment. Under this scenario (of globalized capital), environmental training for low/medium level workers in a factory becomes reduced to knowledge of the rules and safety/health and environmental management systems, and neglects more complex forms of environmental knowledge which in many instances is simply ignored or not engaged with in training at all, as shown in some empirical studies (e.g. Mohanoe, 2013). Understanding the presence of such a glaring absence in a national system context is, as argued in this chapter, important for mapping out possibilities for transformative praxis.

What is visible at the level of the empirical is that environmental occupations are scattered, extremely poorly differentiated and are not visible across levels of skill. For example, the occupational descriptions and titles listed under environmental science technician (which is the primary environmental occupation in the technical major group) suggest unclear connections to higher and lower skill levels, creating the impossibility of a 'seamless' learning pathway as envisaged by policy. This fragmented occupational differentiation contained in a major policy implementation instrument of the state surprisingly reflects an imprecise picture of feeder occupations (currently the only possible feeder elementary occupations listed are Rubbish Collector, Refuse Sorter, Waste Material Sorter or Environmental Practices Inspector) into the environmental technician occupation and unclear linkages exist into professional occupations e.g. Environmental Scientist. This is despite the fact that these occupations have been identified as being scarce and in high demand in a range of other policy documents such as the national scarce skills list, and the sector skills plan of the Sector Education and Training Authorities.

This reveals lack of engagement with, and poor knowledge of, the systems of practice at intermediary occupation levels in the environmental sector, leading to poor occupational differentiation. This in turn has the effect that the occupations are often mapped incorrectly by Human Resources (HR) personnel, which in turn affects how funding is allocated for training, and access to training. This is in stark contrast to more established economic sectors such as agriculture. For example, the occupation of farmer is highly differentiated (see below), and all occupational differentiations are listed with independent OFO codes, meaning that Sector Education and Training Authorities and HR personnel can allocate training funding to these occupations:

- Agronomy Farmer
- Field and Vegetable Farmer
- Tree and Shrub Farmer
- Aboricultural Farmer
- Ornamental Horticultural Farmer
- Mixed Crop Farmer
- Livestock Farmer
- Dairy Farmer
- Poultry Farmer
- Insect Farmer
- Game Farmer
- Aquaculture Farmer.

All of the above represent only a snapshot of the occupational range of farmer as listed in the OFO (DHET, 2012) (e.g. there is a whole range of subsistence farmers, crop farmers etc.). What is important about this is that this clear differentiation enables a sector to plan and plot clear connections within occupational families; it enables better human resource practices in terms of career pathing and enables collection of coherent labour market information. It also allows for allocation of training funding, and ultimately access to training and expansion of human capabilities.

These disjunctures raise further concerns when viewed on NOPF – the absence of clearly differentiated technical and elementary occupations reflects loose, unfocused occupational progression pathways, e.g. Environmental Practices Inspector (skill level 2 occupation) is mapped to Environmental Manager (skill level 5) without any intermediary options, making a system supported viable learning pathway virtually impossible. Unlike the example depicted above from the agriculture sector (and other sectors such as mining, banking and other sectors that propel the capitalist economy), environmental occupations reflect discontinuous pathways. The DEA (2010a) critiqued the system for privileging private sector interests over the public good.

Under neo-liberal capitalism and the market, one finds the dominance of labour market perspectives and the 'language of occupations'. In skills development and planning, this has become the dominant code of meaning. It holds power in that resources, qualifications development and skills planning are centred on the language of occupations. Identifying absences such as those identified above starts to intensify concern around systemic articulation for public good occupations (e.g. some green jobs) within an occupationally driven system of skills development. These appear to be critical to address if meaningful learning pathways as envisaged by the DHET (2012) are to emerge for this new area of economic development and social-ecological practice in South Africa.

Qualifications (or the absence thereof)

Qualifications remain another key leverage point into sector employment and are thus a critical system element of learning pathways. To understand the qualifications landscape, I studied the SAQA qualifications database (as at 20 December 2012). From this analysis, I identified 1,330 environmentally related qualifications. The qualifications crossed a variety of environmental faculties – mining, forestry, agriculture, nature conservation and environmental science, showing up some aspects of the cross-sectoral nature of environmental education and training provisioning. The data reflected that the NQF level distribution of the registered qualifications is as shown in Figure 7.2.

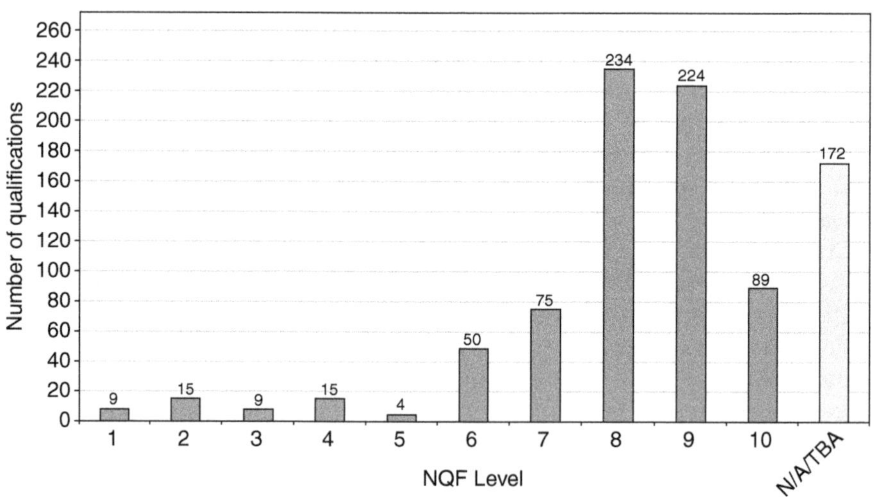

Figure 7.2 Distribution of environmental qualifications across NQF (source: developed by the author).

The data reflects a sector that has overwhelming access points at NQF levels 7 and 8 indicating that the system appears to be viewing environmental qualifications mainly as a specialization – again reflecting a dominant postgraduate entry-point into the sector (also reflected on the OFO categorization of environmental occupations). This data suggests that learners unable to gain access at level 6 will have limited options for access or for alternate pathways reflected by the poor options at levels 4–5. A learner who is unable to gain access to the traditional academic pathway will struggle to shape an alternative or to navigate into an environmental job. The data clearly depicts how qualifications shape who enters the sector, and where and at what levels. This is an issue of concern, as the DEA (2010a) for example identified up to 30,000 entry level employees working in local government alone. Subsequent empirical research has shown that there are almost no learning pathways available for these workers in the environmental sector (Mohanoe, 2013), and is also a concern given the status of scarcity of environmental technical occupations noted above.

Clear, visible and accessible intermediate skilling pathway options

The South African Human Resources Review (HSRC, 2009) notes that three differing labour market pathways to intermediate skilling have emerged in South Africa: the traditional apprenticeship route, the 'learnerships' pathway and further education and training (FET) college programmes (Kraak, 2008, p. 479). However, ineffective functioning of these pathway options have been recognized by several studies (Kraak, 2008; Perold *et al.*, 2012). Kraak (2008, p. 479) argues that these three differing pathways for the 'production of intermediate skills constitute a highly malfunctioning labour market for technically skilled labour'. Graduates of education and training institutions are unable to traverse coherent and visible pathways into intermediate skilled employment (Kraak, 2008). Within this unpromising national scenario the Environmental Sector Skills Plan of South Africa (DEA, 2010a) found that only 2.4 per cent of learnerships in South Africa are 'environmental' learnerships. The Local Government Sector Education and Training Authority's Sector Skills Plan (LGSETA, 2012) reveals that in the 2009–2010 period, 5 per cent of interns used within the constituency were utilized for environmental management purposes across environmental health and pollution control. Both these findings suggest that dominant intermediate skilling pathways are underutilized by the environmental sector. The DEA (2010a) attributed this partially to weak Human Capital Development Planning capacity in the sector, and also to the newness of the sector and a lack of systemic coherence in and for skills planning for the sector. Being primarily an empirical study, the Department of Environmental Affairs (2010a) analysis did not consider additional underlying generative mechanisms as proposed to be possible in the retroductive theory presented above.

Implications of this absence for environmental learning pathways

As a sector with a complex multi-disciplinary and multi-sector engagement and transformation agenda, the environmental sector is dependent on enabling multiple entry points into the sector and multiple exit points that can allow learners entry into occupations from various pathways. For this to be realized, it requires occupations that are differentiated and connected to enable progression and movement through an occupational field.

The chapter so far has tried to illustrate that the absence of clear technical intermediate skill jobs, absence of sound occupational knowledge of these intermediate jobs and labour market information about demand of these jobs constitutes a highly significant 'missing middle' in the skills pyramid of the environmental sector.

This emergent 'missing middle' constitutes a stratum that can enable better learning pathways, and provide articulated access when connecting to artisan occupations. It can also enable movement of entry level skills into intermediate occupations and it can provide the mechanisms to enable parallel pathways. Parallel pathways enable entry via another route (e.g. for a student who does not obtain university entrance) as well as creating the cushion for the student who drops out of traditional 'academic' pathways. A significant observation from analysis is that everything below professional occupations and jobs are conflated, hence prompting inaccurate and unclear labour market information about the sector, negatively affecting the public good role and possibilities of the environmental sector, affecting access into and building of a more benign, 'greener' economy.

The nature of the missing middle, or this substantive absence, that characterizes the environment sector highlights how intermediate skilled workers face a number of distinctive challenges as it shapes who enters, how they move and how their jobs are conceived. There is a poor conception of work roles, tasks and the scope of jobs can result in underutilization of available skills; and low-skilled workplaces tend to have few career development and progression opportunities, poor human resource practices and higher staff turnover (DEA, 2010a).

Two empirical observations attest to how this absence has shaped the environmental sector:

- Because there are few or no intermediary learning pathways, high skill staff are often underutilized and tend to be in positions that could be filled by technicians (e.g. employing masters graduates and using them in technical roles that could be filled by NDip diplomats). This perpetuates practices of underemployment and results in graduates moving into other sectors as their skills are undervalued and poorly utilized (Ramsarup, 2015).

- A study by Mohanoe (2013) found that the lack of attention given to the learning pathways of local government sustainability practice workers at entry skill occupation level responsible for improving quality of life in urban and rural communities through their sustainable development practices was paradoxically running counter to numerous policy statements. According to the worker's experiences there is no mobility or progression on their learning pathways (Lotz-Sisitka *et al.*, 2012). Such workers, with years of experience in environmental practices, could potentially, with recognition of prior learning and additional training, be offered opportunities to enter the intermediate skills level.

Using the concept of emergence (Bhaskar and Parker, 2010) to interpret this, it is possible to see that on the first, higher-order level, a well articulated provisioning system provides the boundary conditions for the lower order or more basic level maintaining the status quo via a lack of access to private and public qualifications and courses in the environmental field. On the second, lower-order or more elementary occupational level, the absence of occupational differentiation provides the boundary conditions of possibility or access into the environmental sector. This affects learning pathways progression, and possibilities for access into the emergent or higher-order level. In the case of South Africa, where access to training is determined by occupational differentiation at this level, without occupational differentiation there can be no formally constituted education and training provisioning, a practice that continues to maintain patterns of exclusion from possible learning pathways and jobs. Thus it can be observed that several distinct mechanisms, at different and potentially emergent levels, are combining to produce a unique result (Bhaskar, 2010, p. 5), as depicted by the emergence of a high skilled thesis within the environmental sector.

Using an example of engineering provisioning to illustrate options for transformative praxis

A further aspect of the identification of absence, as outlined in the analysis above, is the possibility of understanding what is at 1M, absence at 2E, possibility at 3L and transformative praxis, or that which is possible and can be enacted at 4D of Bhaskar's MELD schema (Bhaskar, 1993). To consider possibilities and potential transformative praxis, I turned to an analysis of the engineering sector which, unlike the environmental sector described above, has well defined and clearly differentiated occupations, as well as parallel learning pathways. My interest in this learning pathway is also because the occupation of 'environmental engineering' has been identified as a scarce skill in the sector. Besides providing useful insights into why this occupation was defined as 'scarce' in South Africa (Ramsarup,

2015), I found that it provided useful transformative praxis insights for the sector as a whole.

Engineering sector: illustrating transformative praxis

Despite being widely regarded as a high skilled sector, the engineering sector has developed a fairly progressive model of access that embodies multi entry and exit options. In this section I use the pathways framework of the engineering sector to explore Bhaskar's (1993) suggestion that sometimes an absence is present somewhere else. Enabling broad access that caters beyond the 'privileged learner', i.e. the learner who gains access into higher education, is dependent on enabling multiple entry points into the sector and multiple exit points. This allows students entry into occupations from various pathways. It also requires differentiated occupations that are connected to enable progression and movement through an occupational field, as also mentioned above.

The Engineering Council South Africa (ECSA) is the statutory body charged to act in the public interest. The data in the engineering sector provides a stark contrast to the data on environmental learning pathways. Key to this appears to be the presence of a professional association that understands the skills needs of the sector well.[2] This is displayed by well quantified, high quality skills intelligence from a demand side perspective and clear occupational differentiation which allows for balance of skills mix in the sector. ECSA is able to define the 'mix' of skills needed to adequately resource the engineering sector as cited by Case (2006): one graduate engineer; one BTech; four technicians; sixteen artisans.[3] Data from the engineering sector (retrieved from ECSA on 31 May 2012) reflects a broad range of access points into the engineering sector, each with well defined pathways as shown in Table 7.1.

Table 7.1 Pathways in the engineering sector

Engineering	*Qualification route*
Engineers	4-year Bachelor of Science: Engineering (BScEng)/Bachelor of Engineering (BEng)
Engineering technologists	Bachelor of Technology (BTech)
Engineering technicians	National Diploma (NDip)
Certified engineers/ engineering artisans	Government certificate of competence that enables them to work in mining or a particular engineering industry. They are generally artisans who then study at a Further Education and Training (FET, equivalent to Technical and Vocational Education and Training (TVET)) College, and take on industry roles

Source: Developed by the author.

In South Africa, national diplomas are offered across ten higher education institutions across nine engineering disciplines. The Bachelor of Technology (BTech) is offered across ten higher education institutions across seven engineering disciplines. There are five reflected environmental specializations at BTech, BSc and BEng qualifications across eight institutions across 14 engineering disciplines.

This data reflects an intention to create complementary sectors – with 'equal missions but differentiated missions' (Case, 2006). Case (ibid.) reports that in the engineering sector, students who drop out of university engineering degrees can progress very successfully through National Diploma into B.Tech. Engineering reviews have also found this engineering artisans group to have developed high levels of competence – in some instances equivalent to technologists (Hanrahan, 2000). This qualification differentiation clearly gives different access routes and opportunities for 'second-chance' pathways and provides recognized mechanisms for learners to change from one learning track to another. This provides a framework that contains 'bridges and ladders' which enable young people to move through the system in accordance with their needs (Hoppers, 2009).

The challenge faced by the sector, as reported by Case (2006), was that engineering output was equal to output of technicians, i.e. one engineer: one technician. This resulted in engineers doing work that a technician should have been doing (a problem not different to that identified above in the environmental sector). Consequently, in recent years, an expanded effort has been made to attract and increase artisan and technician practitioners. Case (2006) reports that although BScEng numbers have remained fairly steady there has been a significant increase in University of Technology students since the introduction of the BTech qualification in 1995. The sector has additionally also introduced strategic interventions that are now strengthening BTech programmes to enable crossovers into Masters in Engineering.

Outlining possibilities for transformative praxis

As can be seen from the above, the dialectical critical realist methodology adopted in this study involves identifying 'what is' or 'being' using a depth ontological perspective; it also involves identifying absences, possibilities for transformative praxis, and ethical action (the actualizing of transformative praxis). I have given attention to all of the above in the chapter, except the actualization of transformative praxis. In a chapter like this is it not possible to report on how such transformative praxis will play out; thus the best I can offer is to re-imagine such transformative praxis for the sector. I conclude the chapter with this contribution.

In offering this conclusion, I conclude the dialectical movement that is required from concrete and structured experience of being (e.g. the experience and realities shaping progression in a learning pathway) to the possibility for transformative praxis, or a new way of 'becoming' where such realities can be transformed (I have shown that this is possible by the presence of the absence I have identified somewhere else). This has helped understand how it can be possible to absent the absences through giving attention to systemic elements of learning pathway construction.

Re-imagining transformative pathways for the environmental sector: absenting the absence of intermediate skills

To absent the absences identified above, I propose the following possible actions through which, I argue, it is possible to re-imagine systemically articulated and transformative learning pathways development possibilities for the environmental sector.

Pro-active skills planning from a base of better labour market information

This analysis has highlighted the absence of a clear understanding of the complex multifaceted nature of the environmental sector and the nature of the different jobs needed across the sector (now and in the future) coupled with an understanding of the system that produces and uses environmental professionals. The Environmental Sector Skills Plan for South Africa (DEA, 2010a) identified the need for better skills information, as without this labour market information it is not possible to enable more inclusive skills planning. Having systemic capacity to provide for better labour market information can potentially foster improved understanding of the skills mix (across levels of skills) needed in the sector. This recommendation is strongly supported by the Environmental Sector Skills Plan for South Africa (DEA, 2010a). It also supports a social justice argument for greater access into the sector via a range of pathways and chances. Additionally, it also supports a sustainability argument that more people need to be engaged in environmental concerns in order to counter existing hegemonies of resource management and control proposed in my retroductive argument.

This would be inclusive of a more *reflexive and open systems approach* to labour market analysis, as it is understood that the environmental skills needed today may not be the same as those needed in future given the rapidly changing social-ecological context. This may require innovation in labour market intelligence system designs.

Skills pyramid resource and job needs of sector

Better labour market information has potential to enable a better understanding of resource needs across levels of skill – and could enable a better understanding of the nature of our jobs making more explicit what is entailed in jobs, job descriptions, occupational differentiation, scope and level of expected tasks and so on. All of this can enable a better understanding of occupational progression, making learning pathway possibilities more transparent and accessible to learners and workers and managers alike. This understanding of skills mix can potentially also foster better co-ordinated operations across the sector amongst stakeholder groups such as employers, government agencies, professional bodies and employees.

Prioritizing multiple pathways

In this chapter I have made a strong access argument for multiple pathways. At the same time the chapter highlights how these pathway links are structured and how they relate to the production of different types of skill which cannot be disconnected in the emergence of a high skills sector, if wider transformation goals are to be attained. It is interesting that the environmental sector to date has prioritized high level skilling, with very little attention being given to those systems of skills provisioning that are essential for longer-term support and expansion of the high skills base for transformation purposes (HR related), but also wider social-ecological transformation objectives and widespread sustainable development which are high on the agenda of the environmental sector. This is perhaps also related to inadequate education and training or human capital development capacity in the country, the newness and rapid growth of the sector, and may also be related to the way in which power over natural resources is structured in society.

Pathways with destinations

As we approach the artisan revolution set to hit South Africa, a multi-pronged approach to skills development with a focus on joined-up processes in the environmental sector is advocated with clearly articulated and programmed attempts to connect low skill programmes to intermediate skills development programmes. The issue 'is not the creation of low-skills work in itself, but how routes out of such work can be created and protected' (Lauder and Brown, as cited in Kraak, 2008). Except for the waste sub-sector of the wider environmental sector (DEA, 2010b) and the Expanded Public Works Programmes (EPWP) programmes, there is very little knowledge of artisan, entry level occupations, and these too are poorly defined in occupational systems and learning pathways development terms. In any discussion around entry level provisioning, upskilling and routes out of low-skills work, the sector needs to learn from the histories of countries that have faced the problem of a low-skills equilibrium in which institutional barriers trap generations of workers in low skills work, as has been found to be the case in, for example, local government environmental practitioner employment (Mohanoe, 2013), due also in part to the problem of wider skills planning neglect as outlined in this chapter and in the Environmental Sector Skills Plan for South Africa (DEA, 2010a).

Role of the professional body

The engineering example has illustrated the role that a professional body can play in enabling transformative potential in a sector – the co-ordinating role can enable the sector to keep their eye on the 'big picture' (sector-wide view). The professional body can also play a key role in maintaining synergy between the systems of practice and the systems of provisioning – raise disjuncture, enable bridging when needed to maintain quality and standards and enable a proactive approach to skills planning. Here it should be noted that the new NQF Act No. 67 of 2008 foregrounds and gives new credence to the role of professional bodies in skills planning and development. The sector would need to consider the requirements for establishment and recognition of professional bodies under this act. The professional body, its position and role can become the 'bridge' between the 'different pathways' enabling better systemic articulation. In this process care should be taken to not only foreground the green economy (economic development) aspects of environmental skills and qualifications development, but also the natural resource management and longer-term sustainability aspects of environmental skills and qualifications development (i.e. the public good aspects of the sector's work). A Green Economy Skills Planning Forum has recently been established in South Africa considering the former, while an Environmental Sector Skills Planning Forum also exists (considering mainly the latter). These may need to 'come together' for a more co-ordinated approach to environmental sector skills planning in South Africa in future.

Role of workplace

Workplaces have a critical role to play in mediating young graduates into the development of professional competency. Enabling progression along pathways, especially at low to intermediate skills, is very reliant on the workplace and diverse and creative mechanisms for workplace learning and human resource development. The study by Mohanoe (2013) identified that aspects such as power relations and access to power and resources in the workplace, high quality and futures-oriented internal skills planning and Human Resources Management systems, workplace mentoring systems and visionary management are critical factors for enabling a workplace where learning pathways are opened up, and not simply left absent or neglected.

Focusing on the whole pipeline

For a mixed skill sector to emerge there is a need to focus on the whole pipeline – starting at schools. Career guidance around non-academic based careers is critical. Generating public knowledge and interest around diversity of technical jobs is also critical, as is a recognition of their importance to society and sustainable development. The attention given to training technicians associated with South Africa's Green Drop programme is perhaps an example of such a process where the technical skills are seen to be highly valuable to society as a whole.

Conclusion

I have explored connections and disconnections between systems of provisioning (such as institutional arrangements) and systems of practice (such as the actual job roles) and have argued that the absence of clear labour market information, coupled with institutional and structural constraints, has opposed the emergence of clear occupational differentiation for environmental occupations. This perpetuates ongoing problems of education and training system provisioning for a sector that is struggling to differentiate occupational roles (professional, technical and elementary), and leaves it unable to clearly depict occupational progression and career paths.

This scenario constrains the emergence of a well-articulated provisioning system within the environmental sector as it hinders the emergence of new or parallel learning pathways from technical into professional pathways and by not providing connecting points for artisan pathways. The emerging learning pathway picture is thus one that favours single, traditional academic pathway trajectories constraining broad access, mobility and progression with negative social justice, and social-ecological constellation impacts as outlined above.

Through the critical realist analysis, the chapter provides for transformative praxis possibilities within a systems wide approach to learning pathways development and actualization. I have discussed a number of key leverages of change above, e.g. occupational differentiation; better labour market information; improved skills planning; effective coherent provisioning that can all enable multiple entry multiple exit and support an improved pathways model. The final caveat is related to the nature of social-ecological systems change. It is rapidly advancing, affecting societies and economies everywhere and rapid, yet reflexive, engagement with these changes is necessary. Responding within this system with clear social justice objectives is necessary for a view of sustainability that encompasses the flourishing of all, and may also be necessary for shifting the power$_2$ relations of the elite in managing South Africa's environment and gaining access to green economy opportunities. As shown in this study, an open and emerging system can learn internally from other parts of a system through identifying absences that may be present elsewhere.

Acknowledgements

This chapter draws on research conducted for a PhD study conducted by the author in the Environmental Learning Research Centre at Rhodes University. The research was funded by the South African Qualifications Authority, via a research partnership with Rhodes University.

Notes

1 It is noted here that environment and sustainable development is actually cross-cutting across various economic and other sectors, but for purposes of understanding the problems related to the emergence of education and training system responses, we use the term 'sector' here.
2 The environmental sector does not have an equivalent professional body. It has a number of affiliated professional bodies that deal with some aspects of the sector's supply and demand issues. This is proving to be inadequate for substantive analysis of supply and demand, and for in-depth knowledge of skills needs and planning.
3 The environmental sector currently has no such understanding of demand.

References

Allais, S. (2009). *The changing faces of the South African National Qualifications Framework.* Employment Working Paper 45: Learning from the first qualifications frameworks. Switzerland: International Labour Organisation (ILO).

Archer, M. (2000). *Being human: The problem of agency.* Cambridge: Cambridge University Press.

Archer, M. S., Bhaskar, R., Collier, A., Lawson, T., and Norrie, A. (1998). *Critical realism: Essential readings.* London: Routledge.

Bhaskar, R. (1975). *A realist theory of science.* London and New York: Routledge.

Bhaskar, R. (1991). *Philosophy and the idea of freedom.* London: Verso.

Bhaskar, R. (1993). *Dialectic: The pulse of freedom.* London: Verso.

Bhaskar, R. (2010). Contexts of interdisciplinarity: Interdisciplinarity and climate change. In R. Bhaskar, C. Frank, G. Hoyer, P. Naess, and J. Parker (Eds), *Interdisciplinarity and climate change: Transforming knowledge and practice for our global future* (pp. vii–xiii). London: Routledge.

Bhaskar, R., and Parker, J. (2010). Introduction. In R. Bhaskar, C. Frank, G. Hoyer, P. Naess, and J. Parker (Eds), *Interdisciplinarity and climate change: Transforming knowledge and practice for our global future* (pp. 1–24). London: Routledge.

Case, J. (2006). *Issues facing the engineering profession and engineering education in South Africa: Briefing paper to inform further research.* Paper commissioned by the Human Sciences Research Council, Pretoria.

CATHSSETA (Culture, Arts, Tourism, Hospitality and Sport Sector Education and Training Authority). (2012). *Culture, Arts, Tourism, Hospitality and Sport Sector Education and Training Authority sector skills plan 2012/13.* Johannesburg.

Cloete, N., and Butler-Adam, J. (2012). Introduction. In H. Perold, N. Cloete, and J. Papier (Eds), *Shaping the future of South Africa's youth: Rethinking post-school education and skills training.* Cape Town: African Minds.

DEA (Department of Environmental Affairs, South Africa). (2010a). *Environmental sector skills plan for South Africa: A systems approach to skills planning and human capital development. Summary Document (second draft).* Pretoria.

DEA (Department of Environmental Affairs, South Africa.). (2010b). *Comment on the Revised Seta Landscape and NSDS iii: Environmental Sector Submission* (first draft). Pretoria.

Death, C. (2014). The green economy in South Africa: Global discourses and local politics. *South African Journal of Political Studies, 41*(1), 1–22.

DED (Department of Economic Development, South Africa). (2011). *New Growth Path: Accord 4 – Green Economy Accord.* Pretoria.

DHET (Department of Higher Education and Training, South Africa). (2010). *Strategic Plan 2010–2015.* Pretoria.

DHET (Department of Higher Education and Training, South Africa). (2012). *Guidelines: Organising framework for occupations.* Pretoria.

EWSETA (Energy and Water Sector Education and Training Authority). (2012). *Energy and water sector skills plan (2011–2016).* Johannesburg.

Hanrahan, H. (2000). *A nationally integrated model for qualifications, registration and licencing in the engineering profession.* Paper presented at the South African Conference on Engineering Education, Vanderbijlpark, 28–29 September 2000.

Hoppers, W. (Ed.). (2009). *Post primary education in Africa: Challenges and approaches for expanding learning opportunities.* Synthesis prepared for the 2008 ADEA Biennale on education in Africa, Maputo, Mozambique, 5–9 May 2008. Tunis: Association for the Development in Africa (ADEA).

HSRC (Human Sciences Research Council). (2009). *Guidelines towards a Human Capital Development Strategy in the biodiversity sector.* Research report. Lewis Foundation and the South African National Biodiversity Institute (SANBI), October 2009. Pretoria.

Kraak, A. (2008). Incoherence in the South African labour market for intermediate skills. *Journal of Education and Work, 21*(3), 197–215.

LGSETA (2012). *Local Government Sector Education and Training Authority sector skills plan (2012–2017).* Johannesburg.

Lotz-Sisitka, H. (2011). *Change-oriented workplace learning and sustainable development practices. Phase 2: Understanding learning pathways and systems of work and learning across the NQF landscape.* Unpublished research programme proposal, Rhodes University Environmental Learning Research Centre, Grahamstown.

Lotz-Sisitka, H. and Ramsarup, P. (2012). Absenting absences in learning pathways: A methodological exploration. SAQA/Rhodes University discussion paper.

Lotz-Sisitka, H., Mohanoe, N., Ramsarup, P., and Olvitt, L. (2012). Boundary making and boundary crossing in learning pathways access and progression: Voices from the workplace. Rhodes University, Grahamstown.

Maia, J., Giordano, T., Kelder, N., Bardien, G., Bodie, M., Du Plooy, P., Jafta, X., Jarms, D., Kruger-Cloete, E., Kuhn, G., Lepelle, R., Makaulule, L., Mosoma, K., Neoh, S., Netshtomboni, N., Ngozo, T., and Swanepoel, J. (2011). *Green jobs: An estimate of the direct employment potential of a greening South African economy.* Pretoria: Industrial Development Corporation, Development Bank of Southern Africa, Trade and Industrial Policy Strategies.

Martinez-Alier, J. (2002). *The environmentalism of the poor: A study of ecological conflicts and valuation.* Cheltenham: Edward Elgar Publishing.

Mohanoe, N. (2013). *Learning pathways of key occupations relevant to sustainable development in Makana municipality.* Unpublished MEd thesis, Rhodes University, Grahamstown.

Norrie, A. (2010). *Dialectic and difference: Dialectical critical realism and the grounds of justice.* London: Routledge.

Perold, H., Cloete, N., Papier, J. (2012). *Shaping the future of South Africa's youth: Rethinking post-school education and skills training.* Cape Town: African Minds.

Ramsarup, P. (2015). *A dialectical understanding of environmental learning pathways: Articulation and access.* Draft PhD thesis, Rhodes University Grahamstown.

RSA (Republic of South Africa). (2009). *National Qualifications Framework Act No. 67 of 2008.* Pretoria: Government Printers.

Shipway, B. (2011). *A critical realist perspective of education.* London: Routledge.

8 The emergence of environmental ethics discourses in laminated, open systems

Some educational considerations

Lausanne Laura Olvitt

Introduction

The broad intention of this chapter is to stimulate deeper, more critical awareness of environmental ethics discourses circulating in the field of environmental education. It proposes that a critical realist understanding of the emergence of discourses in open, laminated social systems can develop our reflexive practice as environmental educators regarding the language, images and other discursive elements of educational practice. The chapter focuses on how environmental discourses convey both tacit and explicit messages about what is (or is not) 'of value' in the world, and messages about which actions can be regarded as 'good' in relation to those values. Taking a critical realist approach, the chapter explores how discourses emerge relationally within complex, laminated, open social systems. When understood in this way, discourses come to life as dynamic and generative mechanisms able to shape environmental learning processes in powerful ways, ranging from ideological imposition to reflexive ethical deliberation.

It is important for environmental educators to be able to identify such discursive patterns in their own educational work and consider their potential implications for learning in contexts of social-ecological risk and change. To this end, in the second part of the chapter, I propose an analytical framework informed by critical realist concerns for relationality and causality that might be used by other environmental educators seeking to engage with environmental ethics discourses.

Environmental ethics discourses and learning

Discourses are shared ways of understanding and sharing ideas about the world. They allow their subscribers to develop meaningful 'stories' from the fragmented experiences and bits of information in the world around them, thereby helping to define common sense and legitimize particular forms of knowledge or perspectives on some aspect of the world (Dryzek, 2005; Fairclough, 2003).

Environmental discourses, in particular, are the ways in which groups of people talk about their environments and articulate arguments about past, present or future people-nature relationships (Mülhäusler and Peace, 2006). The intersecting complexity of the biophysical world and human social systems has led to a proliferation of perspectives on – and hence ways of talking and writing about – environmental issues, particularly since the modern environmental movement emerged in the 1960s. Despite these varying perspectives that make claims about the natural world, 'there is no single uninterpreted "nature" capable of putting an end to political dispute' (Dryzek, 2005, p. 12).

Consider, for example, some of the ways people talk about tanzanite, the rare blue-violet gemstone mined from the foothills of Mount Kilimanjaro, Tanzania. The Tanzanite Foundation (2014) describes the gemstone thus: 'A thousand times rarer than diamonds and with a little over a decade of mine life remaining, tanzanite is the fashion gem of the millennium' (www.tanzanitefoundation.org/tanzanite_history). Elsewhere, tanzanite is described as 'the gemstone that brings visions. It has a special ability to connect one with the spiritual world and bring that connection into one's life in the form of knowledge' (www.stephenkris.com/the-mystery-of-tanzanite). However, Dutch journalist Adriana Stuijt asks:

> Do the American women adorned with these stunning and unique gems even know that most of these were torn out by African children's hands, digging and hacking away at the Tanzanian volcanic rock – often forced to live in deep, unsafe mineshafts where many have already drowned horribly?
>
> (https://humangoods.wordpress.com/page/5/?option=com_content&view=article&id=7038)

Each of these representations of tanzanite reflects different values and concerns which, when circulated in informal or formal learning situations, can influence people to endorse, oppose or remain indifferent to the tanzanite mining industry. In the textual examples given here, we can identify (1) a discourse of socio-economic elitism (portraying tanzanite as a rare 'fashion gem' that people should buy now before the limited source is depleted); (2) a discourse of spiritual connection to the earth (because tanzanite 'brings visions' and connects its wearers to the spiritual world); and (3) a discourse of social justice (that portrays the wealthy tanzanite mining industry and those who buy the gems as conspirators in child slave labour and the exploitation of poor people in Africa). There is a close relationship between the knowledge that people access about tanzanite, the values and ethical responses that they develop in relation to the gemstone, the diversity of discourses they encounter, and the power ascribed to each discourse in its socio-cultural setting.

Similarly, multiple and often conflicting discourses related to all manner of environmental topics are circulated and taken up in environmental education programmes and activities. When educators make decisions about what texts[1] to include in their teaching, some discourses get privileged over others and shape the potential of teaching-learning interactions to stimulate critical thinking, agency and ethico-moral reflexivity. Educators like myself (consciously or otherwise) select, draw on and advance discourses about waste management practices, genetically modified organisms, energy use, water access, carbon trading, poaching, ecotourism, fracking, organic food production, logging, vegetarianism, dam construction and so on. Children's and adults' developing understandings of what is 'of value' and what counts as a 'good' response in relation to these topics can be substantially influenced by the choice of words, examples, metaphors and images used when talking about them.

Consider how people might learn differently about forests, depending on their choice of which of the following training programmes[2] to attend:

- 'Forest Finance and Valuation: What every forester should know';
- 'Caring for your Woodlands';
- 'Governance for Forest, Nature and People'.

Despite being only a few words long, each course's name reflects a different discourse about forests and people's relationships with them. The discourses range from valuing forests as *economic commodities* (implied through terms such as 'forest finance' and 'valuation'), to regarding forests and woodlands as needing *stewardship* (implied through the phrase 'caring for'), to recognizing forests as part of *broad, social-ecological systems* (implied through the combination of the words 'forest', 'nature' and 'people') which, it would seem, all require 'governance'.

These brief examples illustrate how each discourse is embedded in language and symbol which, as noted by Harré *et al.* (1999, p. ix) 'not only reflects and records but also shapes, distorts and even creates realities'. Because discourses function as sets of rules that determine what should and shouldn't be said in particular social settings (Elder-Vass, 2010), they are imbued with political power, able to influence definitions, perceptions and values, attract funding, and advance or suppress certain ideas. This view is supported by Harper (2001, p. 101) who notes that environmental discourses often get deployed by governments, scientific organizations and big corporations 'in ways that co-opt grassroots concerns and marginalise public participation in decision-making'.

Fairclough (1985, p. 750) describes how social institutions have their own ideological norms and discourses which 'impose ideological and discursive constraints on [their subjects] as a condition for qualifying them to act as subjects'. For example, to be or become an environmental educator, a person must align generally with the discursive and ideological norms that the broad field of environmental education and, more specifically, that the employing institution endorse as appropriate. Over time, an institution's or community's particular people-nature relationships and the ideologically-shaped practices associated with them become normalized and discursively distinct. Left undisputed or unexamined, these discursive norms wield power, becoming common-sensical and taken-for-granted within the institution. If an environmental education organization's commitment to engage vibrantly and reflexively with the diversity of environmental ethics positions diminishes, it risks presenting ideologically narrow(ing) educational programmes that uncritically uphold certain environmental perspectives and practices (or groups of people associated with them) as more morally acceptable than others.

I argue that the inclination and ability to analyse discourses is educationally significant because it allows educators (such as curriculum developers, textbook authors, classroom teachers, community liaison officers, agricultural, forestry and conservation extension officers, education officers in museums, botanical gardens and national parks, and so on) to understand and critique the power of the words and images they use to represent people's relationships with the natural world. In earlier work, I have noted that the nature of ethics-oriented learning remains poorly understood and that people's values associated with environmental practices are 'commonly taken for granted, under-examined or contradictory' (Olvitt, 2010, p. 73). These perspectives resonate with the work of Jickling *et al.* (2006, p. 3) who state:

> How we approach the world, how we come to know it, and how we make meaning of our experiences are all shaped by our values, even if we don't notice this on a day-to-day basis. These values are sometimes called assumptions and are at the core of everything we do; and, they are at the core of all subject areas – even science.

Environmental education activities, programmes and publications (not unlike all human activities) are all, in some way, values-based and conceptually laden; they mediate learning through environmental and ethical discourses which may be at odds with learners' deeply embedded histories, cultures and practices. Gonzalez-Gaudiano and Buenfil-Burgos (2009, p. 101) express a similar concern that:

> the majority of environmental educators do not examine the ideological discourses underlying the educational projects they are putting into practice and, as a consequence, their daily work is governed by political inconsistency and immediatism.... Such a constitutive bias shows an environmental education with little capacity to contribute efficiently to the complex challenge of environmental deterioration and to have its intrinsic social and political dimensions articulated.

Symbol systems, language and orders of discourse are central, causally efficacious features of the social world which, if taken for granted or overlooked, may result in oversimplified or incomplete analysis. In the sections to follow, I explore how a critical realist approach to analysis of environmental ethics discourses can allow for more complete analysis. Such a position builds on the work of other critical realist researchers (Parker, 2004; Fairclough, 2005; O'Regan, 2006; Sims-Schouten *et al.*, 2007; Price, 2007) who agree that critical discourse analysis as a research methodology is highly compatible with a critical realist ontology.

A critical realist account of environmental ethics discourses

In this section, I draw on four critical realist propositions that I consider fundamental to understanding the nature and causal powers of environmental ethics discourses in learning processes. These interlinked propositions conceptualize discourses as (1) 'uniquely laminated structurata' (Bhaskar, 1993) that are hence (2) relational and (3) emergent in (4) open social systems. Due to the inseparability of these theoretical propositions, I discuss them below as intertwined and mutually defining rather than sequential concepts.

I begin by making a case for discourses to be understood as 'uniquely laminated structurata' (Bhaskar, 1993, p. 404). Most basically, a structuratum is 'any structured thing' (Hartwig, 2007, p. 442) which is more fully defined by Bhaskar (1993, p. 50) as:

> a concrete individual or singular, which will normally be the condensate of, or the effects of, a multiplicity of disjoint, and even contradictory, structures or of their ways of acting.... [A structuratum] will characteristically remain heteronomously conditioned, dependent upon and influenced by the levels out of which it has emerged, even where it is causally efficacious on them.

In the case of discourses, this 'multiplicity of disjoint, and even contradictory, structures' refers to the discourses' constituent parts that range from the *micro* level (for example, internal relations between words, punctuation, grammatical forms, imagery, typesetting, tone and so on) and the *macro* level (cultural context, economic structures, socio-political conditions and so on). As structurata, discourses are understood to be 'uniquely laminated' (Bhaskar, 1993) because distinct mechanisms at multiple levels interrelate in open social systems to produce unique (that is, contextually emergent) and contingent results that cannot, as such, be predicted.

The social systems in which discourses emerge are described here as 'open' because, drawing on the work of Archer (1995), human agents – who have personal emergent properties (PEPs) – interact with structural emergent properties (SEPs), such as economic frameworks or post-colonial governance structures, and cultural emergent properties (CEPs), such as people-animal relationships, gender roles, or food preferences. The outcomes of these interactions are always contingent, thereby characterizing social systems as change-oriented and ever-open to possibility. As such, discourses – like other social entities – are understood to emerge in stratified, open social systems. They are structured and enduring wholes that interrelate in indeterminate ways at any given moment to have an effect on the world (Elder-Vass, 2004).

For this reason, Fairclough (2005) advocates for a relational view of discourses, noting that discourses cannot be analysed effectively as self-contained entities but rather in terms of their complex relations with non-discursive elements of the social world. He draws on a critical realist 'dialectical-relational social ontology' to propose that discourse analysis involves examining the *textual elements* of 'the relationship and tension between pre-constructed social structures, practices, identities, orders of discourse, organizations on the one hand, and processes, actions, events on the other' (2005, p. 923). To return illustratively to the mining of tanzanite: the conflicting discourses associated with extraction of these gems are best analysed dialectically and relationally. How might the persuasiveness of the phrase 'fashion gem of the millennium' be more fully understood by considering it in relation to the dominant socio-economic structures in Tanzania (the source of the gemstone) and America (its primary market)? What cultural norms, tensions and silences are reinforced here? Who reads this text, and what else might they have read and experienced before that will influence the way they understand and respond to this representation?

Critical realism's dialectical-relational, open systems ontology enables us to understand that environmental ethics discourses do not occur only at the conceptual or ideational level but are bound up in the socio-material world, in particular its social-ecological dimensions. Conversely, people's material practices in relation to the natural world are affected by the causal powers of circulating environmental ethics discourses. Material practices (such as how we treat animals, how we manage our waste, how we grow our food) are independent of, yet relationally implicated in, discursive practices (that is, how we represent and talk to one another about these things). Similarly, concept-dependent discursive practices always have a material dimension (Bhaskar, 1989, p. 4). Fairclough *et al.* (2004, p. 27) argue that social scientists need to locate discourses (and other meaning-making processes) 'within their necessary dialectical relations with persons (hence minds, intentions, desires, bodies), social relations, and the material world – locating them within the practical engagement of embodied and socially organised persons with the material world'.

As noted earlier, Bhaskar explains that each structuratum (such as an environmental ethics discourse) is 'emergent', that is, 'dependent upon and influenced by the levels out of which it has emerged, even where it is causally efficacious on them' (1993, p. 50). Emergence is an important concept for understanding how environmental ethics discourses come to exist and have causal efficacy in social systems, in particular through environmental education organizations, programmes and publications. Philosophically, the concept of emergence is based on the premise that 'a whole has properties or powers that are not possessed by its parts' (Elder-Vass, 2005, p. 316). It presupposes (1) that the existence of a certain whole (a substance, entity, property or system) which we will call β is dependent on some other substance, entity, property or system which we will call α; (2) that interrelations between these mean that fundamental changes in α lead to fundamental changes in β; and (3) that the existence of β (that is, its form, operation and consequences) cannot be reduced to α (Hartwig, 2007).

That causal powers emerge (contingently) from their relations with other causal powers in a complexly interacting, laminated world is the essence of a critical realist understanding of emergence. For Bhaskar (1993, p. 73), emergence is

> the relationship between two terms such that one diachronically, or perhaps synchronically,[3] arises out of the other, but is capable of reacting back on the first and is in any event causally and taxonomically irreducible to it, as society is to nature or mind to matter.

Elder-Vass (2005) argues that emergence can be understood best by identifying the synchronic relations between a whole and its parts:

> The point of emergence is that it is that way that a set of entities is related to each other at a given point of time that determines the joint effect they have on the world at that moment. Emergence, then, is a synchronic relation amongst the parts of an entity that gives the entity as a whole the ability to have a particular (diachronic) causal impact. The relation between a whole and its parts is thus a relation of composition, and not of causation.
>
> (p. 231)

Understanding these diachronic (occurring over time) and synchronic (occurring at one point in time) relationships is important in analysing the emergence of discourses. Where diachronic analysis of the relations between the parts of a discourse enables researchers to trace *change over time*, synchronic analysis enables researchers to identify and describe the unique combinations of the entity's parts *at the particular moment* of social elaboration (that is, to understand the 'knot' of events and mechanisms that enabled the change to occur in the particular way that it did). Researchers must be able to describe, for example, textual elements such as grammatical forms, vocabulary, tone and imagery and understand them in their own terms before providing any synchronic or diachronic explanatory reduction to whatever causal powers the emergent discourses may hold at a societal level. This causality is not linear and cannot be predicted because, according to Bhaskar and Norrie (1998, p. 601), 'events are not determined before they are caused'. The causal powers of discourses in social systems are emergent from (but not reducible to) the textual entities on the printed page or digital screen, and are capable of reacting back on them, although such relations cannot be predicted before they are actualized as events.

Having given a critical realist account of environmental ethics discourses as relational and uniquely laminated structurata that are emergent in open social systems, I now consider how these ideas might usefully guide environmental educators to examine discourses in their own educational settings.

A framework to analyse the emergence of environmental ethics discourses

As explained in the previous section, synchronic analysis of discourse allows researchers to scrutinise the 'knot' of a particular discourse's multi-levelled, relationally bound components. Understanding a particular discourse's causal impact over time (achieved through diachronic analysis) is dependent on the researcher being able to give an account of the synchronic relations amongst the parts of the discourse. To this end, I draw primarily on the work of Fairclough (1989) who emphasizes that the relationship between texts and social structures is complex, relational, indirect and mediated by human agents. His concern is for understanding not only what discourses are like (that is, providing a descriptive account), but also 'why' and 'how' they develop (an explanatory account). This concern reflects a retroductive process that is consistent with critical realism's commitment to pursuing causal explanations by looking 'behind' phenomena and their regularities to consider what produces them. Fairclough's (1989) framework for critical discourse analysis analyses the relationships between texts, processes and their social conditions; what he calls analysis at the levels of text, interaction and context.

The resultant synchronic analysis of an environmental ethics discourse would entail attending to:

- *the textual level:* the particular words, language, format, grammatical structures, images, examples and socio-cultural references that have been used;
- *the level of interaction:* mediating interactions between the text's particular readers, educators and their resources;
- *the macro social-ecological level:*[4] the particular historically constituted forms of knowledge, politico-economic and socio-material circumstances, socio-cultural norms and social-ecological dynamics that constrain or enable interactions with the text.

These multi-levelled and multi-disciplinary components interrelate 'in the knot of the moment' to produce a contextually particular way of thinking and talking about people-nature relations and the kinds of responses that are 'good' or 'right'. In other words, their unique, synchronic relations enable the emergence of environmental ethics discourses. From this vantage point, the diachronic relations of the discourse can be considered by asking what structures and mechanisms have preceded this moment, and what future possibilities might shape/be shaped by it.

Below, I present a three-stage analytical tool based on Fairclough's framework of text, interaction and context. I apply it, for illustrative purposes, to a textual extract from a South African teacher professional development programme in environmental education. The analysis has been guided and inspired by a range of analytical questions and processes, including Fairclough's seminal work of 1989, O'Regan's (2006) 'text as critical object' approach, Price's (2007) critical realist 'explanatory critique' framework, and the author's earlier development of an analytical tool focused on environmental ethics discourses (Olvitt, 2012) which drew heavily on Kronlid and Öhman's (2013) environmental ethics framework in order to bring to the fore a concern for people-nature relations and values. These researchers share a concern for examining how language conveys broad, historically and socio-culturally embedded meanings and thereby validates knowledge and influences social power gradients.

As a research methodology, critical discourse analysis is thus generally associated with deconstructive critique towards emancipatory or transformative ends. For example, researchers may analyse texts to reveal how political and economic hegemonies or patterns of sexist, racist, classist social norms are reproduced or transformed through discourse. Although the immediate goal of the analytical tool shared in this chapter is perhaps more modest (to identify environmental ethics discourses circulating in an environmental education project, institution or community), its transformative potential may well be significant if the discourse markers identified through the tool are brought into wider reflexive processes at the institutional or communal level.

Fairclough's (1989) procedure for analysing discourse begins with a *descriptive* stage in which textual features significant to critical discourse analysis are identified and described, for example, the interrelationships of words, metaphors and other imagery; tone; and grammatical features such as the use of punctuation, tense, imperatives, nominalizations and active and passive voice.

The *interpretive* stage is important because 'the values of textual features only become real, socially operative, if they are embedded in social interaction' (ibid., p. 140). This position is reflected in Fairclough's later work, following his uptake of a critical realist ontology, where he explains: 'a critical realist discourse analysis is not merely concerned with languages and orders of discourse; it is equally concerned with texts as (elements of) processes, and with the relations of tension between the two' (Fairclough, 2005, p. 923). The interpretive stage thus involves identifying which textual features resonate with the individual cognitive resources of the text producers and text interpreters, and identifying the dialectical interplay of textual 'cues' and the interpreters' ideological and common sense assumptions and background knowledge.

The objective of the *explanatory* stage of critical discourse analysis is to give an explanatory account of 'the relationships between transitory social events (interactions), and more durable social structures' (Fairclough, 1989, p. 27). The researcher's objective is to gain insight into the complex relations between structures, events, social practices and the causal powers of social agents that together constitute discourses as uniquely laminated and causally efficacious structurata.

Table 8.1 presents a developmental analytical tool for educators who are interested in the emergence of environmental ethics discourses in their contexts but who do not have much experience with textual or discursive analysis. The tool is structured to guide the researcher to examine a text, its interactions, and the way its social-ecological context has provided the preconditions for those interactions.

To illustrate how the analytical tool presented in Table 8.1 can be applied to a piece of text, I now present an analytical discussion of a short extract from the introductory section to an environmental education teacher professional development course[5] (see Block 1). The text was selected because of its function as an orientating text in the introductory section of the course handbook. This extract aims merely to illustrate one way an analytical tool can guide detailed analysis of text, interaction and context. Discussion is limited to features associated (more or less directly) with environmental ethics discourse as this is the focus of this chapter. Extension to analyse distinct political, educational or scientific discourses go beyond the scope of this chapter and would require an adapted analytical tool. Brief textual extracts such as the one analysed here provide useful starting points to examine environmental ethics discourses. However, researchers should recognize that a more substantial enquiry would involve, first, analysis of a much wider selection of texts from the environmental education programme under consideration and, second, extension of the multi-dimensional analysis of the production, interpretation and use of the text than is modelled here. The discussion that follows is guided by Fairclough's (1989) recommended sequence of description, interpretation and explanation.

Table 8.1 Analytical tool to guide a critical analysis of environmental ethics discourses

Level	*Orientating questions*
TEXT	*1.1* What is the topic of this text? *1.2* How is the topic presented (formal, informal, persuasive, aggressive, humorous, sympathetic etc.)? *1.3* What is the preferred reading of this text (i.e. the meaning that corresponds most closely to the author's intention)? *1.4* What vocabulary seems significant in this text (e.g. words that are especially emotional, distant, nostalgic; words that are repeated; use of modal verbs [should, must, could, can]; use of personal pronouns [you, we, I])? *1.5* Are any words capitalized, italicized, underlined or put in inverted commas? What effect does this have on the meaning? *1.6* What tenses are used and do these have an effect on the tone and meaning of the text? *1.7* Who or what is the focus of concern in the text? How do the textual components establish this? *1.8* What information is put first, or otherwise prioritized in the text? *1.9* What metaphors, symbols and examples are used? What kinds of ideas or feelings do they evoke? *1.10* How is the text organized visually? (e.g. words only? words and pictures? Are there pictures, cartoons or photographs? What are the effects of these visual devices?) *1.11* When are active and passive constructions used? What is generally backgrounded/foregrounded in these constructions? Who are the 'do-ers' – people, animals, objects, places?
INTERACTION	*2.1* Are there any tensions or contradictions between the components of the discourse, as described above? *2.2* What assumptions are evident in the text regarding the kind of society people aspire to live in? *2.3* Who or what is the moral object of this text (i.e. what is presented as the focus of moral concern and responsibility? Individual organisms? Habitats? Global or local humans? Past/present/future human generations? Poor humans? Wild/domestic animals? etc.) *2.4* Overall, how does the text represent people-nature relationships? (e.g. is the relationship exploitative, mutualistic, managerialist, spiritual, oppositional, restorative etc.?) *2.5* How does the text represent the value of nature? (e.g. nature has intrinsic worth which is dependent/independent on how people construct it; nature has value in so far as it meets human needs; or the value of nature is indistinguishable from the value of humans etc.) *2.6* How are the causes of the environmental crisis constituted through this text? *2.7* How are solutions to the environmental crisis constituted through this text? *2.8* Do any of the textual components echo/resonate with/contrast other relevant and influential texts such as local or international policies and strategic frameworks? 2.9 What is *not* said in this text? (i.e. are there any omissions, silences, absences? Might they be strategic?)
EXPLANATION IN SOCIAL-ECOLOGICAL CONTEXT	*3.1* What power relations at situational, institutional and societal levels help shape this discourse (e.g. political, economic, gendered, cultural, racial, religious, geographical power relations)? *3.2* Is there evidence of ideological bias in this discourse? To what/whom can these ideologies be traced? *3.3* Does this discourse contribute to reproducing or transforming social and cultural structures associated with environmental problems? Are these contributions overt or covert?

Source: Developed by the author.

Block 1 Textual extract from the introductory section of the teacher professional development programme's handbook

Building a better world

Since the Industrial Revolution began in the late 18th century, 'development' has provided humankind with numerous benefits, such as modern medicine, housing, transport and communication systems. However, progress has also caused problems, as non-renewable resources have been over-extracted, huge amounts of waste have been created, and pollution has impacted on the health of people and the environment. Most people are now aware that human actions are changing the climate in unpredictable ways. Societies around the world must 'adapt' and change their practices for a 'low carbon' future. But do most people actually understand what this means?

Massive over-consumption of resources and continued environmental degradation are undermining the natural systems we depend on, and impacting most severely on the poor and marginalized people in our society. How should teachers respond to these challenges?

One thing we can do is to give children hope that together we can 'build a better world'. We can use our imaginations and problem-solving skills to creatively develop diverse solutions to our problems, and to change how we live for the better. But do most teachers understand the issues well enough to be able to guide such useful processes?

What would 'a better world' look like to you? Some of the words people are using to describe this 'better world' – and the ways in which we might build it together – include:

- Equity
- Sustainability
- Resilience
- Adaptation
- Transformation

Stage one: description of the text

Analysing discourse at the textual level requires close attention to the interrelationships of words, metaphors and other imagery, and grammatical features such as use of tense, imperatives, and active and passive voice (Fairclough, 2003). In the extract in Block 1, these elements combine to produce a general discourse of social care and optimism in the face of pressing social-ecological problems. The focus of the text is on 'building a better world'. Besides the title, the word 'better' is used four times in six lines, thereby establishing that positive societal transformation is the central concern of this text and, by implication, of this professional development course. The authors' assumption appears to be that this 'better world' is to be achieved through changed lifestyles, low carbon practices, reduced consumption, creative and problem-solving teachers, equity and sustainability. Beyond these 'cues' as to what might constitute a better world, the authors leave a more precise definition of a better world open to exploration and invention; teachers are invited to think about: 'What would a better world look like to you?'.

In terms of tone and intention, the extract seems to operate in two parts: the first two paragraphs describe how industrial development and patterns of unsustainable production and consumption have resulted in an environmental crisis that threatens the very systems on which people depend for survival. Equity-oriented social change is needed to turn the situation around. Here, there is a predominance of negative vocabulary and descriptions ('problems'; 'over-extracted'; 'huge amounts of waste'; 'pollution'; 'changing the climate in unpredictable ways'; 'massive over-consumption'; 'continued environmental degradation'). This is in stark contrast with the second part of the extract (paragraphs three and four) which takes a more personal, direct and optimistic approach, encouraging teachers to develop a concern for equity and social-ecological sustainability and to teach in creative, problem-orientated ways to build a better world with and for the children they teach. Here, the vocabulary is substantially more positive and encouraging than the preceding paragraphs. The word 'can' is used three times in paragraph three. The focus is on 'hope', building a 'better world', and finding 'diverse solutions' through imagination, creativity and problem-solving skills.

The shift in position between parts one and two is further established through the use of tense. In part one, mostly past and present tense are used to describe current environmental degradation and its causes. Part two, however, uses present and future tense, for example: 'We can use our imaginations and problem-solving skills to creatively develop diverse solutions to our problems, and to change how we live for the better' (paragraph three). This shift in tense suggests that unsustainable practices happened in the past but teachers are now associated with brighter prospects into the future.

In the first part of the text, the authors have used a grammatical construction known as 'nominalization' which obscures the identity of the action-takers within a sentence. In paragraph one, 'development' is presented as the subject of the sentence that has provided modern medicine, housing, transport and so on and it is 'progress' that has caused serious environmental problems. 'Development' and 'progress' – here seemingly disassociated from people – are identified as the cause of people's problems. Similarly, in paragraph two, the sentence structure distances people (us) from responsibility for the degradation of natural systems. The subject of the sentence (what or who is responsible for 'undermining' natural systems) is obscure. There is no 'us', 'you' or even 'them' and the subject is instead presented vaguely as 'massive over-consumption of resources and continued environmental degradation'.

Specialists in eco-linguistics such as Goatly (2001) and Schleppegrell (2001) agree that nominalizations such as these are a common feature of environmental reporting and environmental education texts. Apart from shifting environmental responsibility to an undefined subject, this grammatical form also blurs the distinction between an environmental *process* (such as how some people's patterns of consumption drive unsustainable production) and an environmental *thing* (such as 'massive overconsumption'). A possible outcome of the nominalizations used in this text is to reduce teachers' feelings of guilt or complicity in the shadow-side of development; the main purpose of the text is to gain the support of teachers and to inspire creative and caring responsiveness through education, not to burden them with guilt for global or national environmental problems.

In line with this, the focus shifts in the second part of the extract from describing a crisis-filled world to calling for a 'better world'. Here, nominalizations are no longer used and the authors' approach to using personal pronouns changes distinctly. The personal pronoun 'we' is used five times in six lines, 'our' is used twice and 'you' once. This direct and active sentence construction, together with positive vocabulary such as 'hope', 'better world', 'imaginations', 'problem-solving skills', 'solutions' and 'useful' establishes an inspiring, optimistic tone which is in stark contrast to the detached and generalized problem-statement of the first section. This section places responsibility clearly with teachers; it positions them as creative problem-solvers and willing agents of change who simply need more knowledge to unlock their potential to make the world a better place.

To understand the emergence of structurata such as environmental ethics discourses, researchers need to understand the synchronic relations between the discourse and its parts. For this, it is necessary to consider the textual elements such as those described in this section in relation to the *situational* level (that is, the people who produced, received or otherwise interacted with that text, the relational dynamics and socio-material conditions of those interactions) as well as the *macro social-ecological* level (that is, the interactions between broad social, cultural, political, ecological and economic systems and structures). Stage two below sketches an interpretation of the text at the situational level, and Stage three thereafter provides a starting point for social explanation through consideration of the macro level.

Stage two: interpretation of the text

As noted previously, the purpose of the interpretation stage is to identify the dialectical interplay of textual features such as the above-mentioned words, images and grammatical features, and the ideological and common sense assumptions and background knowledge of those who produce and interpret them. In this discussion, I focus in particular on those elements that relate to environmental ethics discourses, as guided by Questions 2.1–2.9 in Table 8.1.

One main assumption in the text is that teachers, as the interpreters of this text, aspire to live 'in a better world', and that such a world is characterized by 'equity, sustainability, resilience, adaptation and transformation'. Teachers are invited to participate in elaborating this notion of a better world ('What would a "better world" look like to you?') and the starting point and tone of such an exploration is guided strongly by these listed characteristics.

Listing 'equity' as the first characteristic of a better, more sustainable world may be unusual for many teachers interpreting the text. The emphasis on equity disrupts traditional approaches to environmental education which have more commonly foregrounded moral imperatives such as 'respect for nature', 'self-restraint' and 'nature conservation'. The symbolic power of this subtle shift, from what Cock (2007, p. 2) describes as 'the "victim blaming" conservation discourse' of the apartheid regime, is especially pertinent in the post-apartheid context in which this text circulates.

Kronlid and Öhman (2013) explain that the moral object of a text or social process is who or whatever holds intrinsic value or has moral relevance. The moral object of this text appears to be people and their long term well-being (see, for example: 'undermining the natural resources *we depend on*' (paragraph one); 'impacting most severely on the poor and marginalised people in our society' (paragraph two); 'change how we live for the better' (paragraph three). Several values are evident in this position: natural systems are valued insofar as they meet human needs; and teachers are valued as people who can 'give children hope', contribute to environmental problem-solving and create a better world.

In relation to these values, the text advances a cluster of preferred ethical responses: we should seek ways to reduce our impact on natural systems; 'adapt and change [our] practices for a low carbon future'; oppose pollution which 'has impacted the health of people'; ensure that access to the benefits of natural resources is equitable; be better informed about environmental concerns; work with children in optimistic, creative ways to build a better world; and 'use our imaginations and problem-solving skills to creatively develop diverse solutions to our problems, and to change how we live for the better'.

Within these preferred ethical responses, the text implies an instrumentalist view of people-nature relationships, that is, the natural world is represented in a way that foregrounds its benefit to people. Natural 'resources' are recognized as vital for human development, human health and the well-being of 'poor and marginalised people'. The benefits of this relationship are focused on basic human needs such as health, food, housing and transport, but no mention is made, for example, of people's cultural, spiritual or aesthetic relationships with the natural world. Nor is moral consideration extended beyond humans to include other realms of concern such as plant and animal species, ecosystems, landscapes or the biosphere.

Explanation of the text

At the explanatory stage, analysis goes beyond an examination of the internal features and relations of the text to consider their interactions with extra-discursive entities such as social structures, economic systems, social-ecological conditions, political processes and so on – all of which are historically constituted. Price (2007) describes this stage of explanation as being concerned with how socio-historical processes provide the preconditions for the interactions described in the previous stages.

The text makes a connection to some of its own socio-historical preconditions by tracing the origins of the contemporary environmental crisis to developments traceable to the Industrial Revolution in Britain and Europe in the late eighteenth century. According to this (fairly standard) narrative, development associated with industrialization has brought benefits to human society ('such as modern medicine, housing, transport and communication systems') but has also caused unprecedented environmental decline and an uncertain future. For this reason, teachers are called upon to teach children in ways that build a better world. The text does not refer to more recent history in South and southern Africa when colonial and apartheid structures established unsustainable environmental management practices and entrenched patterns of what Khan (2000, p. 156) describes as 'the gradual but relentless alienation of blacks from the environmental sphere, and the growth of hostility to conservation issues as defined by the mainstream'.

There are no explicit references in the text to South Africa's segregationist and unequal past. This may be an indicator of the authors' reluctance to stimulate politicized interactions in the teacher professional development course, or it may reflect the extent to which environmental injustices of colonial and apartheid governance are so widely and profoundly understood and experienced by South African teachers – the text's primary interpreters – that the historico-political links do not need to be belaboured, especially in an uplifting and futures-oriented professional development course. Whatever the reason, the text's overarching concern for 'building a better world', attending to 'equity' and the well-being of 'the poor and marginalised people in our society' needs to be understood in relation to South Africa's colonial legacy and more recent history of apartheid and post-apartheid governance. This may involve, for instance, referring explicitly to current tensions in South Africa around development, industrialization, governance, energy provision, land distribution, food security, water security and so on.

Through this text, responsibility for positive environmental change is placed in the hands of teachers and children: 'We can use our imaginations and problem-solving skills to creatively develop diverse solutions to our problems, and to change how we live for the better.' Such a view reflects a voluntaristic view of human agency and social-ecological transformation. Voluntarism, which regards human will as the dominant principle or agency in the world, has an important place in the emergence of social possibilities. However, knowledge, experiences, ethico-moral motivations and behaviours – no matter how reflexive – are alone not enough to transform life's circumstances.

The text foregrounds voluntaristic actions and assumes that ethical action in and for the world (for example, ensuring 'equity, sustainability, resilience, adaptation and transformation') is possible merely through applying our minds and actions to it. This upward conflationist view of human agency (Archer, 1995) portrays social and cultural structures as passive and malleable in the hands of human agents; it denies the real powers and constraints to achieving equity, sustainability and social-ecological resilience imposed by, for example, failing urban infrastructure, government corruption, energy shortages, limited economic models, and the politico-spatial apartheid legacies of land use and distribution.

Textual analysis in summary

This short textual extract was written as part of an orientation handbook for South African school teachers participating in an environmental education course. The text aims to convey the urgency of addressing environmental problems, and the potential role of teachers in re-orientating education for a better world. The textual analysis summarized here has highlighted features of the text that collectively convey perspectives and priorities regarding what is 'wrong' in the world and what can be done to make it 'better'. It has enabled a defensible account of the text's predominant environmental ethics discourse as being a transformation- and social justice-oriented discourse that imbues the natural world with value only insofar as it meets people's needs, in particular the needs of poor and marginalized people. Within this general discourse is a voluntaristic discourse that represents transformation towards a better world as dependent on the (good) will, imagination and commitment of human agents – in this case teachers.

It is important to note that this analytical process and the presuppositions it makes about environmental ethics discourses are based on a single and very brief illustrative example. Claims about the dominant environmental ethics discourses of the overall course cannot be made based on such a limited example. A much richer and more compelling discussion of the course's environmental discourses would include similar textual analyses of, for example, other extracts from the handbook, transcripts of course facilitators' presentations during the course's implementation, extracts from course reports, promotional material and so on.

An educational synthesis

This chapter started by describing the diversity of environmental ethics discourses circulating in society and, more especially, in the field of environmental education. Each discourse advances selected environmental values, represents people-nature relationships in specific ways, and promotes particular ethical actions. Environmental education organizations, programmes or activities may consciously and strategically advance one particular environmental ethics discourse, aiming for it to permeate all teaching sessions, publications and informal interactions. This approach is common in environmental education initiatives associated with specific religious, political or cultural affiliations, or in programmes whose activities are guided closely by corporate funders. Whilst this approach of aligning with a single environmental ethics discourse may achieve coherence and continuity within educational processes, it risks narrowing learners' scope for wider and more critical ethico-moral engagement with the natural environment and is, as such, at risk of being morally impositional.

By contrast, some environmental education initiatives seem not to align with any particular environmental ethics discourse. This may be because they aim to widen participation in environmental learning and action through an all-embracing response to the question of what constitutes 'good conduct' on and for the Earth. It may also occur when an organization's own ethical stance in relation to people-nature relationships is unexamined, indistinct or inconsistent. Within the same organization, educators may hold radically or subtly different ethical stances to one another, thereby introducing different environmental ethics discourses to their educational practices. Learners have to navigate not only the discursive variations of their educators but also those advanced in educational texts (course handbooks, textbooks, posters, audio-visual resources, diagrams, models and so on) and those in wider social texts (advertisements, internet sites, television, newspapers, magazines, books, movies and so on).

The consequent challenge for environmental educators is to enable educational responses that help learners to avoid the pitfalls of ethical relativism and to progress instead towards reality-congruent, ethical attentiveness that is both reasoned and reasonable. Ethical relativism is 'the idea that what is good is simply relative to one's point of view' (Sayer, 2011, p. 135). If people's ethico-moral judgements are determined by their culture and social circumstances then there can be no standard of right or wrong outside of each particular situation. A relativist approach to the diversity of environmental ethics discourses would therefore establish each discourse as valid on the grounds that its constituent values, beliefs and ethical standpoints are beyond rational evaluation.

Critical realism regards ethical relativism to be an invalid proposition. Where purely interpretive or hermeneutic accounts of the world are likely to conflate content and context to conclude that 'there are no good or bad interpretations of the world, just different ones' (Sayer, 2000, p. 43), a critical realist approach allows us to anchor their interpretive acts in practical contexts, recognizing the causal powers and liabilities of both social structures and human agents.

The risk of relativistic engagement with environmental ethics discourses is acute in educational processes founded on the principles of participation, democracy and equality but which fall short of enabling authentic, knowledge-led critique of ethical action. Consider the following example of an educational activity that aimed to open up democratic discussion about diverse environmental ethics positions (Olvitt, 2012). The activity involved presenting learners with seven ethical responses to a proposed titanium mine along South Africa's east coast. Each response reflected a particular environmental ethics discourse ranging from: 'The titanium mine proposal should be rejected because it violates the intrinsic value of the pristine ecosystems in the area' to: 'Titanium should be mined in the area as it is a natural resource whose exploitation will stimulate much-needed economic activity and employment'. Learners had to align themselves with one response and were given five minutes to brainstorm a defence of their position before presenting it to the rest of the class. With limited time for the activity, limited prior knowledge or experience of titanium mining, and no new knowledge introduced to support the activity, learners took up the discourse of their chosen group in superficial, emotive ways. The educators supporting the activity assumed neutral positions, focusing on critiquing and 'playing devil's advocate' with each group's ethical stance. Once all seven groups had made a case for why their response to the proposed titanium mine was the 'right' response, tutors commended all the groups for their enthusiasm and hard work on the activity, and noted that the activity reminds us of the complexity and contestation of environmental issues. The activity thus ended on a relativistic note; the pedagogy guided learners to identify a range of conflicting ethical responses but fell short of giving them conceptual or practical tools to assess the validity and long-term consequences of each position. The educational activity was indeed participatory and deliberative, but it was dominated by the politics and posturing of participation and deliberation instead of by knowledge-based, contextually anchored reasoning processes that enable judgemental rationality.

This example and the others outlined earlier in this chapter illustrate that environmental ethics discourses are pervasive and powerful in environmental education organizations, programmes and activities, yet they are frequently taken-for-granted and under-examined. Consequently, this chapter has argued that educators need to be, first, more individually and collectively reflexive regarding the environmental ethics discourses circulating in their educational programmes and, second, explore ways of mediating these discourses more explicitly and critically in care-filled ways with learners. The contribution of critical realism to this educational challenge is the power of explanatory critique and judgemental rationality, made possible by recognizing environmental ethics discourses as relational, contingent and uniquely laminated structurata that are emergent in open social-ecological systems.

Notes

1 Here, in line with Fairclough (2005, p. 925), I refer to 'texts' in the generalized sense of 'written text but also spoken interaction, multi-semiotic televisual text' and so on.

2 These training programmes were identified through a search on the worldwide web in July 2014 (http://extension.psu.edu/natural-resources/forests/courses/caring-for-your-woodlands; www.forestry.co.za/uploads/File/home/notices/1090.pdf; www.wageningenur.nl/en/show/CDIcourse_Governance_ for_forest.htm). It is interesting to note that, to find such diverse examples, the author needed to use different keyword searches because, for example, a search for 'learning to sustain forests' yielded very different results from a search for 'forest management' + 'training'. This suggests that the form and function of electronic search results might contribute to the durability of discourses by directing the online searcher to websites whose discourses most resemble the one she used to initiate the search.

3 *Diachronic* analysis allows us to describe the interplay of antecedent social structures and the actions of autonomous human agents **over time** and report that the discourse was like *that*, but now it is like *this*, under the influence of *these* structures, events and mechanisms. *Synchronic* analysis adds to this important but often generalized account of the constituent parts of the change by enabling researchers to give a time-bound and specific 'snapshot' of what parts of the discourse needed to be arranged **at a given point in time** with a specific group of people in that specific context so that a particular set of other social interactions, learning and change occurred as they did, and not otherwise.

4 Given this chapter's specific interest in environmental ethics discourses, I have foregrounded the social-ecological dimension of a discourse's context to make explicit the dialectical relations of people and nature within geo-political, economic, cultural and ecological systems.

5 The professional development text analysed here is from the Fundisa for Change project (www.fundisaforchange.co.za), co-ordinated and implemented from the Rhodes University Environmental Learning Research Centre where I work. The extract that is being analysed was produced by Heila Lotz-Sisitka who wrote the first version of the Core Texts, working with Charles Chikunda in a piloting programme for the BEd Honours teachers. At the time of writing this chapter, the interpretations of the text as reflected in this chapter had not been deliberated with the author of the text or the co-ordinators of the programme in any depth. The text was, however, analysed with their knowledge, and my interpretations will be deliberated and reflexively used for informing the programme's future development, since I am a close colleague of the authors and co-ordinators of the Fundisa for Change Programme.

References

Archer, M. (1995). *Realist social theory: The morphogenetic approach.* Cambridge: Cambridge University Press.

Bhaskar, R. (1989). *The possibility of naturalism.* Hemel Hempstead: Harvester.

Bhaskar, R. (1993). Dialectical critical realism and ethics. In M. Archer, R. Bhaskar, A. Collier, T. Lawson, and A. Norrie, A. (Eds) (1998), *Critical realism: Essential readings* (pp. 641–687). Abingdon, UK: Routledge.

Bhaskar, R., and Norrie, A. (1998). Dialectic and dialectical critical realism. In M. Archer, R. Bhaskar, A. Collier, T. Lawson, and A. Norrie (Eds), *Critical realism: Essential readings* (pp. 561–574). Abingdon, UK: Routledge.

Cock, J. (2007). *The war against ourselves: Nature, power and justice.* Johannesburg: Wits University Press.

Dryzek, J. (2005). *The politics of the earth* (2nd edn). Oxford: Oxford University Press.

Elder-Vass, D. (2004). *Re-examining Bhaskar's three ontological domains: The lessons from emergence.* Paper presented at the International Association of Critical Realism Conference, Cambridge, UK, August 2004.

Elder-Vass, D. (2005). Emergence and the realist account of cause. *Journal of Critical Realism, 4*(2), 315–338.

Elder-Vass, D. (2010). The causal power of discourse. *Journal for the Theory of Social Behaviour, 41*(2), 143–160.

Fairclough, N. (1985). Critical and descriptive goals in discourse analysis. *Journal of Pragmatics, 9,* 739–763.

Fairclough, N. (1989). *Language and power.* London: Longman.

Fairclough, N. (2003). *Analysing discourse: Textual analysis for social research.* Abingdon, UK: Routledge.

Fairclough, N. (2005). Peripheral vision: Discourse analysis in organisation studies – The case for critical realism. *Organization Studies, 26,* 915–939.

Fairclough, N., Jessop, B., and Sayer, A. (2004). Critical realism and semiosis. In J. Roberts and J. Joseph (Eds), *Realism, discourse and deconstruction* (pp. 23–42). London: Routledge.

Goatly, A. (2001). Green grammar and grammatical metaphor, or language and myth of power, or metaphors we die by. In A. Fill and P. Mühlhäusler (Eds), *The ecolinguistics reader: Language, ecology and environment* (pp. 203–225). London: Continuum.

Gonzalez-Gaudiano, E., and Buenfil-Burgos, R. (2009). The impossible identity of environmental education. In M. McKenzie, P. Hart, H. Bai and B. Jickling (Eds), *Fields of green: Restorying culture, environment, and education* (pp. 97–108). Cresskill, NJ: Hampton Press.

Harper, K. (2001). *Environment as master narrative: Discourse and identity in environmental conflicts (Special Issue Introduction).* Anthropology Department Faculty Publication Series, Paper 75 (pp. 101–103).

Harré, R., Brockmeier, J., and Muhlhäusler, P. (1999). *Greenspeak: A study of environmental discourse.* Thousand Oaks: Sage.

Hartwig, M. (Ed.). (2007). *Dictionary of critical realism.* Abingdon, UK: Routledge.

Jickling, B., Lotz-Sisitka, H., O'Donoghue, R., and Ogbuigwe, A. (2006). *Environmental education, ethics and action: A workbook to get started.* Nairobi: United Nations Environment Programme.

Khan, F. (2000). Environmentalism in South Africa: A socio-political perspective. *Macalester International, 9*(1), 156–181.

Kronlid, D., and Öhman, J. (2013). An environmental ethical framework for research on sustainability and environmental education. *Environmental Education Research, 19*(1), 21–44.

Mülhäusler, P., and Peace, A. (2006). Environmental discourses. *Annual Review of Anthropology, 35*, 457–479.

Olvitt, L. (2010). Ethics-oriented learning in environmental education workplaces: An activity theory approach. *Southern African Journal of Environmental Education, 27*, 71–90.

Olvitt, L. (2012). *Deciding and doing what's right for people and planet: An investigation of the ethics-oriented learning of novice environmental educators.* Unpublished PhD dissertation, Rhodes University, Grahamstown.

O'Regan, T. (2006). The text as a critical object: On theorising exegetic procedure in classroom-based critical discourse analysis. *Critical Discourse Studies, 3*(2), 179–209.

Parker, J. (2004). How might the inclusion of discursive approaches enrich critical realist analysis? The case of environmentalisms. In J. Joseph and J. Roberts (Eds), *Realism discourse and deconstruction* (pp. 68–85). London: Routledge.

Price, L. (2007). *A transdisciplinary explanatory critique of environmental education.* Unpublished PhD dissertation, Rhodes University, Grahamstown.

Sayer, A. (2000). *Realism and social science.* London: Sage.

Sayer, A. (2011). *Why things matter to people: Social science, values and ethical life.* Cambridge: Cambridge University Press.

Sims-Schouten, W., Riley, S., and Willig, C. (2007). Critical realism in discourse analysis: A presentation of a systematic method of analysis using women's talk of motherhood, childcare and female employment as an example. *Theory and Psychology, 17*(1), 101–124.

Schleppergrell, M. (2001). What makes a grammar green? A reply to Goatly. In A. Fill and P. Mühlhäusler (Eds), *The ecolinguistics reader: Language, ecology and environment* (pp. 226–228). London: Continuum.

9 Working with critical realist perspective and tools at the interface of indigenous and scientific knowledge in a science curriculum setting

Rob O'Donoghue

Introduction

In this chapter I report work with critical realist perspective and tools to examine and review prevailing dispositions on indigenous and institutional knowledge (Western science) in environmental learning. I open with a review of some of the macro social processes that have come to inscribe assumptions of incommensurable difference between the two kinds of knowledge. Whilst the previous hegemony of positivism would have resulted in the dismissal of much indigenous knowledge as mere superstition, contemporary intellectual perspectives (poststructural and hermeneutical) have shaped a proliferation of worldview modelling that has resulted in a macro-level exemplifying of indigenous knowledge as different from and opposing Western science (Cobern and Aikenhead, 1998; Aikenhead, 2006). Here, the lack of adequate mediating tools has given rise to a problematic inscription of assumed difference between the knowledge of indigenous peoples and that of scientific institutions. Furthermore, despite an overt emancipatory intention in worldview discourses, the marginalization of indigenous peoples and knowledge remains.

I then move into the micro arena with a case study of learning interactions in the South African science curriculum. Specifically, I explore some patterns of exclusion in relation to the manner in which students are able to gain access to the knowledge of scientific institutions. The experience and evidence reported is of a preliminary nature but the insights and emerging models of process provide a useful perspective on how assumed incommensurability of knowledge can be tenuous. The study suggests that a critical engagement with both indigenous knowledge and Western science can reveal integrative synergies.

An unexpected outcome of the study was insights into how the co-engaged learning process evident here detoured much of the ontological terror and uncertainty that has come to characterize problem-centred approaches to environment and sustainability education. At least two beneficial social outcomes of a plural and better situated approach are becoming apparent: enhanced epistemological access to the knowledge by the students (those with non-Western backgrounds will be less susceptible to exclusionary processes); and better contextualization of the knowledge, making it more relevant to the students' daily, real world issues.

Context of the study

At a meeting to develop a core curriculum for a teacher education course, I was advised that environmental education was dealt with as a field in its own right with its own 'awareness creating' experiential and problem-solving methodologies that do not lend themselves to being included as core. This response brought me to the realization that despite the centrality of new environment and sustainability knowledge in the education project of the twenty-first century, disciplinary conventions and paradigmatic bracketing commonly exclude a deeper and more integrative engagement with some of the pressing questions of our times.

The experience had me undertake this review as a ground-clearing enquiry and a case study to clarify situated learning in relation to the environment and sustainability concerns in a science education setting. My aim was to explore some of the barriers to relevance and knowledge access where abstraction and an inscription of assumed incommensurability is commonly sustained in preference to critical engagement with knowledge in a real world context. The issues and research evidence examined here emerged around questions of knowledge access and relevance in a small pedagogy and learning mediation project involving student teachers enrolled in the Rhodes University Post Graduate Certificate in Education (PGCE) in Natural Science.

Early approaches to teaching about the environment, particularly during the apartheid era in South Africa, simply excluded indigenous knowledge as of little merit or painted much of it as myth and superstition in need of correction. More recently the prior knowledge of learners became a concern as a foundation for meaningful learning where participants might correct any misconceptions and come to have meaningful knowledge that relates to the social-ecological context and associated matters of concern. Within the transition from early exclusionary practices to more inclusionary approaches, the problem of indigenous ways of knowing being posed against the ways of knowing reflected in the curriculum emerged. Here it became clear that a comparative approach of assumed difference in knowledge system discourse was producing another form of exclusion that merited careful review by looking into assumptions about paradigms and worldviews that had shaped the emerging exclusionary distinctions in question.

Constituting the research process

The chapter opens with a process of immanent critique that is developed as an analytical history of the emergence of the common inclination to oppose knowledge systems that was shaped within particular ways of working with paradigms and worldviews. I had noted in work with indigenous people that, whereas they commonly see themselves as different to others, their knowledge practices are represented in relation to the achieving of certain ends in the world. Seldom is what they know positioned against the knowledge of others unless they are in the company of researchers. This insight suggested that inscriptive assumptions of difference may primarily be constituted through intellectual discourse of ethnolinguistic differentiation that are carried over into abstractions in relation to knowledge systems and worldviews.

I approached the enquiry to initially probe the emergence of incommensurable difference at a macro level of paradigmatic cultural worldviews and then to explore assumed differences applied in the planning of learning programmes at the micro level. Here I used the open model for scientific processes (RREIC schema) of critical realism (Bhaskar *et al.*, 2010) to probe environmental learning interactions across an indigenous knowledge practice and scientific concepts at a micro level. An underlying intention was, on the one hand, to examine those substantive theories that result in discrimination against indigenous knowledge and, by association, indigenous peoples, whilst, on the other, to explore what might be done in education practice to resolve exclusions.

In framing the review, I found that Archer's (1995) morphogenesis and the distinction she makes between 'persons, agents and actors' enabled me to uncover and to track the emergence of intellectual narratives that inscribed difference. I noted an interesting irony that, on the one hand, discourses on knowledge difference can produce alienating separation but in the mediating of learning sequences, difference (dialectic) most often produces the discontinuities necessary for learning. This enabled me to contemplate the prospect of less exclusionary and more open-ended learning progressions. Archer (1995) notes that, as persons engage in a process of narrative agency (they become 'agents') to emerge as citizens ('actors') in the company of others. These actors come to hold particular ways of knowing and doing things in real world contexts of everyday life that they share. Thus I assumed that the same antecedent conditions (processes of individual and collective agency) that reproduce questionably irreconcilable difference (indigenous/Western) can be traced and probed to transform (reconcile) many of the inscriptions of difference in question. Here, seen more widely, through learning, agents have the potential to become actors who can shape their perspectives and practices in relation to the possibilities offered by the social-ecological state of affairs (social realism) in which we interact to make meaning in the world. Concluding the research process involved suggesting a more integrative way of working with knowledge (paradigm/cultural world view) that is centred on testing knowledge in a social-ecological context, in ways that avoid a priori assumptions of differences between indigenous knowledge and the knowledge reflected in the school science curriculum. This is discussed further below.

Critical realism ground clearing to situate the study

In this chapter, I approach education research as a social practice that engages persons as agents to become actors in their social-ecological surroundings. As noted earlier, I aim to probe and disrupt conventional wisdom that assumes incommensurable differences that are overly dichotomizing, with indigenous knowledge and Western science commonly being treated as worlds apart. In particular, I aim to explore how this conventional wisdom may stand in the way of learners acquiring the knowledge reflected in the school curriculum (epistemological access). I also ask how, in the planning and enacting of this science curriculum, we might enable the mediated acquisition of both indigenous and Western scientific knowledge in meaningful ways with the systems thinking necessary to ameliorate the problems of sustainability in a globalizing world of and at risk (Wiek *et al.*, 2011).

In approaching this task, I disrupted the assumption that science is always an expert mediated disciplinary field. In so doing I moved beyond a legacy of institutional appropriation and mediation of knowledge and the insistence that knowledge is valid only when developed in scientific institutions and by a scientific elite. Everyday ways of knowing and the differentiation of knowledge in situated practices can also be scientifically sound. I note also that indigenous knowers have been sidelined from actual and potential knowledge, initially through the appropriation or invalidation of their everyday ways of knowing and then in being unable to access subsequent advances in knowledge. This is notable in the failed induction of students into knowledge and systems thinking in science curriculum settings today. Indigenous knowers therefore suffer a double blow to their ability to access knowledge to enable their agency: their everyday knowledge is most often assumed to be redundant; and access to the information in the sciences of the curriculum is commonly cut off to them owing to reification and abstraction that are barriers to learning.

Informed by these perspectives, the study developed as a process of immanent critique. This examined how historical processes came to shape the precepts of a general ontology of assumed difference. The emergent, domain specific landscape of theory inscribing difference is probed for an alternative and potentially commensurable perspective to support both our being in (acting) and referencing of the world in how we go about our science teaching in the current curriculum setting.

To develop the review, I open with an illustration of a process where Bhaskar (1989) differentiated his ontological *œuvre* (1). This illustration opens the way to a disruption of inscriptions of dichotomizing and simplistic difference that is then deepened within cosmopolitan perspectives on paradigms and cultural worldviews (2). Finally, the beginning of a possible reorientation is developed drawing on the social epistemology of Norbert Elias (1987) and Bhaskar's realist theory of science (3).

1 A historicized ontology of being in a real world

At a recent seminar on Critical Realism, Leigh Price[1] reported how Bhaskar started out in economics, attempting a PhD dissertation in economic theory on the problem of underdevelopment. The thesis proved impossible since there was no direct relation to be found between the formulae of modern economic theory and the real world issues of underdevelopment amongst the poor. It became clear that modern economic theory after Milton Friedman (and before Sen) was immunized from any direct reference to the socio-economic and ecological conditions of the poor in the world, developing as a self-referential cycle of predictive equations to inform its theoretical formulations. Simply put, being in the world of economics was not a worldly ontological process that related to the conditions of the poor. Work with economic theory was an interpretative theoretical act of modelling to predict trends and to inform capital economic gain within which the poor were excluded or at margins where their changing conditions bore little relation to the predictive modelling of capital market growth in the global arena.

Price described how through this experience Bhaskar (1989) came to probe the question of knowledge reference in a real world. His realist critical project uncovered how science cannot be reduced to empiricism and that there are three dimensions in reality: the real, the actual and the empirical (a layered ontology). This *œuvre* revindicated ontology and redescribed epistemology as a historicized, purposeful and contextual practice for reorientation and producing emancipatory agency in a changing world.

2 Disrupting prevailing orthodoxy of paradigms/world views

Working from this position, it is of note that during the modern period of living in what Beck (1992 [1986]) termed a Risk Society, new social movements shaped pedagogy and research to foster awareness and change. Here, a failing modernist project gave rise to and shaped environmental education as one of many responses initiating education activities in relation to emerging concerns like environmental degradation, peace, the plight of the poor, human rights and HIV for example.

The education research culture of the times and an ecological sciences focus in early environmentalism initially demanded a positivist, quasi-experimental (empirical analytical) design in environmental education research. The structural functionalist approach of the times was centred on measuring states of awareness and how these produced changed attitudes and practices (behaviour). The parallel emergence of more interpretative conventions soon shaped the so-called 'paradigm wars' in environmental education research during the early 1990s (Mrazek, 1993). An uneasy truce was variously achieved with the mediating hand of a particular reading of Thomas Kuhn's *paradigms* and the *knowledge-constitutive interests* of Jürgen Habermas (Robottom and Hart, 1993). Popkewitz (2008) notes how the mediating inscription of constitutive differences here produces the distinctions, differentiations and divisions through which the objects of social life come to be ordered for and governed with mediating research.

Positivist, interpretative and critical research paradigms are now inscribed in most research methodology texts so that the co-opted notion of parallel paradigms and associated worldviews is now taught to science education researchers wishing to investigate aspects of schooling and disciplinary knowledge acquisition, for example. Here paradigms have become 'regimes of truth' (Foucault, 1972: 131) ascribed to differing worldviews. These provide security and identity in a particular disciplinary convention, often acting more as a fortress to insulate upholders of the faith from the assaults of outsiders than a framing of conventions that are constitutive of a particular disciplinary field. Setting aside this take on paradigms and their use to uphold the crumbling ramparts of earlier times is not a trivial matter as the emergent and flimsy demarcations of territory and dominance in education research are insulated from the prospect of blind spots and barriers to change being reconciled. In review, and in this case, a more expansive terrain can emerge so that what had been differentiated as paradigms of different but equal merit is exposed as uncertain and constitutive of existing knowledge separated by flawed assumptions and blind spots that could not be differentiated at the time of earlier paradigmatic inscription.

Reconstituting some contours of the paradigmatic conventions by which the objects in research, schooling and pedagogy have been produced, ordered and classified is no easy matter. Few have critically read the substance and qualifying detail of Kuhn's scientific revolutions (or the communicative rationality of Habermas (1994) for that matter) to probe for possible contradictions and to note how many of the competing understandings were incommensurable at that time. Put simply, the inability to reconcile competing cultural orientations had research fall headlong into the blinkered calm of plural perspectives and an accompanying relativism of assumed equal value. An inability to adequately mediate amongst competing positions within the individualizing trajectories of the times had shaped a populist muddle of parallel research paradigms and cultural worldviews (different but of equal merit) that persists in liberal humanism today (O'Donoghue, 2014).

Critical realism after Bhaskar (1993) elegantly surfaces how the transitive dimension in rival positions must most often refer to the same intransitive dimension of reality or the practices would not be rivals. The perspective also notes that when theory or perspective changes there is little likelihood that the phenomenon in question also changes. These realist perspectives allow competing propositions to be reconciled with a mediating weight of evidence that resolves impasse, effectively consigning many of the inscribed differences in parallel paradigmatic worlds to historical artefact. Much of the differentiation in practice and perspective here is an outcome of inscription and is very different from diversity in culture and practice in a plural world where possibilities are relatively open within the bounds of the social conventions and the scope for diverse possibilities with the earth's biophysical systems and processes.

3 Re-orientating within a primacy of practice

Norbert Elias (2011) was relentless in his critique of analytical philosophy for blind spots that obscure a historicized and emergent interplay between a close ontological engagement in the world (involvement) and a standing back that accompanies more formal symbolic representation (detachment). He documented this as an epistemology of the modern sciences, a 'detour via detachment' (Elias, 1987) that encompasses an emergence of an increasing 'object congruence' that remains open to refinement and reorientation (Elias, 2011).

Bhaskar (1993) notably found a flaw in the detaching empiricism of science (positivism), identifying a hidden ontology implicitly 'smuggled in', yet left out in the positivist ideal of empiricism after Kant, Hume, Hempel, Popper etc.[2] Without this *œuvre*, philosophy was always about inscribed conventions of theory and practice (sacred/profane) within which modern education science emerged as a hybrid and incomplete orthodoxy of distinctions and differentiations to achieve freedom through a knowledge mediated agency.

As briefly noted earlier, the modern education move to decentre the subject in child-centred-education came with the imperative to ascertain what is already known for meaning-making engagement in learning (prior knowledge after David Ausubel, 1963) but this was little more than a surface marker in a science orthodoxy of expert mediated concept-to-context pedagogy. In a similar way to the economics review by Bhaskar, the conceptual framing of the science curriculum gives lip service to contextuality and cultural sovereignty against service to the scientific method and its ordered hierarchies of concepts. Reading this science as Western in the recent loosening associated with the postmodern condition (Heller and Feher, 1988) has done little but ferment epistemological confusion now often mediated as matters of difference in worldview; the latter (as mentioned earlier) being rooted in the individual without adequate reference to emergence (history) or a real world of objects and processes in and against which ideas were read and tested in the realizing of human purpose.

Bhaskar (1993) locates science as involving both intransitive (being in a human-independent world – ontology) and transitive (with an emergent cultural capital of purposeful propositions – epistemology) where ontology exists and, as noted earlier, the human ontological condition is layered. Here science is a process that cannot be reduced to empiricism and might best be understood as having been 'naturally constituted' over time and in a world of structures and regularities where human purpose and practice has shaped a hierarchical capital of propositions now primarily held and mediated by experts from within modern institutions. Here the modern orthodoxy of paradigms legitimates diverse theories and methodologies of investigation shaping a proliferation of research registers for mediating social life. This change made widening social science research reasonable for the ordering of social life in a cosmopolitan setting (Popkewitz, 2008). A widening proliferation in relation to cultural worldviews emerged within the same social processes and can similarly be read within modernist cosmopolitan trajectories of freedom and individualization.

Popkewitz (2008, p. 7) reviews emerging systems of reason in the modern era, noting how they are orientated to 'simultaneously produce what is inside and outside', functioning to normalize diversity in social life within which the individual is constituted as one who 'uses reason and science to perfect the future'. He concludes that:

> The individuality lives in a world where external verities and certainties no longer seem safe. The spaces of change are stabilized and tamed through the calculations of principles of reflection and action. The joining of uncertainty and certainty confront each other not as opposites but in the same phenomenon and as limits of governing. The modern governing, I argue, is the 'taming' of social contingencies through inscriptions that orders the future by disciplining the conduct of conduct.
>
> (Ibid., p. 185)

In these conditions, critical realism perspective and tools might serve to open the way to education research and pedagogy that is *historically informed and purposefully engaged* in evidence generating practice that informs the co-engaged social mediation of everyday life for the common good. An emergent and situated perspective such as this necessitates a critical review of Worldview Theory (Cobern, 1996) towards a reading of its emergent precepts as *lived world ontology and open to mediation for the common good.*

Positioning the study

Out of an initial ground clearing exercise that surfaced historicized contexts of situated practice within modernist, cosmopolitan inscriptions of reason and science for perfecting the future, a more open perspective on knowledge emerged for work at the micro level of mediated teaching and learning interactions and the science curriculum. *With this came a critical sensitivity to give close attention to the contours of* ***cultural preconditions for mediated learning****, particularly any discontinuities and real world attributes in teaching and learning interactions.*

Taking note of the mediating role of cultural preconditions in situated learning interactions began to reveal how a science curriculum in an African context that primarily ordered knowledge around reified abstractions (concepts) presented learning to understand the African social-ecological context as a process of mediated induction from the abstract to the concrete. This made it difficult for most learners to access without expert mediation and tangible, culturally relevant examples. Catherine Odora-Hoppers (2002) coherently pointed to this problem when she noted from a cultural practices perspective, that 'what had been theirs as Africans' and remains meaningful to them in lived world experiences and for providing orientation in the world is not recognizable as belonging to their world or of carrying any meaning with which they can easily identify. This insightful perspective is notable in research evidence of classroom performance in the sciences and recent research after Hanisi (2006), Kota (2006) and Mandikonza (2007), to mention a few local research projects. Their work is showing some movement towards a resolution of the exclusion of indigenous knowledge, opening the way towards situated, co-engagement approaches to the relating of scientific knowledge and indigenous knowledge practices, an integrative pedagogy that works both ways.

The question of knowledge access and relevance is at times a seemingly insurmountable problem that is currently bedevilled by the empiricism of the scientific method and a muddle of parallel cultural worldviews that most often hold indigenous knowledge to be different from and opposing Western sciences. This problem is, however, surprisingly easily resolved within the ontology of being in a real world and coming to know this world within emerging practices of purposeful engagement in developing context and over time. Here, working with a layered ontology, a research process involves Description, Retroduction, Elimination, Identification and Correction, that is, DREI(C). This is expanded to RRREI(C) for applied work so learning science can involve the open *resolution* (1) of complex events into causal components. These can then be *redescribed* (2) in a context to allow *retrodiction* (3) where relational links are noted and modelled as mediating hypotheses that are deliberated and tested to enable the *elimination* (4) of competing explanations and the *identification*

(5) of causal mechanisms that give us the necessary explanatory grasp to steer what we know and do in ways that must remain open to *correction* (6) in continuing engagements in a real world. The process is reflected by Bhaskar (1989, p. 109) as:

> There is the epistemically progressive movement after (1) Referential detachment – This is when an absence or anomaly or contradiction generated by, say, a relevant incompleteness (or a tedious homology induced by replication of already well established results) leads to the (2) retroduction of explanatory hypothesis.

Here, the hypothesis is not the starting point of science as commonly found in experimental methods but a detaching turn out of emergent ontology of being closely engaged in a real world context. As in the case of unravelling the barrier of contemporary economic theory not being adequately constituted to refer to the problem of marginalized underdevelopment, the same insights can help resolve the problem of relevance and knowledge access in a modern curriculum of reified hierarchical concepts. Bhaskar's DREI(C) and RRREI(C) tools allow us to take up the meaning-making challenge of being in complex world.

Whilst working with the students, I respectfully set aside the political histories of the paradigms and located the question of pedagogical orientation of teachers within the plural interface of lived world experience (for some students this included indigenous knowledge), formal institutional concepts (natural sciences) and the issues of the day (environment and sustainability concerns). I assumed that any a priori assumptions about worldview difference was present as an intergenerational cultural capital (person) for engagement (agent) with institutional propositions and better ways of seeing and doing things (actor).

Case study: accessing scientific knowledge

Over a seven-week module and into teaching practice, the PGCE Intermediate Phase (IP) science students worked to differentiate key propositions in the science curriculum and to align these with the everyday knowledge practices and prior knowledge of school learners. The students worked in groups to unpack the science curriculum in relation to daily life and the issues of the day. Their task was to explore continuities and contradictions using three start-up questions to guide their work:

- What are the big ideas (in this case, the Western science of fermentation chemistry) that we have to teach?
- What are the connecting (indigenous knowledge) practices in the everyday?
- How do these relate to the issues of the day?

Some of the topics that we have worked on together have been particularly interesting, especially probing the scientific propositions that they as learners in school did not fully understand themselves. The emphasis on knowing disciplinary content coincides with one of the PGCE curriculum imperatives of teaching student teachers the knowledge of the discipline. Here a subtext of thinking about 'how I came to learn this content' often plays out to help open up the double hermeneutic necessary for professional practice, i.e. thinking about my learning and scaffolding the learning processes towards curriculum outcomes that are assessed.

Experience and evidence from our science curriculum module on fermentation has been briefly used to develop this case study. A concern for addressing problems of relevance and epistemological access in school curriculum settings is significant as successive waves of institutional appropriation (colonial and modern) and socio-economic marginalization has characterized the colonial era of Western scientific imperialism. This legacy is still with us as barriers to learning continue to exclude young African learners in the sciences today.

A patchwork of fermentation practices were described as we opened up the topic together, with side notes on patterns of how things had changed in ways that shaped emerging environmental issues that are included as topics in the science curriculum. Evidence for the Nguni mastery of fermentation that preceded the colonial era was gathered. We discovered that Nguni knowledge practices were not static. Notable innovations were the making of *amaRewu* with cooked maize meal (introduced by the Portuguese) and later with a wheat ferment starter (wheat was introduced by the British). A similar trajectory of innovation was tracked in the making of dumplings and steam bread. *Isonka* (oven bread) was mastered in Dutch and later African Trekboer kitchens. Dutch oven bread was later used by amaHlubi groups excluded by apartheid from the mainstream economy. Evidence for this was found locally in amaHlubi oral histories in the Herschel District. These issues and ideas were further explored in a mother tongue demonstration of fermented foods by Gladys Tyatya. Here isiXhosa knowledge practices were demonstrated in our science classroom and were discussed in terms of the diverse cultural practices found amongst the participating student teachers. These activities provided a cultural capital of knowledge practices that allowed the commensurable integration of Western chemistry and indigenous knowledge, hopefully improving the epistemological access of the student teachers and providing them with ideas for integrating Western science and indigenous knowledge in their own teaching practice.

The demonstration of Xhosa fermented foods was followed by brainstorming and web search activities to map out ideas and open up information on health risk issues that had arisen with the advent of the modern diet. Obesity, diabetes and gluten allergies were highlighted here, as it was noted that bread is now the staple diet of modern South Africa. The group then did a 'back-analysis' (retroduction) of changing food production practices to identify how changed fermentation practices were, over time, shaping modern health risks. These activities served to open up environmental matters of concern to be included in teaching and learning materials.

The third arena of engagement was the modern science curriculum. The purpose here was to search for how abstractions might enable an additive explanatory logic to either shed light on what was common sense knowledge and what change practices were contributing to emerging issues and risk. Some of the things to surface here were modern baking practices and the industrial appropriation of rural baking as multinational monopolies like Tiger Brands[3] developed, particularly after 1994.

The emergent complex mix of indigenous knowledge practices, modernizing patterns of change, curriculum concepts in modern science and emerging issues of health and sustainability were challenging for the student teachers to work with. We thus mapped these out against a model for the design and assessment of environmental learning programmes in the South African curriculum. The learning programme framework for generating materials and for mediating learning progression and assessment in the practice of science teaching is provided in the slides shown in Figures 9.1 to 9.3.

In Figure 9.1 an active learning framework (Start-up story (DR) leading to finding out and trying out (EI) dimensions and deliberating synthesis (C) was used to scaffold the learning programme assignment.

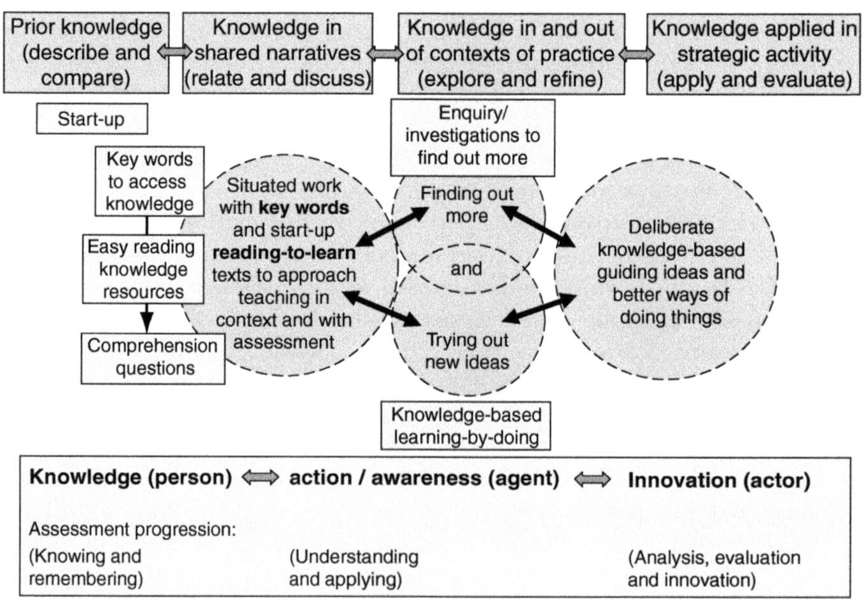

Figure 9.1 A knowledgeable teacher teaching with situated reading-to-learn material, assessment and application (source: developed by the author).

And then, 'how to develop the materials to teach the topic' became the agenda. The task was to develop a learning programme that:

- is of relevance in the everyday, as well as providing
- knowledge access to the explanatory power of modern scientific propositions (concepts) for
- informing future social-ecological sustainability.

In a first attempt at structuring materials using start-up stories, although creative and seemingly meaningful in their own learning, these were not very effective as scaffolding for classroom learning interactions. We thus deepened the analysis by modelling the start-up process from heritage practices to scientific concepts in the curriculum topic. This involved using DREI(C)/RRREI(C) to frame the learning progression as in this case related to *amaRewu* and fermentation below.

Figure 9.2 shows how we worked with a simplified representation of a layered ontology, namely the artefacts used (the empirical – calabash and microscope), what was purposefully done with these (the actual – fermenting process and looking into what went on) and the realizations emerging from these practices (the real – the successful and insightful 'realizings' emerging). This allowed us to landscape the critical nexus of events, relating each nexus in indigenous knowledge practice to the object congruent constituents of the modern sciences. In the developing process the students and I co-validated the knowledge systems and mapped out access routes into the disciplinary field of the modern sciences, as this is represented in the

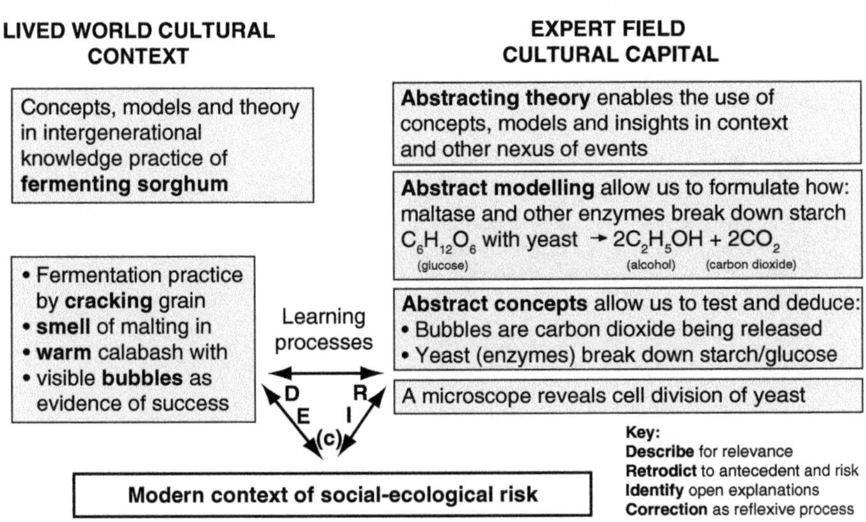

Figure 9.2 Mapping key nexus of events for epistemological access with relevance (source: developed by the author).

current school curriculum. The cultural capital and lines of knowledge access became clearer and this allowed the students to mediate activities and design illustrative experiments to enhance knowledge access.

I then added the critical realist DREI(C) model of scientific processes. This provided a further enhancement that allowed us to contemplate the reconciling, redescription that might enable retrodictive insights (RRR) across the two knowledge-producing practices. It was then possible to chart hypothesis-led processes for eliminating ambiguity and identifying explanations (E-I), with deliberative correction (C) to unlock the explanatory power of causal mechanisms at play. This was followed by an exploration of the health issues of the day and further deliberation towards a creative re-imagining of sustainability practices that might mitigate some of the pressing environmental concerns of our times. This RRREI(C) mapping process has now been used in many other practices and curriculum topics, notably seasonal cycles, food storage and food security.

The key point from the emergent experiences and evidence represented here is that the situated description and retrodiction processes appears to correct a misalignment/gulf between the intergenerational cultures in the everyday and modern empirical science disciplines. It also opens up insights into how modern disciplinary knowledge might be used to more explicitly unlock causal dimensions for a better explanatory grasp of the complex environment and sustainability challenges of the day. A critical realist approach came to be summarized in slide three (Figure 9.3)

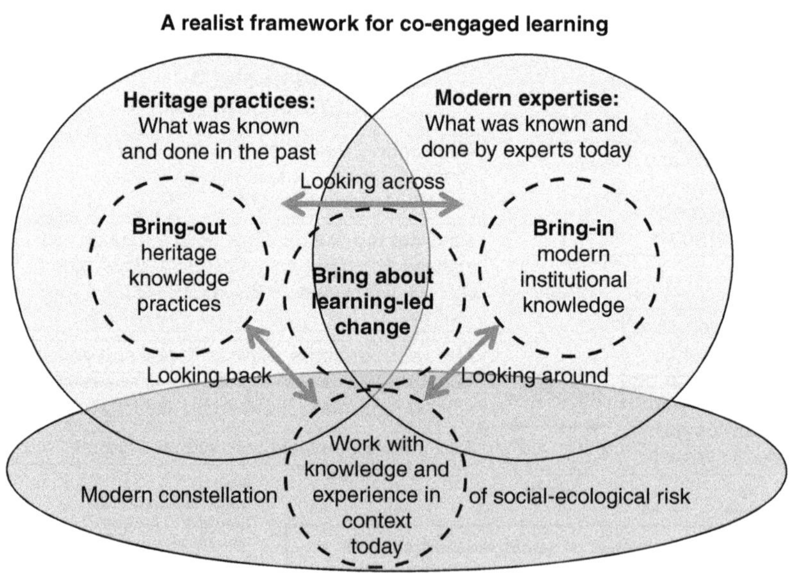

Figure 9.3 A realist framework for co-engaged learning (source: developed by the author).

as a situating and dialecticalization process emergent in 'looking back' to bring out indigenous knowledge heritage and 'looking around' to assess what is now known and we might 'bring together' to 'bring about' a productive disposition to re-imagine better ways of doing things.

I am grateful to a colleague in Japan for pointing to a process of sublation at this nexus, namely an expansion, inclusion or synthesis at a higher level of explanatory grasp. Here this appeared to be reflected for many indigenous participants in the grasp of a truth that had lurked within error/injustice, namely the marginalizing exclusion of indigenous knowledge. Critical realism suggests that through a process of resolution in context and within dialectical learning, sublation (*Aufhebung*, 'transcendence') emerges as a negation of error and as an integrative grasp within a higher level of synthesis. Here Bhaskar notes that error, absence and associated negativity are not preserved[4] amidst a positive open-ended totality that reflects a greater object congruence and coherence in practice. It is learning processes such as this that Bhaskar elegantly describes as 'dialectic pulse of freedom', a process of learning where negativity is lost so positivity and identity are restored within a greater totality of reason. This outcome was notable in how the learners worked from the disciplinary scientific knowledge to explain heritage practices with pride. They were also able to contemplate how *amaRewu* would be a nutritious food for people with immune deficiencies, noting that this might be why it was on sale outside the local clinic where people went to get their treatment for HIV.

The key point from the emergent experiences and evidence represented in the case examined here is that the situated description and retroduction processes appear to allow a historicized process of dialecticalization with boundary crossing that restores reason and identity, unlocking causal dimensions for a better explanatory grasp of the complex environment and sustainability challenges of the day.

The concept maps we developed allowed the students to locate themselves as teachers having to operate at the nexus of intergenerational everyday practices, social ecological risk of the day and the ordered hierarchy of modern scientific objectifications of a real world. The differences/tensions here were not narrated as opposing cultural world views but diversity that opened up dialecticalizing processes in and for learning to emerge. The open framework here avoids the current interpretative paradox of plural cultures and worldviews, providing the prospect of diverse, open processes of co-engaged learning with an integrative gaze. Here it is notable that far from being a cultural process that is unitary in character, the students' ways of working with their students and the science curriculum were creative and diverse. This diversity is key in the continuation of education as a process of the dialecticalization of ontology, processes that are central to learning and change.

Synthesis and discussion

The opening ground clearing work at the macro scale of ambivalence in the conventions of the modern day was both a demanding analysis and an unsettling process. Whereas I was reasonably comfortable with moving beyond paradigmatic inscriptions of difference, I was not prepared for the ripple effect into some of the foundational assumptions of World View perspectives (Kearney, 1984; Cobern, 1991) and as related to indigenous knowledge (Cobern and Aikenhead, 1998). I thus had to initially leave many bigger questions unresolved and give attention to simply working at the interface between knowledge systems governed by seemingly diverse and commonly opposed systems of meaning. As we worked on the knowledge practices and conceptual hierarchies in the curriculum topics using critical realist perspective and tools, the process developed as a co-engagement (immanent critique) towards explanatory grasp and re-imagining that we undertook together. This developed as a co-engaged act of becoming that was centred on the recognition and description of differences that offered learning opportunities towards valued ways of being, knowing and doing things better in a world of and at risk.

Drawing on Deleuze, Paraskeva (2011, p. 173) suggests that curriculum perspectives should not simply be governed by dominant systems of meaning (indigenous and Western sciences for example) but should seek to exemplify and be open to diversity that enables learning interactions towards the production of 'new wor(l)ds'. In line with this, the Vygotskian notion of a 'germ cell' (Edwards, 2011) refers to a common concept (the germ of a shared idea, in our case fermentation) that is open to change in cases of expansion, inclusion and synthesis (integrative sublation). This allows one to contemplate how the descriptive differentiation of commonalities and differences (dialecticalization) is necessary in education so as to enable learning. The engagement across apparent socially inscribed difference and the emergent/integrative insights here are a foil to the common error that the processes of exclusionary appropriation and the marginalization of indigenous cultures can be ascribed to an imperial domination of scientific knowledge (Western/Modern). Clearly a politics of exclusion is evident in imperial histories but, recognizing the ideals inscribed in all human representational systems, this is not intractably rooted in the symbolic representations of sciences that describe and explain workings in and of the world that are open to continuous revisionary review. Here, Edwards (2011) notes that in apparent boundary crossing, open common knowledge can be a font of plurality (dialecticalization) towards shared understanding, opening up an integrative diversity of valued knowings, doings and beings.

Working with critical realism offers tools for understanding causal mechanisms and for teaching in ways that are real world co-engaged. The experience not only enabled us to better chart the historical progression of knowledge access and relevance but it also challenged our creative agency to re-imagine the education process across indigenous knowledge practices and the knowledge in the science curriculum. This opened up education as a process of engaged dialecticalization with learning pathways for narrating new insights and more sustainable practices. It was also notable that we did not foreground the fearful prospect of life in a changing modern world or start our teaching by introducing our students to the imponderables of local and global environmental issues so as to 'create awareness' for resolving the enormity of the challenges faced by humanity in the modern era. The attendant fears and complexities notable here became part of a creative and integrative dialectical engagement with how things used to be, what is now known and what might now best guide what ought to be done for the common good. The emerging insights we experienced in our dialectical struggles to learn in integrative ways that bring hope to the challenges of the day are reflected in how Bhaskar (*The Philosophers' Magazine*, 2013, p. 3) notes that 'the full implications traced through this dialecticalisation of ontology are very radical indeed, and presuppose a vision of the good society viewed as implicit in every human action or remark.' This points the way to ethics-led meaning-making imperatives with guiding principles that might inform valued ways of being, knowing and doing things for the common good. This might take us beyond the current structural functionalist imperative to create awareness and to change values, attitudes and behaviour, a global perspective out of the 1980s that seems to have reached its sell-by date.

Acknowledgements

I am grateful to Lise Westaway and Leigh Price for their support and access to guiding ideas, materials for this project. I also acknowledge the openings on sublation processes that I derived in conversation with Prof. Kensuke Chikamori, Naruto University of Education, Japan. Finally, I must note that all of the conceptual features of the piece are my responsibility for which I alone am accountable.

Notes

1 These orientating narratives were developed with workshop notes reporting on a Realism and Education conference (University of London, 18–20 July 2008) that were produced by Leigh Price, current co-ordinator of the Africa Desk for Critical Realism.

2 Price makes the link between the abstract symbolic field of mediating philosophical expertise as self-referential island and the problem of empiricism revealed in Bhaskar's work. This is a key insight to begin to apprehend the scope of the critical realism project for education to engage a real world and its mediating structures if adaptive change is to be rational and knowledge empowered rather than a matter of competing ideals for sustainability.

3 This company had recently been fined millions of Rand after recent Competition Commission hearings on collusion and price fixing to manipulate and open up rural markets to factory-baked bread.

4 It is here that Bhaskar (1993) departs from an error of sublative inscription in the earlier Hegel dialectic (idealism) to differentiate a negation process where error, absence and negativity are not preserved as in his 'dialectic pulse of freedom'.

References

Aikenhead, G. S. (2006). *Science education for everyday life: Evidence-based practice.* New York: Teachers College Press.

Archer, M. (1995). *Realist social theory: The morphogenetic approach.* Cambridge: Cambridge University Press.

Ausubel, D. (1963). *The psychology of meaningful verbal learning.* New York: Grune & Stratton.

Beck, U. (1992) [1986]. *Risk society: Towards a new modernity.* London: Sage.

Bhaskar, R. (1989). *Reclaiming reality: A critical introduction to contemporary philosophy.* London: Verso.

Bhaskar, R. (1993). *Dialectic: The pulse of freedom.* London: Verso.

Bhaskar, R., Frank, C., Høyer, K. G., Næss, P., and Parker, J. (Eds). (2010). *Interdisciplinarity and climate change: Transforming knowledge and practice for our global future.* London: Routledge.

Cobern, W. W. (1991). Worldview theory and science education research. National Association for Research. In *Science Teaching Monograph, Number 3, 1991,* Manhattan, KS.

Cobern, W. W. (1996). Worldview theory and conceptual change in science education. *Science Education, 80,* 51–59.

Cobern, W., and Aikenhead, G. (1998). Cultural aspects of learning science. In B. Fraser and K. Tobin (Eds), *International handbook of science education* (pp. 39–52). UK: Kluwer Academic Publishers.

Edwards, A. (2011). Building common knowledge at the boundaries between professional practices: Relational agency and relational expertise in systems of distributed expertise. In *International Journal of Educational Research, 50,* 33–39.

Elias, N. (1987). *Involvement and detachment.* Oxford: Basil Blackwell.

Elias, N. (2011). The symbol theory. In *Collected Works, Vol. 13* (R. Kilminster, Ed.). Dublin: UCD Press.

Foucault, M. (1972). *Power/Knowledge: Selected interviews and other writings 1972–77.* Bury St Edmunds: Harvester Press.

Habermas, J. (1994). *The philosophical discourse of modernity* (Trans. Frederick Lawrence). Cambridge: Polity Press.

Hanisi, N. (2006). *Nguni fermented foods: Mobilising indigenous knowledge in the life sciences.* Unpublished MEd study, Department of Education, Rhodes University, Grahamstown.

Heller, A., and Feher, F. (1988). *The postmodern political condition.* London: Polity Press.

Kearney, M. (1984). *World view.* Novato, CA: Chandler and Sharp.

Kota, L. (2006). *Local food choices and nutrition: A case study of amaRewu in the Consumer Studies curriculum.* Unpublished MEd study, Department of Education, Rhodes University, Grahamstown.

Kuhn, T. S. (1970). *The structure of scientific revolutions.* Chicago: Chicago Press.

Mandikonza, C. (2007). *Mobilising indigenous technologies and intergenerational ways of knowing in science curriculum: Case of Mutare Teacher's College, Zimbabwe.* Unpublished MEd study, Department of Education, Rhodes University, Grahamstown.

Mrazek, R. (Ed.). (1993). *Alternative paradigms in environmental education research.* Troy, OH: North American Association for Environmental Education.

O'Donoghue, R. B. (2014). ThinkPiece: Re-thinking education for sustainable development as transgressive processes of educational engagement with human conduct, emerging matters of concern and the common good. *Southern African Journal of Environmental Education, 30,* 7–26.

Odora-Hoppers, C. (Ed.). (2002). *Indigenous knowledge and the integration of knowledge systems: Towards a philosophy of articulation.* Cape Town: New Africa Books.

Paraskeva, J. M. (2011). *Conflicts in curriculum theory: Challenging hegemonic epistemologies.* New York: Palgrave Macmillan.

Popkewitz, T. (2008). *Cosmopolitanism and the age of school reform: Science, education, and making society by making the child.* London and New York: Routledge.

Robottom, I., and Hart, P. (1993). *Research in environmental education: Engaging the debate.* Geelong: Deakin University Press.

The Philosophers' Magazine. (2013). *Roy Bhaskar Interviewed.* Issue 8. Retrieved 10 May 2015, from www.philosophers.co.uk/current/bhaskar.htm.

Wiek, A., Withycombe, L., and Redman, C. (2011). *Key competencies in sustainability: A reference framework for academic program development.* doi10.1007/s11625–011–0132–6.

10 Indigenous knowledge and critical realism on the Eastern Coast of Tanzania

Daniel Sabai

Introduction

In this chapter I demonstrate that indigenous knowledge practice is commensurate with critical realist scientific practice. Critical realism underlabours for Western scientific knowledge, helping to bring its practice in line with its theory. In this paper I similarly underlabour for indigenous knowledge. I use examples from the Eastern Coast of Tanzania to suggest that the kind of knowledge that is generated through indigenous processes is based on retroductive and retrodictive reasoning (as well as inductive and deductive reasoning) and is thus grounded in the theory development principles of DREI(C)/RRREI(C) which, according to Bhaskar (1993), is the basis for all scientific knowledge.

The chapter therefore creates a basis for individuals, groups, organizations, and institutions that are involved in the field of environment and sustainability education and have an indigenous knowledge component, to use the DREI(C)/RRREI(C) for learning and research purposes. In this way, they can assume the commensurablity of both Western scientific knowledge and indigenous knowledge. This work substantiates the significance of indigenous knowledge as science in its own right, which pursues specified scientific principles and procedures to inform practice or praxis in the coastal learning environment in a manner that may enhance social learning. There already exists a body of literature that regards indigenous knowledge as 'local science' (Sillitoe, 2007). However, I hope to expand this view by more closely aligning this 'local science' with the same practices used by science in general, where this science is defined according to critical realist principles.

Indigenous knowledge as a science that has widely enriched Western knowledge

There is evidence that indigenous knowledge has significantly enriched Western science and scientific understanding (Sillitoe, 2007). Turnbull (2000) confirms that from the era of the European global expansion,

Western scientists had drawn on the knowledge of indigenous people to help them understand new plants and animals. It is also documented that in South Asia, Western scientists drew heavily on native classifications and ecology which continue to influence scientific biology to this day (Ellen *et al.*, 2000; Ellen, 2004).

The emergence of Darwin's theory of evolution (one of the grand theories) was also supported by indigenous people (the Galapagos inhabitants). These helped Darwin to confirm the location of tortoise and finches by their markings and shape hence 'micro-evolution' (Sillitoe, 2007, p. 5). Puri (2001) notes that a number of pharmaceutical companies are currently active in using indigenous knowledge in their search for new drugs and advances. Indigenous knowledge has already contributed to over 25 per cent of Western drug prescriptions.

In eastern South Africa, scientific wildlife institutions appropriated traditional ecological knowledge from the Nguni people and used it in creating green nature reserves, and in developing notions of conservation and environmental education (O'Donoghue, 2003). There are also several cases where traditional knowledge has provided corrections to natural science with regard to fish spawning seasons and ecosystem dynamics (Johannes *et al.*, 2000).

In some geographical settings, indigenous groups have developed their own traditional monitoring mechanisms that help them to establish the health status of natural ecosystems (Kofinas and McPherson, 2002). Such knowledge is not only useful in solving problems in local situations but also for classroom activities (Ballantyne and Packer, 2002; Fisher, 2001). Parker (2010, pp. 201 and 211), drawing on Bird-David (1993), indicates that 'many indigenous communities have detailed knowledge of human practices that are conducive to the flourishing of the biotic community as a whole'. She adds that indigenous knowledge 'bequeathed us the seeds that are indispensable for growing all crops today and without which transgenics would be impossible'.

Mangrove as the site of context-based learning

Mangrove ecology, like any other ecological niche, serves as a life support system to a variety of fisheries resources either on a temporary or permanent basis (Berjack *et al.*, 2011). This unique ecosystem has for many decades been used for artisanal fishing on the Eastern coast of Tanzania and has proven to be a reliable source of income, food (protein) and building materials. It has the potential to serve as a conducive environment for context-based learning in the coastal environment where traditional fishers may experiment and engage in reflective practices such as the monitoring of their fisheries (Berkes, 2012). Such reflective practices may lead to the generation of new knowledge. When carried out collectively, such reflective practices may result in the acquisition of greater social skills and lead to the development of relationships and trust. In turn, this may help the fishers to face prevailing challenges more effectively as a group and to agree on reliable collective actions (Muro and Jeffrey, 2008). Furthermore, it is in this kind of environment that experienced mangrove-based fishers may transfer knowledge to the less experienced (especially the youth) by supporting them to internalize daily routine fishing practices as their own, and to become experienced practitioners (Belay, 2012). Knowledge transfer among fishers who practise fishing in the mangrove areas and adjacent sites may also be realized through peers, friends and relatives. These community members engage in reflective practice and interact on a daily basis to share ideas, notions and concepts that are used in the fishing context (ibid.).

Danermark *et al.* (2002) argue that concepts, notions and ideas that people represent may potentially contribute to knowledge production, especially when these are employed to describe objects (like mangroves and fish species) and processes (such as various interactions that take place in the mangrove ecosystem and causal mechanisms which generate them). This assertion suggests that when individuals and social groups who have never been inducted into fishing as a practice are exposed into spaces that allow them to share notions, ideas and concepts about mangrove-based fishing, they are likely to acquire knowledge that relates to the context to which they have been exposed. By understanding the mangrove-fishing context and uncovering the learning space that exists among mangrove-based fishers, we may uncover traditional ecological fishing techniques that are likely to attract wider social learning in the coastal environment, and help to restore the disrupted harmony of ecological systems (Pahl-Wostl, 2006).

Knowledge that emerges from mangrove fishers in their context may be shared and used to describe realities in the social world. This is possible since 'social learning takes place through deliberative processes that involve sustained interactions between individuals, and sharing of knowledge and perspectives in a trusting environment' (Lotz-Sisitka, 2012, p. 18). As explained earlier, realities, or complex situations, that occur in the open system of the mangroves, may be articulated through experiences of involved participants (coastal indigenous communities). Their learning will be enhanced not only through consensus but through discussions in which individuals sometimes have differing opinions, in a process akin to critical realism's *judgemental rationality*. The open system in this context refers to a situation where there are many generative mechanisms operating in combination with each other; as there are many involved mechanisms, it may become difficult for mangrove-based fishers to anticipate the results or outcome (Danermark *et al.*, 2002; Sayer, 2000) of processes that are taking place in the fishing environment, especially in the absence of theories or models which may help them to describe or explain what is going on in the mangroves. This necessitates the need to explore how indigenous theories or models are developed and employed in informing fishing practices in the coastal environment (particularly mangrove-based fishing), and how the same pursue scientific procedures and principles in a manner that justifies consideration of *indigenous knowledge* as a form of science that need not be dismissed or undermined.

Linking indigenous knowledge to the DREI(C)/RRREI(C)

The DREI(C)/RRREI(C) schema is applied in the analysis of a 'complex event or a situation of interest' (Hartwig, 2007, p. 196), where there is a multiplicity of causal mechanisms, structures, fields or processes operating in combination with each other to produce certain outcomes (Bhaskar *et al.*, 2010; Bhaskar and Danermark, 2006). This is the case with indigenous knowledge systems that interact with coastal and marine ecosystems in the eastern coast of Tanzania, to generate events and experiences that may be apparent to fishers and mangrove restorers. According to Bhaskar *et al.* (2010), when more mechanisms are involved in complex systems, it becomes difficult to anticipate the outcome or potentially infer causes, hence a need to use the 'schema' to understand the characteristics or the nature of the multiplicity or plurality of components that produce what may or may not be observable. The schema contains two related models 'operative to the social domain' (Hartwig, 2007, pp. 195–196), namely the DREI(C) and RRREI(C). The latter has an applied nature, and may potentially help us to explain or analyse complex systems for *application* purposes while the former may help us to resolve *analytical* complexity or tensions (Hartwig, 2007).

Bhaskar (1998) and Bhaskar *et al.* (2010, p. 3) describe the DREI(C) as formed by components as follows:

D – description of pattern of events or phenomena
R – retroduction of possible mechanisms or structures
E – elimination of competing alternatives
I – identification (we apply the theory in the real world, often using induction and deduction)
C – correction of earlier explanation or findings.

On the other hand, the second operative model (RRREIC) (see Hartwig, 2007, pp. 195–196) contains the following components:

R1 – resolution
R2 – re-description
R3 – retrodiction
E – elimination
I – identification (using induction and deduction)
C – correction.

The fishers' theoretical knowledge assumes the presence of a 'stratified ontology' in that it uncovers drivers (causal factors) that may potentially actualize ecological, economic and social events, leading to experiences that may or may not be apparent to coastal communities. Indigenous knowledge helps mangrove-based fishers to explain processes that occur in the mangrove ecology by describing their relevant component parts and relations (resolution), re-describing them in a way that makes them relevant or theoretically significant (re-description), explaining natural and social events by making inferences back to mechanisms that are capable of producing them (retrodiction), eliminating alternative expressions and explanations that suggest existence of confounding mechanisms (elimination), identifying correct generative mechanisms from those considered (identification), and making any possible corrections (C) through praxis. Indigenous use of RRREI(C) in this context depicts a form of experiential learning or 'learning by doing' which may potentially pave the way for transformative practices and eventually attract wider social learning in the coastal area, as was shown to be the case in a more substantive study on this process (Sabai, 2014).

In this chapter, I demonstrate the presence of theories arrived at by retroduction and retrodiction in the knowledge-base of the mangrove-fishers (henceforth referred to as 'fishers') on the eastern coast of Tanzania. However, I will start with an explanation of how the fishers also use the better known scientific logics of induction and deduction, as defined by Danermark *et al.* (2002) and associated with the DREI(C)/RRREI(C) schemes at the stage of *identification.*

Induction

Danermark *et al.* (2002) argue that induction as a process occurs when we draw general conclusions from observation of individual cases. According to Gilbert (2008), it is a process of finding a single case and observing a relationship, then observing the same relationship in several more cases and finally constructing a general theory to cover all the cases. Induction helps us to generate theories by moving from a set of observations to a theory (Jago, 2013).

Induction may be presented in the following example:

1 In the first occasion, when we experience torrential rain, juvenile fishes escape from mangroves to the ocean.
2 In the second occasion, when we experience torrential rain, juvenile fishes escape from mangroves to the ocean.
3 In the third occasion, when we experience torrential rain, juvenile fishes escape from mangroves to the ocean.

We may conclude that: always, when we experience torrential rain, juvenile fishes escape from mangroves (nursery grounds) to the ocean. Likewise, mangrove fishers begin the process of theory making by observing specific or multiple events in the mangrove ecology. Reoccurrence of similar situations are observed as fishers engage themselves in routine fishing practices. These are later confirmed and considered as general premises that may potentially inform fishing practice or praxis in the coastal area.

When fishers, for example, observe a constant conjunction of events in the mangrove area where they always see one particular fish arriving in summer, and they finally conclude that that particular fish will always arrive in summer, their reasoning is regarded as 'inductive'. This is the case in the following fisher's illustration of fish dynamics in the mangroves: 'Mullet fish and other fisheries that are not permanently residing in the mangroves tend to visit mangroves in large companies in the rainy season than other seasons of the year' (Fishers' Focus Group Discussion held at Moa Village, March 2012).

There are several observations that constitute fishers' inductive reasoning in the above example that are worth noting. First, fishers seem to be aware that there are fish species that live permanently in the mangroves and also some which visit mangrove sites on a temporary basis. Second, the fishers also seem to be aware that in other seasons of the year, mullet species and other unnamed fish species may be found in the mangroves but not in large or big companies except in the rainy season. Their inductive premise, therefore, establishes a clear correlation between rainfall and fish abundance in the mangroves.

Another example which indicates how fishers develop theory by induction is the following illustration: 'In the northerly monsoon there are more fish in the mangroves than in the southerly monsoon (Fishers' Focus Group Discussion, Moa village, March 2012). Fishers in the given inductive illustration establish a correlation between the monsoon winds and abundance or decline of fish in the mangroves. The former is associated with the northerly monsoon while the latter is linked to the southerly monsoon. The above inductive reasoning which serves as a theory among fishers on the eastern coast of Tanzania is widely trusted in the area and used to guide local fishing plans and projections. There is evidence that wind direction or variation affects fish movements (Ayub, 2010), hence indigenous inductive reasoning creates the basis for theory advancement, improvement or falsification.

Deduction

While induction helps us to develop shallow theories of causal constant conjunctions, deduction on the contrary helps us to apply theories (Gilbert, 2008). Deduction is defined as 'the process of starting with a theory and using it to explain particular observation' (ibid.). This process helps us to take the data about a particular case and apply the general theory in order to deduce an explanation from the data. According to Danermark *et al.* (2002), the process of deduction is characterized by two main logics, namely propositional and predictive. The former pursues a relational route (marked with words that examine the validity of a statement or argument such as '*if*', '*not*' and '*or* ... *then*'), while the latter (predictive logic) usually employs inferences that use words like '*all*' and '*no*'. For example we may use an illustration from the fishing context to explain how 'predictive logic' may be realized:

Rule: All mullet fishes visit mangrove sites.
Case: These are mullet fishes.
Result: These mullet fishes visit mangrove sites.

We may see from the above example that deductive reasoning deduces the 'particular' from the 'general'. In other words, it starts with the general to arrive at explaining the 'particular'.

Deduction is a *propositional* or *predictive logic.* Fishers deduct specific premises from existing (known) theories in the coastal environment. For example, Fisher A has worked in traditional artisanal fishing for over 30 years. He knows some theories that explain species availability in the coastal systems; and he uses deduction to predict the best times to fish. Thus he will prefer to go fishing when the northerly monsoon blows as he will predict that at this time there will be more fish. Here he uses deductive logic: *if* fish are attracted to the warm northerly monsoon, *then* when the northerly monsoon is present, there will be more fish for me to catch. When he goes out fishing and as he expects, finds plenty of fish, then he will have his theory confirmed.

In another example, Fisher A provides a theory to explain turtle reproductive behaviour in the following account:

> If you happen to see turtle's spoors trailing from the ocean to offshore or near mangroves, then know that that's a female sea-turtle that had just gone to lay eggs. Male turtles tend to stay in the ocean, and do not go out as females do.

The above theory leads to the deductive prediction that since there are many threats currently being reported in the Eastern Coast of Tanzania involving the killing of sea-turtles on the shore (URTSER, 2006), then the most affected turtles are likely to be 'females'. A testable deductive conclusion that emerges from the above two premises is that the population of male turtles may be higher than that of female turtles, since the former spend most of their time in the ocean and thus protect themselves from human threats. That is: if most sea turtles are killed when they leave the sea to come onto the shore, and if all turtles who come to the shore are females, then more female turtles than males will be killed. Such deductive predictions create opportunities for learning and provide Western scientific research with possible research questions.

Retroduction

Sometimes, fishers' explanations go to the level of analysing the underlying mechanisms that result in certain events or constant conjunctions. For example, Fisher J (below) uses retroduction to explain why the presence of some fish like mullet (*Liza/Mugil spp*) is correlated with certain times. He says that at these times, certain fruits (*mpia*) have fallen into the water and have begun to decompose, which attracts fish: 'Some fishes visit mangroves in search for food. Mkizi (mullet species) for example, prefers feeding on decomposed Mpia (Sonneratia alba) fruits.' Berjack *et al.* (2011) confirm that fish are frequent foragers in the mangroves eating 'mature fruits that drop from the trees at high tide' (p. 71).

Retroduction is defined as the process of explaining the events by postulating (and identifying) 'mechanisms which are capable of producing them' (Sayer, 1992, p. 107). Fishers also tend to attach various explanations to events by specifying particular causal factors that produce them or suggesting conditions that necessitate or determine involved processes. Such explanations set the stage for deduction and thus may also serve as initial platforms for Western scientific investigations.

Another example of retroduction is as follows:

> In the past, fish had no fear at all, you could find them in groups and it was easier to catch them; today, juvenile fishes run away from fishers in fear; they sense and think like human beings and would definitely ascribe one's presence to danger … such fear is caused by frequent dynamite fishing near mangrove sites.
>
> (Fisher J)

Here the fisher uses retroduction to arrive at the conclusion that the relatively recent use of dynamite fishing by illegal fishers has generated fear among fishes in the area. This is not induction because he has not lived in several places, some of which had dynamite fishing and some not, thus allowing him to see a constant conjunction between dynamite fishing and frightened fish. Instead, he has asked the retroductive question, 'What mechanism is present that could explain the development of fear amongst the fish in these mangroves?'

Retrodiction

Retrodiction is simply a process of explaining or interpreting present information or ideas by making an assertion about the past using already available theories. Let us consider the following theory, given by Fisher X: 'You can easily identify fish that have been fished using dynamite; their skin becomes too soft, fins too loose; some parts of their bodies also appear abnormal as compared to those that are fished using traditional or acceptable gears'. Fisher X thus has a theory that fish caught by using dynamite look abnormal. If Fisher X, for example, visits the nearby fish market to obtain some fish for his family and finds a fish with the indicators he has established, he will retrodictively conclude that the condition of that particular fish is due to dynamite fishing. However, a fisher who is not familiar with the theory of dynamite blasting and its effect on fish, on being faced with an abnormal fish, will not be able to retrodictively apply this explanation to understand its strange appearance (he will perhaps have to use retroduction to arrive at a new theory of his own).

Practical advice on applying the RRREI(C)

As alluded to earlier in the introductory section, this chapter also serves as a basis for practitioners in the field of environment and sustainability education to use the DREI(C)/RRREI(C) for learning and research purposes. It is necessary to recall that the schema contains two related models 'operative to the social domain' (Hartwig, 2007, pp. 195–196). The DREI(C), as explained earlier, helps us to resolve *analytical* complexity or tensions while the RRREI(C) fits well in the applied context. Since this section is primarily aimed at addressing practical aspects of indigenous knowledge generation rather than analytical aspects (which are already covered in previous sections), I will focus on the application of the RRREI(C) in knowledge generation.

Adopting the RRREI(C) as a tool for capturing and applying indigenous knowledge requires a participatory space where indigenous communities such as fishers and mangrove restorers may meet, under the researchers' facilitation, depending on the nature of the context, and engage in collective activities that steer learning amongst themselves (Sabai, 2014). Workshops are often used as an opportunity to generate knowledge in research. They differ from focus group discussions in that they are more interactive and take a longer period of time. They often follow a structured or semi-structured form of interaction. In expansive learning research, for example, Engeström (2001) and others such as Virkkunen and Ahonen (2011) have developed a sophisticated workshop methodology called the 'Change Laboratory Workshop' which engages participants in a mutual learning process in order to expand their learning and practice. This allows workshop participants to 'share' or 'mirror back' knowledge that was generated from previous fieldwork for further dialogue and discussion. Likewise, other forms of workshops such as the Experiential Learning Intervention Workshop (ELIW), which draws on the Change Laboratory Workshop may also be employed in capturing indigenous or Traditional Ecological Knowledge (Sabai, 2014).

Carrying out the ELIW requires specified moments or sessions. In the first moment of resolving the problem into its component parts, the researcher as a key facilitator is advised to help the indigenous groups to participate actively in the specified learning activities. The session may include discussion of previous events, processes and/or data, depending on the intended objective. He or she will afterwards lead them in describing relevant component parts of the processes they have specified, in a manner that allows them to compare or relate them, up to a level that may potentially allow them to infer causal factors that trigger those processes or predict possible outcomes. By taking them through this level, they will have engaged themselves in the 'Resolution' stage (the first stage of the model). They will have also participated in a learning process that not only yields traditional ecological knowledge but also learning from each other.

In the second stage (re-description), component parts of the events, processes or data may be compared, contrasted or related by workshop participants at a deeper level. This may yield some significant theoretical insight which may help them to contextualize the nature and extent of discussed issues and pave a way to make inferences back to causal factors that might have produced the events, data or processes (retrodiction).

In the course of applying the RRREI(C) model, it may happen that there are alternative causal factors, explanations or expressions. The facilitator is advised to lead the workshop participants in discussions that will disregard minor factors and focus only on those that truly explain, trigger or reflect prevailing conditions. If this is carefully and wisely executed, the process of 'elimination' (fourth stage of the model) will have been reached. As workshop participants deeply analyse events, processes or data, they may also realize that some generative mechanisms may not directly trigger prevailing situations. In the event such observations are made, the facilitator is advised to help them identify those that are truly correct (identification). The entire process of applying the RRREI(C) model requires thorough checking of presented explanations and that possible corrections are made at every stage, to ensure that the final input truly reflects the intended objectives.

Conclusion

It is interesting to see how the fishers' descriptions of seasonal variations and temperature characteristics closely match those described in the Tanzanian State of Environment Report (URTSER, 2006). They achieve this knowledge using retroduction, retrodiction, induction and deduction. Their practices are guided by theories that are developed and improved over time through an epistemology accurately described by the DREI(C)/RRREI(C) schemes. It is thus justifiable to suggest that indigenous knowledge is in many ways equivalent to Western scientific knowledge. Indigenous knowledge achieved through the DREI(C)/RRREI(C) schemes can potentially be used to fill knowledge gaps. For example, the most up-to-date field guide for marine species in the Eastern Coast of Tanzania was prepared by FAO in 1985 and has many gaps that could be filled from the indigenous knowledge base.

A more substantive study undertaken by the author (Sabai, 2014) showed that this approach can also be used practically, following the process outlines above, to develop and refine coastal zone monitoring indicators where indigenous knowledge can correct indicators produced by scientists, and also lead to wider social learning processes.

Acknowledgements

This chapter draws on data produced for a PhD study conducted by the author in the Environmental Learning Research Centre at Rhodes University (Sabai, 2014). The study was supported by the World Bank, and its purpose was to develop indigenous knowledge monitoring approaches to coastal zone management on the East Coast of Tanzania.

References

Ayub, Z. (2010). Effects of temperature and rainfall as components of climate change on fish and shrimp catch in Pakistan. *Journal of Transdisciplinary Environmental Studies, 9,* 1–9.

Ballantyne, R., and Packer, J. (2002). Nature-based excursions: School students' perceptions of learning in natural environments. *Geographical and Environmental Education, 11,* 218–236.

Belay, M. (2012). *Participatory mapping, learning and change in the context of biocultural diversity and resilience.* Unpublished doctoral thesis, Rhodes University, Grahamstown.

Berjack, P., Campbell, G., Huckett, B., and Pammenter, N. (2011). *In the mangroves of South Africa.* KwaZulu-Natal: WESSA.

Berkes, F. (2012). *Sacred Ecology* (3rd Edn). New York: Routledge.

Bhaskar, R. (1993). *Dialectic: The pulse of freedom.* London: Verso.

Bhaskar, R. (1998). Critical realism: Essential readings (Critical realism: Interventions). In M. Archer, R. Bhaskar, A. Collier, T. Lawson, and A. Norrie (Eds) (2002), *Explaining society: Critical realism in the social sciences.* New York: Routledge.

Bhaskar, R., and Danermark, B. (2006). Metatheory, interdisciplinarity and disability research: A critical realist perspective. *Scandinavian Journal of Disability Research, 8,* 278–297.

Bhaskar, R., Frank, C., Hoyer, K. G., Naess, P., and Parker, J. (2010). *Interdisciplinarity and climate change: Transforming knowledge and practice for our global future.* London and New York: Routledge.

Bird-David, N. (1993). Tribal metaphorization of human-nature relatedness: A comparative analysis. In K. Milton (Ed.), *Environmentalism: The view from anthropology.* London: Routledge.

Danermark, B., Ekstrom, M., and Jacobsen, L. (2002). *Explaining society: Critical realism in the social sciences.* London: Routledge.

Ellen, R. (2004). From ethno-science to science, or 'What the indigenous knowledge debate tells us about how scientists define their project'. *Journal of Cognition and Culture, 4,* 409–450.

Ellen, R., Parkes, P., and Bicker, A. (Eds). (2000). *Introduction in indigenous environmental knowledge and its transformations.* Amsterdam: Harwood.

Engeström, Y. (2001). Expansive learning at work: Toward an activity theoretical reconceptualization. *Journal of Education and Work, 14*(1), 133–156.

Fisher, J. A. (2001). The demise of fieldwork as an integral part of science education in United Kingdom schools: A victim of cultural change and political pressure? *Pedagogy, Culture and Society, 9*, 75–96.

Gilbert, G. (Ed.). (2008). *Researching social life* (3rd edn). London: Sage.

Hartwig, M. (Ed.) (2007). *Dictionary of critical realism.* London: Routledge.

Jago, M. (2013) The content of deduction. *Journal of Philosophical Logic, 42*(2), 1–18.

Johannes, R. E., Freeman, M. M. R., and Hamilton, R. J. (2000). Ignore fishers' knowledge and miss the boat. *Fish and Fisheries, 1*, 257–271.

Kofinas, G. P., and McPherson, F. (2002). Community contributions to ecological monitoring: Knowledge co-production in the U.S.-Canada Arctic Borderlands. In I. Krupnik, and D. Jolly (Eds), *The earth is faster now* (pp. 54–91). Fairbanks Arctic Research Consortium of the United States.

Lotz-Sisitka, H. B. (Ed.). (2012). *(Re)views on social learning literature: A monograph for social learning researchers in natural resources management and environmental education.* Grahamstown/Howick: Environmental Learning Research Centre, Rhodes University/EEASA/SADC REEP.

Muro, M., and Jeffrey, P. (2008). A critical review of the theory and application of social learning in participatory natural resources management. *Journal of Environmental Planning and Management, 51*, 325–344.

O'Donoghue, R. (2003). Indigenous knowledge: Towards learning materials and methodologies that respond to social processes of marginalisation and appropriation in Eastern Southern Africa. *Australian Journal of Environmental Education, 19*, 57–67.

Pahl-Wostl, C. (2006). The importance of social learning in restoring the multifunctionality of rivers and floodplains. *Ecology and Society, 11*(5), 10.

Parker, J. (2010). Dialectics of knowledge and care. In R. Bhaskar, C. Frank, K. G. Hoyer, P. Naess, and J. Parker (Eds), *Interdisciplinarity and climate change: Transforming knowledge and practice for our global future.* London and New York: Routledge.

Puri, K. (2001). *Legal protection of traditional knowledge and cultural expressions of indigenous peoples.* Lecture given at Corpus Christi College, Cambridge.

Sabai, D. (2014). *Mobilising processes of abstraction, experiential learning and representation of traditional ecological knowledge, in participatory monitoring of mangroves and fisheries.* Unpublished PhD study, Rhodes University, Grahamstown.

Sayer, A. (1992). *Method in social science: A realist approach* (2nd edn). London: Routledge.

Sayer, A. (2000). *Realism and social science.* London: Sage.

Sillitoe, P. (Ed.). (2007). *Local science vs global science: Approaches to indigenous knowledge in international development.* New York: Berghahn Books.

Turnbull, D. (2000). *Masons, tricksters and cartographers: Comparative studies in the sociology of scientific and indigenous knowledge.* Amsterdam: Harwood Academic.

URTSER (United Republic of Tanzania State of the Environment Report) (2006). Vice-President's Office, Division of Environment.

Virkkunen, J., and Ahonen, H. (2011). Supporting expansive learning through theoretical–genetic reflection in the Change Laboratory, *Journal of Organizational Change Management, 24*(2), 229–243.

11 Dialectical critical realism and Cultural Historical Activity Theory (CHAT)

Exploring and expanding learning processes in sustainable agriculture workplace contexts

Mutizwa Mukute

Introduction

This chapter seeks to illuminate the use of dialectical critical realism in a PhD study that was conducted to explore and expand learning processes in sustainable agriculture workplace contexts (Mukute, 2010). The study was conducted in three case study sites in Lesotho, South Africa and Zimbabwe. About 80 people in three study sites participated in the multiple case study research. In Zimbabwe the study was located in Hwedza District, St Margaret Primary School and its community that learns, practises and facilitates the learning of Permaculture within the Schools and Colleges Permaculture Programme (SCOPE). In South Africa the study took place in Durban urban and peri-urban areas where a community of organic farmers, facilitators and entrepreneurs co-ordinated the marketing of their produce through Isidore Farm and Earth Mother Organic and support each other to learn and practise organic farming. In Lesotho the study took place in the Mafeteng and Mohale's Hoek districts where the focus was on farmers who learn and practise the Machobane Farming System (MFS) and were supported in this by the Rural Self Development Association (RSDA) and the Machobane Agricultural Development Foundation (MADF). The other research participants were agricultural extension workers who tend to promote high external input agriculture, sustainable agriculture researchers and organic produce marketers.

The analysis aimed to show that dialectical critical realism:

- is an important ontological theory that enables an in-depth understanding of contemporary agricultural issues that have historical and structural causal explanations;
- provides the necessary methodological tools to enable processes of understanding and absenting the absence of sustainability in agricultural practices;

- provides the essential systems thinking that is essential for approaching and dealing with issues that have a wide spatio-temporal character such as sustainability in agriculture through the concept of *constellations*, which are collections of related ideas, things and people. This concept is closely tied to that of interacting activity systems in Cultural Historical Activity Theory (CHAT) (Engeström, 1987; Engeström and Sannino, 2010).
- effectively underlabours CHAT, a constructivist and epistemological theory that was worked with in the study, so as to provide ontological depth, addressing the problem that Nunez (2013) raises related to CHAT functioning with an 'implicit ontology'.

The analysis is achieved through identifying the elements of dialectical critical realism that are evident in: the focus of the study as encapsulated in its title and research questions; the features that it shares with CHAT and the associated complementarities; and the application of its four elements in the interventionist research process (methodology) and outcomes. The four elements being referred to are: non-identity (first movement); real negativity and contradiction (second edge); unfinished whole in motion as totality (third level); and practical human agency (fourth dimension) (Bhaskar, 1998).

The research context and thrust, and its connection with dialectical critical realism

Dialectical critical realism

Before analysing the import of dialectical critical realism in the focus of the study, it is necessary to outline how dialectics enriched original or basic critical realism. It was the infusion of the notion of absence that created the 'second edge', which moved the original critical realism from its main interest in layers of reality, that is being, to the spatio-temporal character of *becoming* as well. The process of becoming is a process of change as explained below:

> [A]nd change involves the passing away, the negation of what is in favour of what has come to be in its place – a thing becoming something else, and in the process ceases to be what it was. What is was is no more, has passed into non-being, and has been absented through the process of change.... Becoming is the absenting of what was in favour of what now is, for every becoming is also, in some part at least, a 'begoing' and absenting of what was there.
>
> (Norrie, 2010, pp. 14–15)

Since critical realism is a philosophy concerned with ontological matters, absences being discussed within this context should essentially be differentiated and stratified so that in addition to covering the empirical layer and the actual, they should also include the real layer of structures and mechanisms – the invisible part that has causal powers and liabilities.

Study justification

The study was informed by three main factors: the purpose of the Rhodes University and South African Qualifications Authority (SAQA) Research Programme, within which it was conceptualized; the main sustainable development and sustainable agricultural issues that were being faced in southern Africa; and the experiences and interests of the researcher. The objectives of the Rhodes University and SAQA programme were:

- developing conceptual understandings of how change and adaptation processes occur in and through workplace learning, thus deepening knowledge of workplace learning;
- developing new knowledge of sustainable development practices and how they 'play out' in workplaces and contribute to a critical understanding of such practices; and
- enhancing understanding of the learning-development practices relationship.

This suggested the need for *stratified* and *differentiated* knowledge about learning needs and processes in particular places in time, which is consistent with the first movement. There was also a growing understanding that ambivalence, ambiguities, tensions, contradictions and risk – different forms of *absences* – emerge from environmental issues and can produce learning and development opportunities (Wigley, 2006; Wals, 2007; Pesanayi, 2008). The study also drew on the growing need for solidarity and co-operation among different actors of the world in order to address risks. For example, Beck (2000) notes that global dangers result in global mutualities in which contours of a global public sphere begin to take shape; the dangers trigger an impulse towards the development of co-operative international institutions; *constellations* of sub-politics appear, which are at once global and direct, which may lead to a global alliance of alliances of mutually exclusive beliefs (p. 20).

The study was interested in sub-politics around sustainability and resilience in agriculture. In line with critical realism's position that our knowledge of reality is inadequate and therefore has absences, Visvanathan (2006, p. 169), discussing cognitive justice, argues that modern science should dialogue and negotiate with other forms of knowledge 'to create a pluralist world of cognitive possibilities where *emergence* rather than reduction was emphasized' [my emphasis]. Cognitive justice appears to be an important part of the fourth ethical dimension in Bhaskar's dialectics for 'what it means to ... enter into relations of solidarity with others' (Norrie, 2010, p. 17) because learning and agency in agriculture is a joint project between farmers and others.

Aim of the study

The study being discussed in this chapter was entitled 'Exploring and expanding learning processes in sustainable agriculture contexts' (Mukute, 2010), which was conducted in 2008 and 2009. The study sought to understand how learning in agricultural workplace contexts takes place in relation to sustainability issues and how such learning can be enhanced. This was based on the informed understanding that sustainability in sustainable agriculture practices was inadequate and desirable; but more importantly on the growing realization that modern and conventional agriculture had historically and culturally absented sustainability in agriculture. The research objectives were to: investigate learning processes that expand the current scope of sustainable agriculture practices; and develop mediation tools that support expansive learning of sustainable agricultural practices. The study sought to answer the following four questions:

1 Why and how do farmers learn sustainable agriculture in their workplaces?
2 What are the current limitations and contradictions of sustainable agriculture learning processes among farmers?
3 How can sustainability be better learnt and more reflexively practised in the farmer's workplace?
4 What conceptual artefacts can the study develop to support expansive learning for sustainability in farmers' workplaces?

The transformative part of the research was concerned with the development of reflexivity, resilience and agency among farmers and their development facilitators through absenting contracted or degenerative activity (by making expansive learning present), vulnerabilities and inaction.

The process of exploration as discussed in the study was concerned with seeking to understand the state of sustainable agriculture workplace contexts in three research sites – their *being*. The sustainable agricultural practices that were explored (in order to be expanded) were: Organic Farming, Permaculture and the MFS in South Africa, Zimbabwe and Lesotho respectively. Organic farming involves the production of safe and nutritious food using production methods that respect the principles of fairness, care, ecology and health. Permaculture is a land-use design system that seeks to create the most beneficial and productive relations between elements in a system while respecting and copying nature. The Machobane Farming System is a practice that seeks to improve land productivity, agro-biodiversity on farms, progressive economic development, solidarity and joint learning among farmers and care for the land. Sustainable agriculture is an umbrella term for agricultural practices that seek to achieve economic viability, ecological sustainability and social justice, the achievement of which requires a *dialectical* approach.

Constraints to the practices were examined at various levels during the exploration and ahead of the expansion, from the experiences of the research participants and the historical and contemporary events that shaped and shape sustainable agriculture, right down to the *contradictions*. The contradictions were viewed as the 'structural causes and generative mechanisms that render the actual and events present and observable' (Norrie, 2010, p. 10). This was an attempt at taking an otherwise constructivist study to ontological depth through 'taking realism beyond the empirical and the actual ... to the underlying domain of the real' (ibid., p. 9). These contradictions also provided the energy for *movement*, because Bhaskar's 'fertility of contradiction' creates the necessary tension for learning and action (Parker, 2001).

Expansion was concerned with processes of improving the sustainable agriculture practices by *absenting* some of the constraints to the achievement of the three-dimensional and complex object. It involved the construction of solutions to absent the *absences*. While the construction of the solutions involved joint learning – human *mind*, it also involved the use of human *agency* and intentionality in the formation of new human activity. In order to ground and enable the development of their agency beyond reasons given for why people choose or avoid practising sustainable agriculture, the study drew on the theory of habitus/practice (Bourdieu, 1990) to help reveal unconscious motivations. Through implementation of solutions to sustainable agriculture challenges that were identified in each study site and across them, the process of expansion put the focus on the dialectical critical realism space between *being* and *becoming*.

Theoretical and methodological framing

Key features of dialectical critical realism that underpinned the study

Bhaskar proposes that mechanisms and structures have powers and liabilities that cause the actual occurring of events; and that such structures and mechanisms are distinct from the events. Both nature and our knowledge of it are stratified and differentiated. The main structure of Dialectical Critical Realism (DCR) is called MELD, which stands for first **M**ovement, second **E**dge, third **L**evel and fourth **D**imension (Bhaskar, 2008). These elements are elaborated as follows.

First Movement – non-identity

First Movement refers to non-identity: things are not all the same but involve some stratification and differentiation. This argument forms the foundation of basic critical realism, which captures Bhaskar's early philosophical thought. The key words (motifs) include: transitive and intransitive; the real, actual and empirical; and epistemic fallacy. Bhaskar noted that reality is stratified with the empirical, the actual and the real respectively (Sayer, 2000, p. 2; Benton and Craib, 2001, p. 125). Empirical reality is that which can be observed; actual reality is the second layer of reality and is what happens when events are activated. The real is whatever exists, whether people know it or not, and can be social or natural. It is associated with causal powers and structures. This layered reality can further be divided into two groups, the transitive, which can be changed, and the intransitive, which is nearly impossible to change. Epistemic fallacy occurs when reality is reduced to what is known about it. The potential value of the First Movement resided in its ability to provide an explanatory critique, with ontological depth, that goes beyond the actual and the observed to the causal mechanisms that are invisible (Bhaskar, 1998).

Second Edge – absence

Second Edge is concerned with the notion of absence, and consequently, of absenting. It is based on the understanding that the world is made up of things that are present as well as those that are not. Change and development is therefore concerned with the need to fill an absence; as much as it may be concerned about the need to absent a constraint. Given this argument, absence has a causal significance (Norrie, 2010). The key words are: absences and ills; axiology of freedom; contradiction and constraint (Mingers, 2010); omissions and liabilities. Bhaskar (2008) identifies several kinds of absences that are worth noting here as they were important in the study: absence of an entity in a particular place or time because it exists in others or because it does not exist at all; because it is finite and has perished; because it never existed before. Norrie (2010, p. 14) points out: 'Natural necessity involves change, and change involves the passing away, the negation, of what is in favour of what has come to be in its place.' The process of expanding learning processes and improving sustainable agriculture practices was essentially about removing absence through what Bhaskar (2008) called 'presences, commissions and powers'.

Third Level – totality

Third Level is concerned with *systemic* constructs by emphasizing interconnectedness between parts and wholes; interrelations and interactivity; and 'recursive embeddings and reflexivity' (Mingers, 2010, p. 320). This focuses on totality or holism (or the broader system), which is essentially real, open and incomplete or unfinished and in motion thus providing an emergent sense of open totality (Norrie, 2010, p. 16). The key words are: holistic causality and reflexivity. Lotz-Sisitka *et al.* (2012, p. 63), writing on how to absent ontological collapse in social learning research, note:

> Critical realist social theorists ... see learning processes and interactions as emergent and reflexive, shaped by structural emergent properties, cultural emergent properties, and personal emergent properties that interact resulting in new meaning making possibilities and possibilities for agents to act which always occur in the 'concrete universal'.... They explicitly recognise the socially and ontologically located nature of learning and agency.

The notion of totality resonates with CHAT's concept of third generation activity systems, which was worked with in the study (Mukute, 2010) and provided an important theory of change of both the practices and the agricultural communities in the study sites. The layered relationships that exist within and between such communities, that is, activity systems, is well captured in the following statement on dialectical critical realism:

> It is the idea that different entities or layers of being may be constellationally embedded in each other to constitute a totality. A constellation, involving the interlocking of distinct entities within a whole, provides a sense of both the autonomy of the parts and of their dependence on the broader relations.
>
> (Norrie, 2010, p. 17)

The sustainable agriculture farmers, agricultural extension workers, sustainable agriculture facilitators, policy makers and marketers exemplified activity systems that interacted as distinct entities that were part of a whole in the study.

Fourth Dimension – human agency

Fourth Dimension is concerned with human agency by recognizing that people are causative agents that can cause change in a purposeful way. At this point the concern moves to using practical human agency and intentionality to change the world so that it becomes more just. Such agency is underpinned by the quest for freedom with others, truth and solidarity. The key word for such agency is transformative praxis (Mingers, 2010). The actual absenting of ills such as hunger, soil erosion, agro-biodiversity loss and unfair market prices through engagement with others and with causal structures (such as government and the corporate sector), and mechanisms (such as agricultural policies and unfair practices) happens through the exercising of intentional and ethical human agency.

Key features of CHAT

Cultural Historical Activity Theory is a theory of learning and development (Vygotsky, 1978) built on contradictions, which are a form of dialectics, as well as on reflexivity and agency (Engeström, 1987). It offers an explanation of learning through activity that helps to develop understandings of workplace learning. Vygotsky drew on Marxist materialism which was concerned with the use of tools to mediate the transformation of nature.

The concept of knowledge in CHAT

In CHAT knowledge is viewed as contextual and transformative. It is generated through processes of reflexive investigation and learning. CHAT posits that learning takes place through collective activities that are purposefully conducted towards a common object. Edwards (2005, p. 50) defined learning as 'concerned with within-person changes, which modify the way in which we interpret and may act on our world ... and in turn change it by our actions'. The learning is facilitated by the use of conceptual and material tools, which help the learners to understand the object better. The incorporation of new knowledge and concepts into the individual happens first at the interface of the community and the individual through internalization – inter-mental and, second, within the individual, intra-mental. The individual externalizes acquired knowledge through applying it to the object towards an intended outcome. Billett (1994) pointed out that the Vygotskian school uses the concept of knowledge appropriation in learning to refer to what happens when inter-psychological processes happen and before the knowledge becomes an intra-personal attribute, because the knowledge is not absorbed unaltered. 'Appropriation refers to a personally active – and at the same time – multidimensional process; it indicates that new knowledge and competence are actively transformed rather than simply internalised by the learner' (Simovska, 2008, p. 64).

The unit of analysis in CHAT

The unit of analysis in CHAT has been evolving over time, from the first to the third generations of CHAT as explained by Engeström and Sannino (2010). The study worked with the third generation unit of analysis, which is called interacting activity systems. It uses systems-based thinking in order to gain insights about the real world. It is based on the proposition that learning is a social and cultural process, which benefits from historical achievements. Engeström (2001) identifies the following five principles which guide third generation activity theory:

a The prime unit of analysis is a collective, artefact-mediated and object-oriented activity system seen in its network relation to other activity systems.
b Activity systems are multi-voiced and are a nexus of many points of view, traditions and interests. Multiple layers and strands of history are embedded in the rules and division of labour. The multi-voicedness of the activity systems is a source of both tension and innovation.
c Activity systems take shape and are developed over long periods of time and need to be analysed in terms of local history of the activity, and the genealogy of conceptual tools that have shaped it over time.
d Contradictions between and within activity systems are potential sources of change and development. They are *historically accumulated structural tensions* between and within systems. Activity systems are also seen as *open-ended* learning systems.
e Activity systems have the potential for expansive transformations, which occur through relatively long cycles of qualitative transformations.

Research methodology in CHAT

CHAT is built on and addresses matters of contradictions, reflexivity and agency. Recent work in CHAT is evolving a methodology called Developmental Work Research, and a method called Change Laboratory (Engeström, 1987; Engeström and Sannino, 2010), which resonate with MELD, and have potential to:

- identify and analyse issues, constraints and contradictions – absences;
- model and examine solutions to address problematic situations using expansive learning processes that draw on the distributed cognition and reflexivity of those concerned;
- mobilize individual, relational and collective human agency and intentionality to overcome the constraints and enable their situations to improve; and
- ensure that the process of improving (making more advanced) the activity systems is continuous and never-ending as there is always room for improvement.

Justification for working with dialectical critical realism and CHAT

Critical realism provides an ontological framing that permits one to delve beyond the current and the surface into the history and the underlying to find 'real reality' that lies beneath the empirical and the actual. Price (2008) identifies the critical features of critical realism as: being an underlabouring science; being aimed at human well-being and emancipation; seeking theory-practice congruence; and based on immanent critique of what it examines by looking for internal inconsistencies – contradictions.

Shared purpose and origins

One thing that connects CHAT to critical realism is emancipatory politics, in particular those associated with Marxism and neo-Marxism as can be inferred from this statement:

> Critical realism was developed during the 1970s at a time when Marxism was strongly represented among social scientists. Marxism was one of the few approaches to social science whose explicit philosophical commitments coincided with the main outlines of critical realism.
>
> (Benton and Craib, 2001, p. 136)

The roots of CHAT also lie in Marxism and this makes the theories potentially compatible. Critical realism provided ontological depth to the empirical and historical analysis that I was able to develop using the constructivist expansive learning CHAT framework. Besides providing for depth analysis of ontology, critical realism also theorizes agency, which was also central in this study. The idea of using contradictions as potential sites of learning and development is found in both critical realism and CHAT. A central interest of the research was to understand why things were the way they were or appeared to be. This interest is captured in the notion of explanatory principle found in both CHAT and critical realism. In CHAT the explanatory principle lies in the historical-genetic nature of inner contradictions in the systems under review to explain daily problems (Pihlaja, 2005, p. 190). In the ontology of critical realism it is found by analysing the causal generative mechanisms and tendencies, going beyond the empirical.

Complementarities in change processes in dialectical critical realism and CHAT

Table 11.1 summarizes the processes of change in both CHAT and dialectical critical realism with a view to illuminating the benefits of using the latter to underlabour the former.

Results of the study and how dialectical critical realism shaped them

This section of the chapter discusses the main outcomes of the study and how critical realism informed and shaped them. The outcomes are discussed

Table 11.1 Comparison of dialectical critical realism and CHAT research processes

Dialectical Critical Realism (DCR)	*CHAT (Engeström, 1987; Engeström and Sannino, 2010)*	*Added value of DCR (Bhaskar, 1998, 2008)*
Stratification and differentiation of the world	Context of appreciation (Mukute, in press): ethnographic data generation	Provision of ontological depth – the real layer of reality that lies beneath the empirical and the actual
Identifying absences	Context of criticism: identifying contradictions	Paying attention to deeper causal mechanisms and structures and explanatory critique in order to explicitly identify absences
Making systemic constructs to address absences	Context of construction: modelling and examining solutions	Change that is differentiated and stratified, and conscious of the underlying matters, and therefore laden with transformative intent
Exercising human agency	Context of implementation: implementing and reviewing model solutions	A more deliberate engagement with causal structures and mechanisms as part of addressing the invisible parts of problematic situations. Moral agency which seeks to absent societal *ills* and absences such as social ecological and/or cognitive injustice

Source: Mukute, 2010.

according to the twin objectives of the research – exploration and expansion of learning processes in sustainable agriculture workplace contexts.

Dialectical critical realism and first movement (multiple layers of reality)

The study revealed that farmer learning of sustainable agriculture is influenced by both intrinsic and extrinsic factors. Extrinsic factors include the needs: to produce adequate safe and nutritious food and surplus to generate income; to improve their resource base for their own good and for the benefit of future generations; and to generate ecological services, such as recharging the water table and increasing species diversity. At the same time some farmers have ended up taking up the 'trade' because they have a passion for it; a concern for the future of the people and the Earth; and a disposition to farm. Farmers use different ways to learn. They learn through both vertical and horizontal movement of knowledge, from and with experts and promoters as well as among themselves. Much of their learning has a practical orientation and includes learning by doing, observation, trial and innovation. However, they do not appear to have systematic and joint mechanisms for continuous and strategic learning around their emerging needs and interests. This observation resulted in the development of a conceptual tool to support such continuous learning and improvement (Farmer Learning and Innovation Forum and how they could be implemented) (Mukute, 2010).

Using the ontological lenses of critical realism, some explanatory principles were sought to understand how learning is shaped by other forces that are beyond the surface. This revealed that:

- Women's limited social power in families undermines their choices to attend training.
- The past land tenure system resulted in the uneven distribution of land which created tensions between those who owned huge tracts of land, and those who owned little or none.
- Past land tenure systems also resulted in some farmers who lease land for limited durations of time (in Lesotho) not interested in investing in its improvement but more in its 'mining', as they can lose it back to the landowner at any time.
- The need for labour in South African mines resulted in the stigmatization of agriculture in order to create labour reserves and the promotion of low external input agriculture such as the MFS in Lesotho was stigmatized as backward, resulting in double stigmatization of sustainable agriculture.
- The susceptibility of certain soils to erosion could not only be learnt and understood by looking at them, or at the topography or by the seasonal processes and events of soil erosion but by understanding the

make-up of the soils from which they originated. Critical realism lenses provided this explanatory force.

Dialectical critical realism and second movement (absenting)

The commonly found contradictions in the study sites, which were identified through CHAT lenses, could be summarized as:

- the clash between long-term and short-term needs;
- the clash between the quantity and quality of tools for learning and practising sustainable agriculture and the need for generating social, economic and ecological sustainability in the immediate future;
- ambivalent messages from different forms of sustainable agriculture and the clashes between messages on sustainable agriculture and those of expedient high external agriculture; and
- ecological affordances versus human needs and wants.

Drawing on the history of sustainable agriculture, the study found that agro-business interests and those of ordinary farmers often pull in opposite directions and this is associated with some of the absenting terms used by Norrie (2010, p. 23): contradict, oppose, marginalize, criticize and condemn. The stigmatization of agriculture in general, in South Africa and Lesotho, can be traced back to colonial and apartheid era tactics to discredit agriculture so as to create labour reserves for gold and diamond mines. Baker (1986, pp. 80–81, my emphasis) made a noteworthy and related explanation which has a bearing on all three case studies:

> Another serious weakness of the educational policy of the colonial era, not unconnected to the previous one, was that there were two separate education systems – one for blacks and one for whites. This had the effect of making the blacks covet intellectual excellence as the only ultimate goal.... To be 'educated' was to achieve a position *where one did not soil one's hands*.... Had such a separation between the two systems not existed, perhaps this distorted concept of the true nature of education may not have been so powerful.... So powerful was the myth to take hold that, as with the other educational problems left to newly independent nations, it is proving, indeed, extremely difficult to eradicate.

The second and related explanatory principle is associated with power relations that exist between different actors. The government has the political power to decide on policies, the corporate sector has the economic and cultural power to push high external input agriculture. They hire the bright and learned in the community. Universities and colleges have modern and institutional forms of intellectual or cultural power, which has generally tended to reproduce Western knowledge and agricultural practices. The economic power of government, partly derived from donor aid from pro-conventional agriculture countries, has resulted in some programmes that entice farmers by providing free or subsidised inputs. Within rural communities in Lesotho there are power relations between the landlords and the landless, with the latter being dependent on the former. Those who rent land – often with leases that last a few years – are discouraged from building the ecology of the soil because the lease can be terminated at any time.

Women in all three case studies tend to be the ones who do not have land and generally do not get as much access to agricultural education as men. In Lesotho, the decisions on what to grow, and when to grow it in the fields, are still made by their husbands who are miles away in South African mines. These uneven power relations between men and women in agriculture appear to need further investigation in terms of how they affect sustainable agriculture practice learning and development. Ignorance (a form of absence) about sustainable agriculture explained why government extension workers cannot effectively promote it. This was a common issue in all three case studies as there are limited places to learn about it and no fully accredited courses on sustainable agriculture are being offered at undergraduate levels in the agricultural training colleges and degree programmes under review. Not knowing much about sustainable agriculture also occurs in other places that matter: at government policy-making level; in higher institutes of learning; among curriculum development workers; and bureaucrats who allocate resources. The absence of sustainable agriculture in the mainstream schools and colleges, and the lack of focus on it, sustains the ignorance. On the ground, among farmers, lack of government material and financial support towards sustainable agriculture perpetuates conventional agriculture at the expense of sustainable agriculture.

Using dialectical critical realism to make systemic constructs in open systems that are in motion (totality)

In order to show evidence and traces of critical realism during the expansion, I review and analyse one of the case studies, and later discuss the conceptual tools that were developed in the study and how some of them were shaped by critical realism. In order to show how critical realism thinking was utilized in the study, I draw on two of the three case studies: the SCOPE case study in Zimbabwe and the Isidore case study in South Africa. Figure 11.1 shows how research participants in the SCOPE case study in Zimbabwe 'expanded' their possibilities and practice during and as a result of the interventionist study being reported on.

The case study outlined in brief in Figure 11.1 above, suggests that the expansive learning process can be an effective tool for researching change-oriented learning and sustainability practice where the intention is to

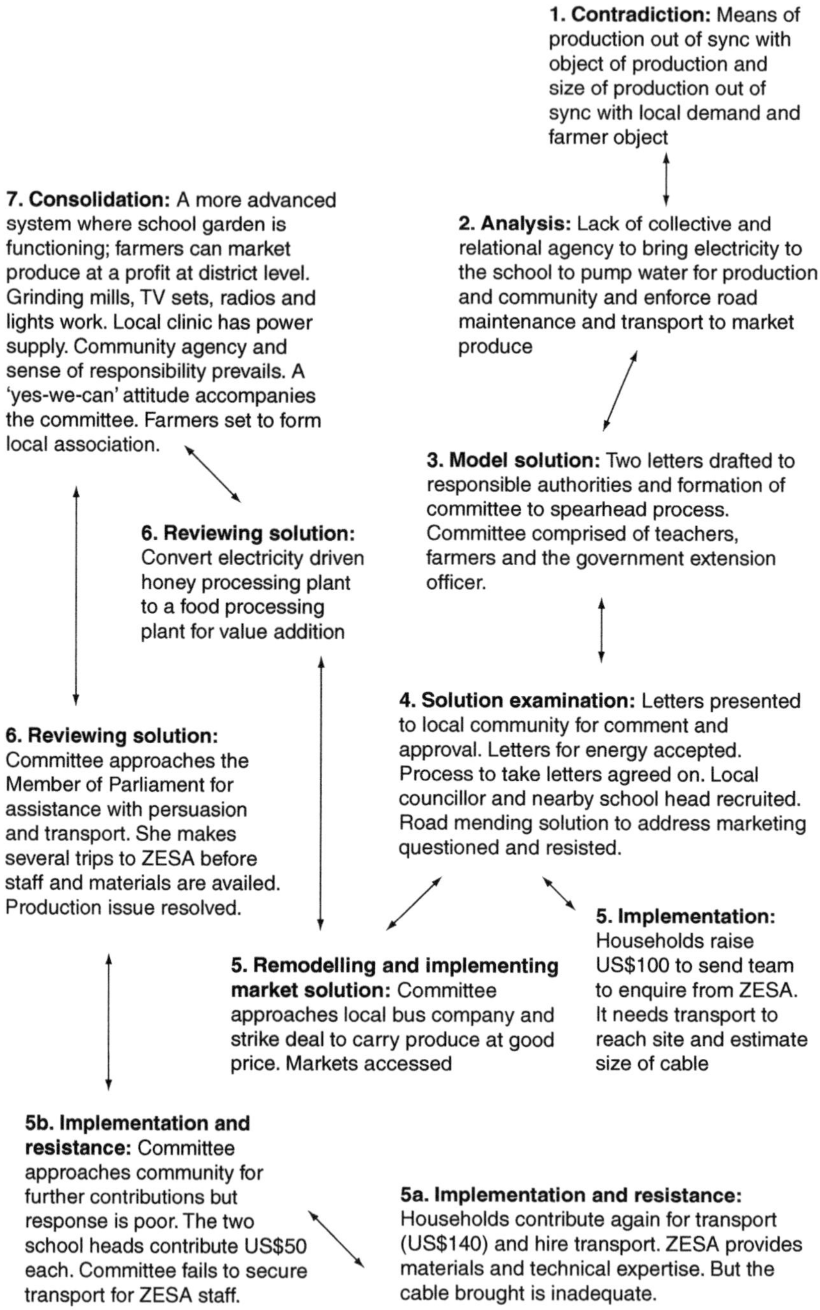

Figure 11.1 Expansive learning process in the SCOPE case study (source: Mukute, 2010, p. 271).

stimulate responsible action and set change in motion. From a critical realism perspective, the research participants developed solutions that were informed by a critical awareness of *interlocking constellations*, these being: structures and systems of local governance (councillors and headmasters); the district political head (Member of Parliament); Zimbabwe Electricity Supply Authority (ZESA) – a parastatal; the NGO which constructed the honey processing plant (civil society); and the corporate sector (bus company).

The process of expansive learning in the Isidore Case Study is captured in Figure 11.2. Using a critical realist perspective the causal mechanism that explains the linkages related to the contradiction being addressed can be defined as indifference as Dean (2009, p. 136) put it:

> In the Marxian account, contradictoriness inheres in the spatio-temporal separation of necessary relations, practices and processes which promote the 'indifference' of these separated elements to one another. 'Indifference' expresses here a kind of 'objective' unawareness of necessary interdependence (i.e. between production, circulation and consumption of goods) which, beyond a certain point, results in crisis.

The solution that was generated in the study, which involved the creation of a Green Growers' Association made up of 'interlocking entities' such as groups of organic producers, organic marketers, trainers, certifiers and inspectors, and donors showed a remarkable awareness of an incomplete and emergent, open totality. It demonstrated how interdependence can be recreated and the absence of elements and people that were necessary for the proper functioning of the organic sector addressed.

Reflexivity and the associated systemic constructs went beyond the individual case studies and, through cross-case analysis, resulted in the development of tools that support learning processes in sustainable agriculture workplace contexts. The tools that had strong features of the third element are:

- *Tool for agentive talk analysis:* Drawing on five categories of agentive talk analysis that were largely derived from the work of Sannino (2008), Engeström (2010) and Lister (2004), the study developed guidelines for analysing agentive talk (Mukute, 2010). The analysis is important for understanding people's preparedness to act. The main contributions of critical realism thinking here was the addition of a category on 'awareness of critical actors' and structures in the talk, which suggests awareness about causal power; and
- *Tool for expanding learning based on identified contradictions:* Largely drawing on CHAT's change laboratory workshops, I developed a tool that described how interventionist research could use contradictions

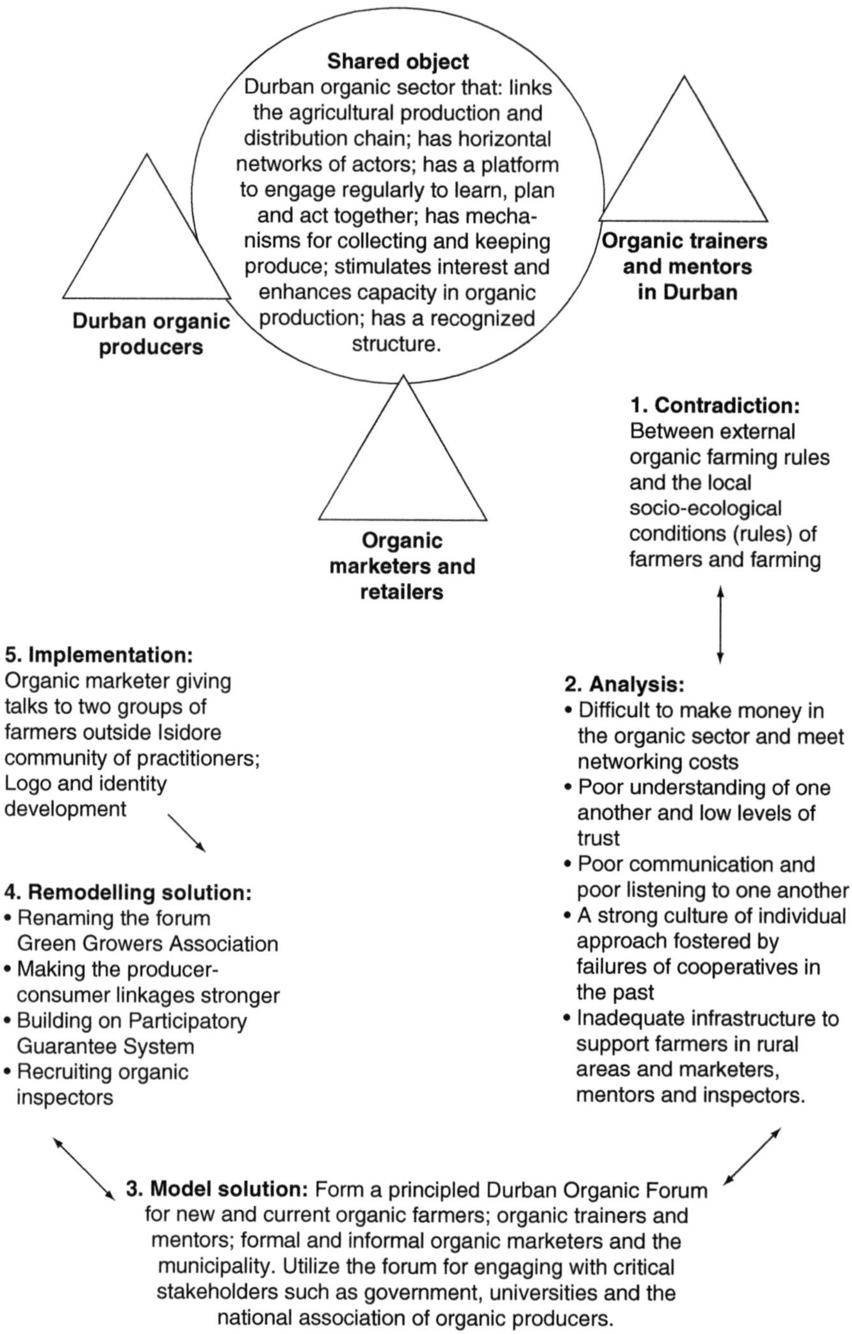

Figure 11.2 Isidore Organic expansive learning (source: Mukute, 2010, p. 290).

for learning and innovation (see Mukute, 2010). The main contribution of dialectical critical realist thinking was identifying the need for 'absenting the absent' (people and their concerns) in the identification and analysis of contradictions as well as in the modelling of solutions, especially in terms of who to work with and how (relational agency).

Dialectical critical realism and human agency (fourth dimension)

The main results of the research in this respect were that it was able to trigger relations of solidarity between and among sustainable agriculture stakeholders in each study site. In the Zimbabwe case study it resulted in local farmers and community members, teachers, pupils, headmasters of two schools, councillors, the local MP, an NGO operating in the area, and the electricity authority joining hands to absent the problem of energy. The research participants also embarked on educational campaigns to address the problem of ignorance that was assumed to have caused the destruction of the electricity facilities in the area. This resulted in the restoration of power, and the revival of the school garden, which benefited orphans among others. The functioning of the grinding mill was restored and villagers no longer needed to travel long distances to process their staple food. Research participants also improved their livelihoods after negotiating for better transport charges to ferry their produce to the nearest market – making them *become* better off.

The structural issue of indifference was tackled to some extent in both the Lesotho and South Africa case studies. In the Lesotho case study this was addressed by working on absenting ignorance as well as motivating local level decision makers and opinion makers in the district in which the studies were conducted. This was achieved through the establishment of an MFS demonstration plot with the joint participation of MFS promoters, MFS farmers, government agricultural extension workers and local government authorities. The demonstration plot was intended not only to show how the MFS should be designed and implemented (and thereby overcome ignorance) but also by illustrating how effective it was in terms of production and productivity, as well as in improving the quality of the land on and around which it was being practised. This way, policy makers and local farmers would, over time, develop a positive attitude towards the sustainable agricultural practice.

In the South African case study, indifference was tackled by drawing on one of the research participant's suggestions: 'So what can help make agriculture work? You need to make it cool. You need to make it sexy. You need to make it socially valuable.' While it was not feasible to make it cool in the 18-month research period, the research participants' efforts in this direction included the establishment of a Green Growers' Association (GGA), with logos that were to be used for labelling organic produce and the establishment of a 'sexy' GGA website. But beyond the immediate actions, the same study recommended the propagation of agrarian consciousness among the youth of the country through changes in the SAQA and Rhodes University work (Mukute, 2010). This was informed by an understanding that SAQA has the causal powers to help improve pupil and students perceptions of sustainable and low carbon agriculture in the country; and Rhodes has the causal powers to generate research and teach sustainability that potentially absences indifference towards sustainability in agriculture and natural resources management. Towards this end, a book on the outputs of the Rhodes-SAQA research programme is being composed (Mukute, in press).

Conclusion

Experience from the study suggests that the process of *becoming* is stratified, from the individual to the institutions, from local to national, to global, from actions to practices. The transformations that occurred in the study focused on those aspects that were within the participants' immediate spheres of influence: the formation of an organic association, the establishment of a committee to overcome challenges, and the joint setting up of an MFS demonstration site in one district. But tackling the deeper layers of the issues at a wide scale needed the co-operation and participation of stakeholders who did not participate in the study. For example, making agriculture, especially sustainable agriculture, 'cool' required the joint efforts of those systems: farming, education, innovation and development, information and marketing and beyond. In tackling this, the study made certain recommendations to SAQA that were likely to influence the education system, which underpins the development of other systems. However, in subsequent post-doctoral studies (in 2011 and 2012) I utilized the tool on establishing farmer learning and innovation forums to enable 32 newly established organic farmer associations in one province of Zimbabwe to *become* sites of learning and innovation (Mukute, 2012). At another level, the study drew on dialectical critical realism's concept of possibility to write about the 'history of the future of sustainable agriculture' beyond the study sites (Mukute 2010). The concept of possibility was built on the foundation of CHAT's Zone of Proximal Development (ZPD), which was extended beyond the individual, the community and the country, to the global level as embodied in the following excerpt of the ZPD description from the study:

> Between the surge of sustainable agriculture and the fall of conventional agriculture, lies the zone of proximal development of the dominant form of agriculture today. Within this space, several, multilayered transformations happened in different places and at different times.... Today, in 2012, local, national and international policies are

> supportive of sustainable agriculture, where the bulk of government budgets on agriculture stimulate *responsible production processes and citizenship*.... School and college *curricula* are inspired by sustainability which combines the ecological, social and economic ... giving birth to new conceptualisations of quality and relevance, generating cognitive justice (Visvanathan, 2006), and reflexive justice (Lotz-Sisitka, 2010).... Isn't it amazing – how the thoughts and deeds of the long dead live in us, in our practices and tools, as will our thoughts and deeds of today live on in the practices and tools of the future?
>
> (Mukute, 2010)

The study shows that dialectical critical realism: (a) is a suitable ontological theory to underlabour the epistemological and constructive theory, CHAT; and (b) provides the necessary conceptual and methodological tools for engaging with social-ecological and manufactured risks that are being faced in sustainable development and associated education and transformation challenges. Underlabouring CHAT with dialectical critical realism offers one greater chances of achieving ontological depth in that it helps one minimize the constructivist limitations in CHAT. The processes of absenting and becoming are consistent with CHAT's project of expansive learning towards shared objects between interacting activity systems across space and time, and its interest in the development of new human activities in ways where learning leads development (Vygotsky, 1978). The study, using dialectical critical realism lenses, suggests that the becoming of sustainability in agriculture is a process that requires structural transformation in government, business and civil society sectors – well beyond the farmer herself.

Acknowledgements

This chapter is based on a PhD study undertaken by the author in the Environmental Learning Research Centre at Rhodes University (Mukute, 2010). The study was supported through a scholarship provided by the South African Qualifications Authority and formed part of the Rhodes University/SAQA research partnership into change-oriented learning and sustainability practices in workplaces.

References

Baker, R. (1986). *Philosophies of education: An introductory course*. Harare: College Press.

Beck, U. (2000). *World risk society*. Cambridge: Polity Press and Blackwell Publishers.

Benton, T., and Craib, I. (2001). Critical realism and the social sciences. In T. Benton, and I. Craib (Eds), *Philosophy of science: The foundations of philosophical thought* (pp. 119–139). Basingstoke: Palgrave.

Bhaskar, R. (1998). Facts and values: Theory and practice. In M. Archer, R. Bhaskar, A. Collier, T. Lawson, and A. Norrie (Eds), *Critical realism essential reading* (pp. 409–417). London and New York: Routledge.

Bhaskar, R. (2008). *Dialectic: The pulse of freedom.* London: Routledge.

Billett, S. (1994). Situated learning: A workplace experience. *Australian Journal of Adult and Community Education, 34*(2), 112–130.

Bourdieu, P. (1990). *The logic of practice.* Stanford: Stanford University Press.

Dean, K. (2009). Agency and dialectics: What critical realism can learn from Althusser's Marxism. In K. Dean, J. Joseph, J. M. Roberts, and C. Wight (Eds), *Realism, philosophy and social science.* Hampshire: Palgrave and Macmillan.

Edwards, A. (2005). Let us get beyond community and practice: The many meanings of learning by participation. *The Curriculum Journal, 16*(1), 49–65.

Engeström, Y. (1987). *Learning by expanding: An activity theoretical approach to developmental research.* Helsinki: Orienta-Konsultit.

Engeström, Y. (2001). Expansive learning at work: Towards an activity theoretical re-conceptualisation. *Journal of Education and Work, 14*(1), 133–156.

Engeström, Y. (2010). From design experiments to formative interventions. Helsinki: University of Helsinki.

Engeström, Y., and Sannino, A. (2010). Studies of expansive learning: Foundations, findings and future challenges. *Education Research Review, 5*(1), 1–24.

Lister, R. (2004). *Poverty.* Cambridge, UK: Polity Press.

Lotz-Sisitka, H., Belay, M., and Mukute, M. (2012). The 'social' and 'learning' in social learning research: Avoiding ontological collapse with antecedent literatures as starting points for research. In H. Lotz-Sisitka (Ed.), *(Re)views on social learning literature: A monograph for social learning in natural resources management and environmental education* (pp. 58–88). Grahamstown/Howick: Environmental Learning Research Centre, Rhodes University/Environmental Education Association of Southern Africa/Southern African Development Community Regional Environmental Education Programme.

Mingers, J. (2010). The contribution of systems thought to critical realism. *Journal of Critical Realism, 10*(3), 303–330.

Mukute, M. (2010). *Exploring and expanding learning processes in sustainable agriculture contexts.* Unpublished PhD thesis, Rhodes University, Grahamstown.

Mukute, M. (2012). Bridging and enriching top-down and participatory learning: The case of smallholder organic conservation agriculture farmers in Zimbabwe. *Southern African Journal of Environmental Education, 28,* 75–93.

Mukute, M. (in press). *Putting developmental work research into practice: A manual for agricultural knowledge and innovation facilitators.* Unpublished draft, Rhodes University, Grahamstown.

Norrie, A. (2010). *Dialectic and difference: Dialectical critical realism and the grounds of justice.* Routledge: New York.

Nunez, I. (2013). *Critical Realist Activity Theory: An engagement with critical realism and cultural-historical activity theory.* London: Routledge.

Parker, J. (2001). Social movements and science: The question of plural knowledge systems. In J. Lopez and G. Potter (Eds), *After postmodernism: An introduction to critical realism* (pp. 251–259). London: Athlone Press.

Pesanayi, T. V. (2008). Investigating learning interactions influencing farmers' choices of cultivated food crops. Unpublished master's thesis, Rhodes University, Grahamstown.

Pihlaja, J. (2005). *Learning in and for production: An activity-theoretical study of the historical development of distributed systems of generalising.* Helsinki: University of Helsinki, Education Department.

Price, L. (2008). Critical Realism and Education Conference: Notes from Bhaskar's plenary presentation. PhD lecture notes, Department of Education, Rhodes University, Grahamstown.

Sannino, A. (2008). From talk to action: Experiencing interlocution in developmental interventions. *Mind, Culture and Activity, 15*, 234–257.

Sayer, A. (2000). *Realism and social science.* London: Sage.

Simovska, V. (2008). Learning in and as participation: A case study from health promoting schools. In A. Reid, B. B. Jensen, J. Nikel, and V. Simovska (Eds), *Participation and Learning: Perspectives on education and the environment, health and sustainability* (pp. 61–80). Bath: Springer.

Visvanathan, S. (2006). Alternative science. *Theory, Culture and Society, 23*, 164–169. Retrieved 16 November 2009, from www.tcs.sagepub.com/cgi/content/refs/23/164.

Vygotsky, L. S. (1978). *Mind in society: The development of higher psychological processes.* Cambridge: Harvard University Press.

Wals, A. J. (2007). Learning in a changing world and changing in a learning world: Reflexively fumbling towards sustainability. *Southern African Journal of Environmental Education, 24*, 34–62.

Wigley, J. J. (2006). *Understanding workplace-based learning contexts to inform curriculum development: The case of a Level 5 Environmental Education, Training and Development Practices Qualification.* Unpublished master's thesis, Rhodes University, Grahamstown.

12 Community learning as a passage through the dialectic?

Engaging with absences in an irrigation scheme in Mozambique

Aristides Baloi

Introduction

In this chapter I examine the social learning and transformation that occurred in a community-based irrigation scheme in Mozambique. The irrigation association in Mozambique that forms of the focus of this applied research case study is called the Massaca Irrigators Association. It was founded in 1996 at Boane district in the Umbeluzi catchment in response to government policies that established a framework for community-based irrigation schemes in Mozambique to strengthen and expand farming practices and food production in response to poverty alleviation in Mozambique.

The river that supplies the Massaca irrigation scheme is a prolific source of water, even in drought seasons, and therefore has the potential to fully cater for the daily needs of farmers. The irrigation association comprises 183 hectares of land and 230 households. On average, each household owns just less than one hectare. The householders practise gravity irrigation to deliver water to their plots. They grow vegetables such as tomatoes, sweet potato, pawpaw, banana and maize. Some maize is consumed by the farmers' households, but most of the crops are sold locally or in the nearby markets of Maputo city. The householder farmers pay a water fee each month to cover the running costs of the irrigation system. The Massaca Irrigator Association (henceforth termed the Association) is run by a general assembly, a management committee and a fiscal council. It is described as being 'participatory', which means that the community members themselves are responsible for running it. The Association's main function is to ensure the adequate supply of water to the farmers for the purpose of irrigating their crops. However, at the start of my research this function was not being adequately achieved. The significance of this from an emancipatory perspective is that successful participatory irrigation scheme management can substantively improve the livelihood opportunities for poor households, improving their food security, as well as their ability to generate cash for survival in the cash economies.

My research within this context had two objectives:

- Objective 1 was an emancipatory objective, looking to increase the well-being of the irrigation project participants by supporting their ability to manage their irrigation project via an expansive learning process that reflexively interrogated the existing situation, and then modelled solutions and alternatives with and by those involved in the scheme's management.
- Objective 2 was an applied research objective, exploring the usefulness of dialectical critical realism, especially the concept of absences within a CHAT methodology to guide thinking about and interpretation of learning and transformation.

Theoretical and methodological framing

One of the issues being addressed in this study was to understand the emancipatory potential of learning for improving well-being, as outlined in Objective 1. There is an expanding body of literature on transformative learning, but it often focuses mainly on the cognitive dynamics of transforming perceptions, cognition or identity (e.g. Mezirow, 1995). In this study I sought to consider the way in which learning could potentially expand the social practice of irrigation scheme management, and I therefore sought theories of learning that could explain changes in the development of human activity, as supported by learning. One of the more developed theories of expanding human activity through learning is found in the cultural historical activity theory (CHAT) tradition, developed out of the earlier work of Vygotsky, Luria and Leont'ev by Yrjo Engeström and colleagues over the past 20 or more years (Engeström, 1987; Engeström and Sannino, 2010). Thus, to address objective 1, I opted to use a post-Vygotskian CHAT approach to expansive learning as developed by Yrjo Engeström and colleagues (Engeström and Sannino, 2010). Nunez (2013) critiques CHAT for lacking adequate ontological depth and she explains that CHAT is based on an 'implicit ontology' that allows activity theorists to work with an implicit form of critical realism. In response to this critique, I sought to underlabour CHAT with critical realism. I not only sought to make the implicit ontology more explicit within CHAT using critical realism, but I also sought methodological guidance from CHAT to facilitate learning through the passage of the dialectic (Bhaskar, 1993; Nunez, 2013). Bhaskar (1993) suggests that transformative learning is a passage through the dialectic and that this occurs as a form of transformative praxis, little pedagogical or methodological guidance is provided within critical realism as to how this process of learning can be facilitated or enabled. CHAT, via recent work produced by Engeström and colleagues to develop the concept of expansive learning (Engeström and Sannino, 2010), provides a sophisticated methodological process that potentially allows for considering learning as passage through the dialectic. Through this, I was able to engage with Objective 2, which was to deepen CHAT expansive learning processes as a form of applied research using Dialectical Critical Realism (DCR). These theoretical and methodological tools allowed me to surface and interpret the learning processes, practices and transformation in the irrigation scheme. Both Vygotsky and Bhaskar appear to share emancipatory objectives and while both share a commitment to historical materialism, Vygotskian and critical realist theorists do not always use the same languages of description. This can be understood from the vantage point of critical realism being an underlabouring philosophy (Bhaskar, 1993), while post-Vygotskian activity theory provides theoretical tools for applied research within developmental psychology and education (Engeström and Sannino, 2010).

In particular, in this research into expanding irrigation system management practices within a transformative learning frame, I explicitly made use of dialectical critical realist concept of *absenting absence.* According to Norrie (2010), absence operates at different levels:

- in *ontology* (the casual change in 'state of affairs'), absence is what engenders change;
- in *epistemology* (the identification and elimination of mistakes), absence is central to the process argument and, by extension, the development of knowledge; and
- in *moral praxis* (the identification and elimination of constraints or ills) absence is the incentive of freedom.

To underlabour CHAT I sought to identify what Bhaskar (1993, p. 392) calls real causes and generative mechanisms influencing the activity of irrigation scheme management the learning of new irrigation scheme management practices in the activity systems. This involved identifying 'real negativity' and contradictions, and things that are absent. I sought to identify things that may be absent from the community irrigation scheme, and especially also to identify if the absence of something may have causal effects on objects in the relevant space-time region. For example, this could be a technology that is not effectively transferred which could affect whether the irrigation scheme works or not (Norrie, 2010). This would be an example of a nominal absence. A real absence associated with this might be the continuities of poverty and marginalization of the poor in development settings meant to service their needs and well-being.

Norrie (2010, p. 14) states that 'The world, including the natural world, contains absences, omissions, liabilities, just as much as presences, commissions and powers.' For transformative social learning research, I considered it important to identify absences, omissions and liabilities affecting the activity systems as these could potentially also be the sources of transformative learning and action. Absence is a key concept in dialectical critical realist research. Dialectic (which has been recommended for participatory practice as noted above) at its most complete, involves 'the absenting of constraints on the absenting of absences, or ills' (Bhaskar, 1993, p. 396). This discussion of absences also indicates that which is present (e.g. an irrigation scheme committee) is a 'tiny, but important ... ripple on the surface of a sea of absence' (ibid.). Absences include those things that are erased, hidden, omitted, forgotten or excluded (amongst others).

The CHAT literature used to inform this research was drawn mainly from a series of publications of Yrjo Engeström (1987, 1999, 2000) and colleagues (e.g. Daniels *et al.*, 2009; Daniels *et al.*, 2007; Engeström and Sannino, 2010). In particular, I used Engeström's third generation activity theory to develop insights into the contradictions and possibilities for expanding the object of irrigation system management as partially shared by different interrelated activity systems (AS) relevant in the irrigation scheme development context.

To expand learning within a process that was also a process of absenting absences, CHAT's change laboratory method (Engeström, 2007; Virkunnen and Newham, 2013) was used as a reflexive framework and methodological process for engaging with the contradictions and absences associated with the object of irrigation scheme management. In this process I was able to engage, as an interventionist researcher with explicit emancipatory research objectives, with the project participants from the interacting activity systems in a dialogical process of mutual learning.

Through the change laboratories, participants from the activity systems and I were able to engage with tensions and contradictions that existed between the main activity system of the farmer irrigators, and other activity systems that were linked to this main activity system. Together we identified contradictions that were arising in the irrigation management practices, and also the absences, and from this we engaged in conciliation and implementation of 'model solutions'. This was done via a series of change laboratory workshops over a period of time. Present at the workshops were all the members of the Association's committee, relevant community members, extension personnel, educators and representatives from relevant commercial organizations such as those that buy the Association's produce.

Change laboratories after Engeström's methodology use what is called 'mirror data' to surface contradictions, providing a means of 'double stimulation' for deliberation and problem engagement (Engeström and Sannino, 2010). Mirror data is obtained by interventionist researchers from various interactions such as examination of the association's status in documents, or individual interview and preliminary focus group data sources. From this initial contradictions between elements of the activity systems, or between different activity systems are identified for discussion with members of activity systems. Contradictions are identified using historical object data (i.e. data on the history of irrigation scheme management) and via actual empirical data (what is visible and known to participants) (Engeström and Sannino, 2010). In this research I also sought to identify underlying mechanisms and structural factors shaping contradictions.

In Engeström's expansive learning process, he emphasizes the development of 'model solutions' to respond to contradictions and tensions that arise in activity systems; he does not explicitly couple this with 'absenting absences', although such a process may be implied. In this research I was curious to see if working more explicitly with the idea of 'absenting absences' through the model solutions may enhance the transformative outcomes of the expansive learning processes in the context of the irrigation scheme development. I was also curious to see if such a process may also help to identify ongoing possibilities for transformative praxis in the irrigation scheme's ongoing development as related to Bhaskar's notion of totality, which involves relating the negativity or absence as identified at 2E to 'totality' at 3L where wider possibilities for how things might be could come into being via transformative agency. This allows for the dialectic to be engaged.

Norrie explains that Bhaskar's work develops a crucial link between absence as ontologically real, and absence as negation (Norrie, 2010). Norrie (2010) argues that this is important as it shows the dynamic nature of the world, and that processes of change are possible. Norrie (2010, 15) states that it is at this level that the 'processural categories of negation, contradiction, development, becoming, emergence [transformation?]' are important. For transformative social learning research, these processes of negation, contradiction, development, becoming and emergence, i.e. transformation, would seem to be important to monitor and understand. The methodology of CHAT encourages identification of contradiction, development and emergence within the expansive learning process. Through critical realist work it would seem possible to add an emphasis on negation and absenting of absence to this process to strengthen its transformative intent.

Absenting of absence requires transformative praxis (4D), or agency, which involves the capacity for practical human agency to change the world, or 'agential ethical praxis' (Norrie, 2010, p. 15). Norrie explains that this involves acting according to ethical positions of 'what it means to be free, and to enter into relations of solidarity with others'; which could also include 'the environment on which we depend' (ibid., p. 17) as is reflected in sustainability thinking. The sustainability and participatory concepts used in this study establish an ethical direction for the irrigation system management practices. Norrie argues, however, that while these concepts set ethical direction, they should be contextualized in practices that matter to people (e.g. poverty alleviation practices). Giving attention to these ethical questions in the expansive social learning processes may also influence the way in which transformative social learning is conceptualized, understood and monitored.

At a broader level, the emphasis on sustainability in Learning for Sustainability in some ways denotes an 'absence' that requires ongoing attention in social learning processes (see also Lotz-Sisitka, Chapter 17 in this volume); signalling transformative intent, hence developments in the field of environmental education that are related to Learning for Sustainability, are also relevant to this study, particularly if sustainability is seen as a general 'absence' in the way that irrigation scheme developments have emerged, and may emerge. However, in interpreting sustainability as an 'absence' it is also important to consider Bhaskar's point that 'natural necessity' and real ethical praxis (Norrie, 2010) requires contextualization of the notion of sustainability to the Mozambican society context at the level of transformative praxis (4D), while understanding its importance at a wider global level (2E).

The research was also guided by what Sobh and Perry (2005) describe as critical realism principles; specifically it assumes that reality is 'real' but only imperfectly and probabilistically apprehensible and so triangulation and retroductive analysis from many sources is required in trying to know it. The researcher is value-aware and therefore needs to triangulate any perceptions that are collected, via a process of self-reflexivity.

Activity systems interacting within the Massaca Irrigators Association

In the case of the Massaca Irrigators Association I identified the following interacting activity systems who were engaged with the partially shared object of irrigation scheme management: (a) community activity system, (b) educational activity system, (c) irrigation management committee activity system and (d) sellers' activity system. These are briefly introduced, mainly to show their diverse interest and engagement with the shared object of irrigation scheme management.

a Community activity system

The main interest of the community activity system in relation to the shared object of irrigation scheme management is to ensure that the community is able to use this scheme for crop management. They need skills for operation and management of the irrigation scheme, and have a range of rules that guide their shared use of the irrigation scheme and their farming practices which includes irrigation policy, guidelines, membership mandates, and the statutes of their association. Farmers are mainly from local households and they obtain mediation tools from courses and training programmes; they undertake different tasks, and work with other community members to implement the activity. All of these dimensions of the activity system have historical roots and are also related to cultures of practice and socio-cultural histories (e.g. levels of education of farmers; the history of poverty that has affected and continues to affect farmers in Mozambique).

b Educational institution activity system (EIAS)

The EIAS aims to provide knowledge at different levels, both formal and informal, in irrigation activities as shown in Figure 12.1. The EIAS subjects are facilitators and lecturers from Agricultural Colleges and extension services. They use manuals, flyers, workshops and other mediating tools, as well as demonstrations. Their activities are undertaken in workshop settings, but also in fieldwork settings, and they too are governed by various rules such as extension policies, educational institution mandates and policies. These also have historical and cultural roots, for example extension training is traditionally undertaken by the local Extension Service, while formal training is offered by colleges due to their historical structuring and mandates.

c Irrigation management committee activity system

This activity system comprises an elected committee and the object is to guarantee the effective functioning of the irrigation system. One of their tasks is dispute resolution, and they use guidelines offered by policy and by agreed mandates to manage the activity system. They also give attention to trade and market-based rules, and interact with buyers from the nearby city and government officials. They are further responsible for maintenance of the irrigation pumps, and for managing the financial affairs of the communal irrigation system. Historically they have been active since 1996, but continuity is one of the problems, especially in terms of participants in the management committee.

d Sellers' activity system

This includes the sellers and the buyers, both the producers selling and supplying the fresh products and buyers, namely the local markets and the supermarkets in Maputo city. Their main object is to obtain trade products supplied by food producers, and it is in this role that they interact with the community irrigation scheme farmers and committee.

e Interactions amongst the activity systems around a shared object

The interrelations among the activity systems can be represented as indicated in Figure 12.1, which shows that each of the activity systems contributes in some way or other to the common object of successful irrigation

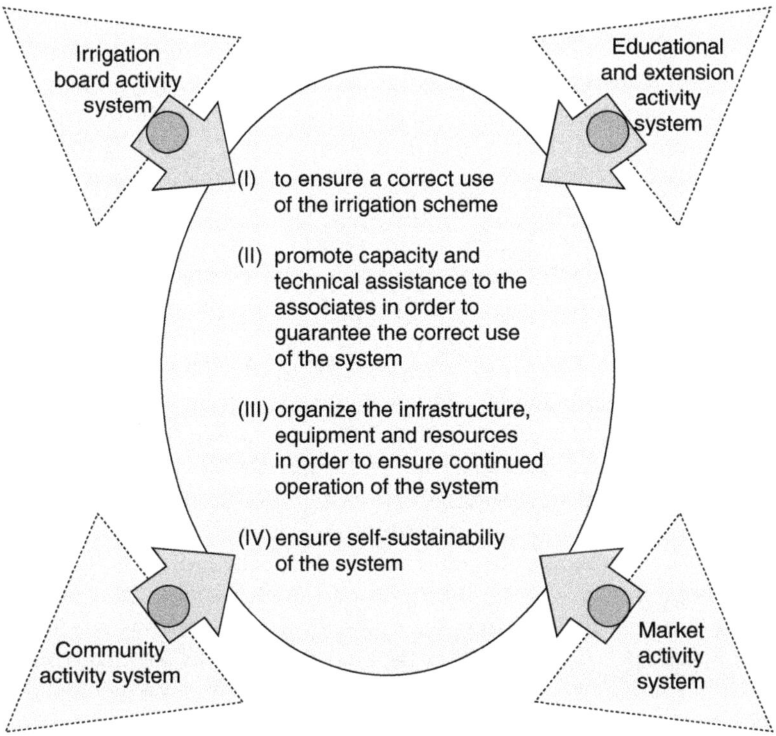

Figure 12.1 Interacting activity systems showing shared concern for the object of irrigation management (source: developed by the author).

management for sustainable food production. Understanding how the different activity systems interact and come together around a shared object 'successful irrigation management for sustainable food production' provides the starting points for learning and transformation. However, to deepen this understanding, it is necessary to analyse the contradictions within the activity systems, and the deep-seated structural tensions or generative mechanisms that shape these contradictions (1M in Bhaskar's DCR). It is also helpful to understand absences (2E in Bhaskar's DCR).

Contradictions and absence

Contradiction typically indicates developmentally significant systemic contradictions and change potentials within the activity. In other words, while the object and motive give actions coherence and continuity, by virtue of being internally contradictory, they also keep the activity system in constant instability (Engeström, 2000). They result in double binds in everyday practices when an individual receives 'two messages or commands which deny each other' (Engeström 1987, p. 174). CHAT contradictions are identified and categorized according to the base structure of activity system, between different activity systems or within an activity system at different stages of development (Engeström, 1987). Contradiction is derived from dialectic and is defined as deep-seated structural tensions that may both hinder and facilitate the development of the activity system (ibid.). In CHAT, contradictions unavoidably exist in activity systems, being the major mechanism for their dynamism and transformation. In Bhaskar (1993), the term 'contradictions' does not always equate to the more common definition used in formal logic, the logical incompatability between two or more propositions. Instead, a contradiction is identified whenever two or more tendencies negate or undermine one another. Hartwig (2007: 82) suggests that 'Instead of fearing contradictions ... we should exploit them for their *dialectical fertility*' (emphasis original). He goes on to say that 'the materialist dialectic differs from the Hegelian in that it is *simultaneously* dialectical and contradictory', involving what Bhaskar (1993, pp. 62–63) describes as 'a simultaneity of grounded (transfactual) presence and (actual) absence, of practical (existential) inclusion and mutual (tendentially transformative) exclusion. It is this that makes it genuinely dialectically contradictory in a stratified ontology that pre-exists the discourse that describes it'. Hartwig (2007, p. 82) explains further that the resolution of such contradictions 'can only be practical ... through what Bhaskar theorizes as a transformed, transformative, totalizing transformist praxis'. What this means in the Massaca Irrigation Association is what I turn to next. I identified the following contradictions within the community activity system (CAS), the irrigation board activity system (CBAS) and within the educational activity system (EIAS).

Contradictions within the community activity system (CAS)

In the CAS, a key contradiction was found to exist between the rules, mechanisms and procedures (rules) meant to be in place for water delivery according to the irrigation scheme policies, and successful participation in food production activities in this activity system (object). This was found to be shaped by historically constituted political, economic, ecological and mental and physical structures associated with tendencies affecting the situation in the form of shortage of water (ecological structure), ownership of land (economic structure), and poor quality education and technical capabilities (mental and physical structures). At a practical level, these manifested as follows.

1 Due to a shortage of water delivery to the plots, some households are left with no option but to practise a less reliable form of rainfed agriculture affecting regularity of cropping and thus capacity to sell to the market.

2 Due to a shortage of water some households situated further from where the water first enters the channels are at times unable to get water, which constitutes a source of conflict within households and affects the capacity of some to produce for market.
3 Poor understandings of the meaning of irrigation system and management, related to a lack of formal education on irrigation agriculture, leads to inadequate responses to the irrigation system management, and consequently to uneven access to the water described above, and to some household plots not being used.
4 Some households do not pay water fees and there is deficient maintenance, which also affects food production.
5 Shared responsibility for the irrigation scheme management is also affected by ownership of the plots, as some household plot workers are not the owners of the plots. The implication of this is that the owners do not participate in the meetings, they let their workers attend and thus they are not able to contribute to the resolution of the wider irrigation scheme management concerns.

Contradictions within the irrigation board activity system (CBAS)

Within the community-based irrigation board activity system, there was a key contradiction that existed between the farmer and irrigation scheme management committee's (subject) abilities to ensure successful management of the irrigation scheme on behalf of the community and the economic and other resources at their disposal to do so (mediating tools). This contradiction was found to be shaped by historically constituted economic structures (market-based economic mechanisms used in communal management schemata), and availability of appropriate physical structures (lack of technical equipment) as well as lack of the forms of knowledge and capacity necessary for the new practice. These manifested as follows at the level of practice.

1 Within community activity system, there is poor understanding of the meaning of irrigation management as this has not been a topic of previous education and/or training.
2 Issues are experienced with the organizational structure of the irrigation management scheme.
3 Failures to pay invoices due to lack of knowledge of the market-based economic system, which was also visible in work patterns that resulted in the CBAS only working when there was an issue to resolve, and a lack of clarity surrounding contract agreements, business principles, and the criteria that were used to determine the cost of water fees. Added to this was the unrealistic economic situation in which the collected fees are not enough to cover power costs, water costs and the maintenance of the irrigation system as a whole (i.e. an unviable economic model).

4 While the water supply to the scheme is adequate, the distribution to individual plots is problematic as some users fail to pay their monthly contribution, and shortages of water are experienced due to a lack of maintenance as broken pipes are not replaced and taps are demolished.
5 There is also poor interaction with and understanding of the broader market activity system, which is meant to provide viability to the communal irrigation scheme practices. Price changes in the market follow a liberal economic model, and fluctuate depending on the product availability and demand, and at times prices are very low. Companies seek more regularly supplied quantities of products, sometimes on a daily basis, and the householder farmers and the association cannot afford the costs of ensuring such a regular supply, especially in the face of other problems outlined above.

Contradiction in the educational activity systems (EIAS)

The main contradiction identified within the educational activity systems was a contradiction within the mediating tools which was constituted by both a lack of resources for offering training to communities and irrigation scheme members, as well as the fact that the available curricula were not relevant to the farmers and were based on more commercial forms of agriculture, and did not include a response to the new policy on community-based irrigation scheme management.

Consequences of the structural contradictions as they manifest in open systems

The structural contradictions manifest in complex open system interactions in which various factors are related and impact on each other. Water provision into the field is inadequate due to the problems with maintaining the pumping equipment and the pipeline system. This results from poor maintenance of the system, as the Association does not have resources either to repair or acquire new pumping systems and the immediate consequence is that the crops lack water. This leads to a vicious cycle whereby poor yields, leading to low sales, mean that the farmers cannot pay their irrigation fees. Some farmers refuse to pay irrigation fees because they do not receive water, which puts the communal system under further pressure. The inadequate water supply thus becomes a source of conflict among the water users. Those farmers near the hydrants and canals tend to have more access to the water and in some cases farmers further away block the supply of water to their neighbours to allow the water to run into their plots. Sometimes is only possible to irrigate parts of a particular plot. If farmers do not plough their plots, invasive plants take hold. These spread quickly to surrounding plots affecting those who are actually producing crops. This affects the overall object of the irrigation association activity system.

Farmers still practise rainfed farming in the irrigation systems because they lack a clear common concept of irrigation management, and have limited knowledge and technical information on irrigation practices, and means to ensure water availability. There are also cultural historical structural factors that shape practices, as some of the members also reject the new knowledge delivered from the extension services. They, for example, say that 'I practise agriculture from a long time ago even before the development of irrigation system, and I know how to do that.' This shows a tension between traditional knowledge embedded in the community that functions to refute new scientific knowledge acquisition, and some of the members are not motivated to learn new practices. This appears to also be linked to a lack of concepts related to the maximization of productivity and cost recovery, i.e. the understanding of the market-based economy and how it works. Most of the community does not have formal knowledge of agriculture, other than that acquired via experience over time.

Many farmers tend to farm according to the seasonal rainfall. This shows a reliance on earlier forms of culturally practised agriculture, which were reliant on rainfed agriculture. The immediate implication is that the market can be flooded with a particular crop at one time, which leads to less cash for farmers (as market prices drop when the market is flooded). This is a waste of the potential of an irrigation system which allows farmers to produce crops throughout the year and therefore benefit from being able to sell crops out of seasons, thus gaining better prices. Another problem is that sometimes the products take a long time to sell and members do not have refrigerators to keep the products fresh.

Farmers are also unclear about the need to be involved in the irrigation management process, and how the Association works; this has implications for their involvement. A curriculum review of the education institution identified the absence of courses on agrarian education, such as courses on irrigation system management and community participation. Overall, the collective production plan for the farmers as outlined by the communal irrigation policy of the country, as initiated by the establishment of the irrigated plots and water resources, whilst present, was not understood by the community. Consequently, most farmers were not following it. This was a key absence that we sought to absent via the expansive learning process.

Absenting absences (solutions to the contradictions)

According to Bhaskar (1993) the category of absence is pivotal to 1M–4D links. Thus a 1M non-identity may generate a 2E absence causing a 3L alienating de-totalization or split-off resulting in a transformed transformative totalizing praxis absenting the split; 2E absence/absenting is the key mediation between 1M non-identity, 3L totality and 4D agency, which has as its prototype the absenting of absence. Dialectics depends upon the positive identification and transformative elimination of absences that means the process of absenting absence. 'Identification of things is a human act, and absents a previous state of non-identification, indicating that absenting is central to a basic human cognitive process' (Bhaskar in Norrie, 2010, pp. 35–36). As a result of the identification of the absences (contradictions) in the analysis of the activity systems, the community members were able to arrive at several solutions to their problems, albeit not all. There is not enough space to report on all their solutions, but I report on two examples below.

One example of a discussion that arrived at a solution (absenting an absence) involved the non-payment of fees and the confiscation of the plots. According to the rules of the Association, the committee has the right to confiscate the plots of those who do not pay their fees. The management committee can then give the plot to another person from the Association or even from outside. However, it was decided that if a plot was confiscated, then that plot would belong to the Association. Profit from cultivating the plots would accrue to Association and would be used to help to pay the maintenance bills, thus reducing the fee requirement for individual members of the Association.

Another example of a discussion that led to action to absent an absence explored the need for a collective production plan. Essentially, the collective production plan was agreed upon and would allow the Association to reduce input costs due to bulk supply of inputs and to ensure profitability by insisting on a focus on cash crop production: crops such as maize, garlic, beans, pimento and tomatoes were most recommended due to their high demand in the cities.

In Bhaskar's conceptualization of learning, he suggests that a learning process, absence (2E) signifies incompleteness, and 'leads to transcendence and a greater totality (3L), in principle reflexively (4D) capable of situating itself and the process whereby it became' (Bhaskar, 1997, p. 144). In the change laboratory workshops, described by Engeström as a mechanism for allowing expansive learning (Engeström and Sannino, 2010), participants were engaged in deliberating on how to absent absences, showing Bhaskar's principle of learning. They said:

> I have learned a lot of things that I did not know its meaning: the advantages and importance of the irrigation, I only knew that I have to irrigate from here to there. Before I did not know the difference between rainfed farming and irrigation farming. I have learned that the Association needs to be well structured and controlled and participation is important.... I have learned many things that we used to hear, but we did not give its value.... I would like to give thanks for the workshop and would like that all we have spoken here gets into

> the mentality of the all and each of us work in order to overcome the problems that were approached here.
>
> Here we have experienced exchange ... from today we will turn the game. I believe that with the water resolutions will allow things to move forward.
>
> We would ask you to monitor the activities in order to have success in this association.... In terms of yield I have started to compare what I use to get as outcome in rainfed and what I get now from irrigation.

These comments are evidence that the community have learned the significance of the irrigation system and have come to better understand some of the issues facing the irrigation system. They have also, as shown by the evidence of these citations, engaged with some of the structurally shaped contradictions and absences. Such reflexivity in irrigation system management might go beyond the given examples. Wals (2007) and Lotz-Sisitka *et al.* (2012) define reflexivity as having a central role in social learning. It involves continuously reflecting on what is learned and taking action to improve what we are doing. An example of this ongoing reflexivity is demonstrated in the last comment where the farmer called for further monitoring of the irrigation system and described an intention to make his own production evaluations. As a result of such reflection, as shown above, further opportunities arise to deal with dissonances and with the deep-seated structural tensions that give rise to the contradictions and absences in the activity systems.

To develop means to resolve or transform these contradictions (instead of merely shifting them elsewhere), thus resulting in a change in the activity system, a new object and motive(s) need to be constructed. Such change is a long-term cyclical and spiral process of internalization and externalization which Engeström (1987) calls learning by expanding. The essence of learning activity is the production of objective, societally new activity structures (including new object, instruments etc.) out of actions manifesting the inner contradictions of the preceding form of the activity in question. Learning new forms of activity involves mastery of expansion from actions to a new activity (Engeström, 1987) and involves reflexive agency (4D) as explained by Bhaskar (1998). In change laboratory interventions, such dialogical and longitudinal relationships form the foundations for practical, material generalization of novel solutions and developmental breakthroughs that have potential to absent absences. These solutions are articulated via the help of the formation of new concepts and models (Engeström, 2007) that become findings that can acquire significant theoretical and practical import. The examples above show absenting the absence of core understanding related to the intentions and processes involved in an irrigation scheme management via an expansive learning process.

Conclusion

A crucial implication of the critical realist ontology is the recognition of the possibility that powers may exist unexercised, and hence that what has happened or been known to have happened does not exhaust what could happen or has happened. The nature of real objects present at a given time constrains and enables what can be achieved but does not predetermine what will happen in the irrigation system. As shown above, the community farmers and irrigators may have ideas and powers that were not activated. Learning for transformation means activation of those unexercised powers that results in the creation of new concepts and resources, reflected in the development of new activity in the activity systems. This case study documents such learning as the power of the community to transform their irrigation system from a non-functioning entity to a (potentially) functioning one. The research shows that this power was always present within the community and merely needed to be expressed. However, further and more deeply embedded structural constraints necessary for enabling community members to express their unexercised powers needed to be removed. In this case, a lack of a framework to understand (learn) about the system was one of these structural factors preventing the efficient management of the system. There are others too, such as the difficulties of aligning with the structural dynamics of the market-based economy and how it functions, and the deep-seated structural influences of poverty and poor quality education, and a lack of access to education and technical and adequate resources for managing the irrigation scheme.

This can also be interpreted within a wider learning for sustainability context, where social learning and particularly transformative approaches to social learning have gained currency in the field of environmental education and Education for Sustainable Development in southern Africa (Mukute, 2010; Masara, 2011; Belay, 2012) and elsewhere. Tàbara and Pahl-Wostl (2007) point out that social learning, particularly within a vision of wider societal change, has been associated with the concept of sustainability. Milbrath (1989) in his work envisioning a sustainable society used the phrase 'Learning Our Way Out', signalling a transformative process of negation. In his view social learning occurs when a dominant institution is replaced by another (or when an absence of a new institution has been absented). This means a change or transformation of societal practices and norms that are shared by a large number of actors. In the Mozambican context of communal irrigation scheme management, this may involve wider learning and change as outlined in the Massaca Irrigation Scheme management case.

Tàbara and Pahl-Wostl (2007), drawing on Milbrath (1989, pp. 85–87) suggest that for social learning to embrace sustainability, people must learn: (1) to become conscious of their ways of knowing; (2) to understand the critical roles played by values and beliefs in the structuring of reality, and that science is not value free (which is reflective of Bhaskar's emphasis on generative structures and mechanisms influencing our actions); (3) to reason together in public debate about their values to redirect scientific and societal development (reflective of Bhaskar's focus on ethics-based agency); (4) to appreciate the complexity and interconnectivity of ecosystems and their implications for social action; (5) to think holistically, systemically and integratively; (6) to not separate human societies from nature, but to live in harmony with nature, rather than dominate it (reflective in some ways of Bhaskar's absences); (7) to avoid interfering with nature's systems and cycles and recognize the limits to growth; and (8) to emphathize with and extend our compassion to people of other lands, other species and future generations to preserve the integrity of the ecosphere and survival of all (which is reflective of Bhaskar's totality). Milbrath (1989) and Tàbara and Pahl-Wostl (2007) point to the need for a 'learning society' that has capacity to move towards a more sustainable world order.

Drawing on Bhaskar (2002) this case study shows that a dialectics of change towards socio-economic-ecological well-being is fostered by generative mechanisms that shape individual and collective learning conditions including absences/contradictions, truth and motivation. The expansive learning process of CHAT allowed the participants in this particular case to develop potential solutions to problems, providing an expansive learning process methodology for enabling irrigation scheme participants to experience learning as a 'passage through the dialectic' (Nunez, 2013). It addresses an issue raised by Nunez which suggests that CHAT is based on what she calls an 'implicit ontology' that allows activity theorists to 'presuppose (aspects of) critical realism in practice' (Nunez, 2013, p. 141), making the ontology shaping expansive learning in CHAT more explicit.

Acknowledgements

This chapter is based on research undertaken for a PhD study (Baloi, 2015) in the Environmental Learning Research Centre at Rhodes University, South Africa. The study was supported by the South Africa–Netherlands Partnership for Academic Development, and by the Murray and Roberts and Gold Fields Scholarship Trust Funds of the Environmental Learning Research Centre.

References

Baloi, A. (2015). *Exploring transformative social learning and sustainability in irrigation scheme contexts in Mozambique.* Draft PhD study, Rhodes University, Grahamstown.

Belay, M. (2012). *Participatory mapping, learning and change in the context of biocultural diversity and resilience.* Unpublished PhD thesis, Rhodes University, Grahamstown.

Bhaskar, R. (1993). *Dialectic: The pulse of freedom.* London: Verso.

Bhaskar, R. (1997). On the ontological status of ideas. *Journal for the Theory of Social Behavior, 27*(2/3), 135–147.

Bhaskar, R. (1998). *The possibility of naturalism.* London: Routledge.

Bhaskar, R. (2002). *Meta-Reality: The philosophy of Meta-Reality. Volume 1: Creativity, Love and Freedom.* London: Sage.

Daniels, H., Leadbetter, J., and Warmington, P., with Edwards, A., Brown, S., Middleton, D., Popova, A., and Apostolov, A. (2007). Learning in and for multi-agency working. *Oxford Review of Education, 33*(4), 521–538.

Daniels, H., Edwards, A., Engeström, Y., Gallagher, T., and Ludvigsen, S. (Eds). (2009). *Activity theory in practice: Promoting learning across boundaries and agencies.* London: Routledge.

Engeström, Y. (1987). *Learning by expanding: An activity theoretical approach to developmental research.* Helsinki: Orienta-Konsultit.

Engeström, Y. (1999). Expansive visibilization of work: An activity-theoretical perspective. *Computer Supported Cognitive Work, 8,* 63–93.

Engeström, Y. (2000). Comment on Bhaskar *et al.* Activity Theory and Social Construction of Knowledge: A story of four umpires. *Organization, 7*(2), 301–310.

Engeström, Y. (2007). Putting Vygotsky to work: The change laboratory as an application of double stimulation. In H. Daniels, M. Cole, and J. V. Wertsch (Eds), *The Cambridge Companion to Vygotsky* (pp. 363–382). Cambridge: Cambridge University Press.

Engeström, Y., and Sannino, A. (2010). Studies of expansive learning: Foundations, findings and future challenges. *Educational Research Review.* doi:10.1016/j.edurv.200212.002.

Hartwig, M. (Ed.). (2007). *Dictionary of critical realism.* London: Routledge.

Lotz-Sisitka, H., Belay, M., and Mukute, M. (2012). The 'social' and 'learning' in social learning research: Avoiding ontological collapse with antecedent literatures as starting points for research. In H. Lotz-Sisitka (Ed.), *(Re)views on social learning literature: A monograph for social learning in natural resources management and environmental education* (pp. 58–88). Grahamstown/Howick. Environmental Learning Research Centre, Rhodes University/Environmental Education Association of Southern Africa/Southern African Development Community Regional Environmental Education Programme.

Masara, C. (2011). Learning commercial beekeeping: Two cases of social learning in southern African community natural resources management contexts. Unpublished master's thesis, Rhodes University, Grahamstown.

Mezirow, J. (1995). Transformation theory of adult learning. In M. Welton (Ed.), *In defense of the lifeworld: Critical perspectives on adult learning.* New York: State University of New York Press.

Milbrath, L. W. (1989). *Envisioning a sustainable society: Learning our way out.* New York: State University of New York Press.

Mukute, M. (2010). *Exploring and expanding learning process in sustainable agricultures workplace learning.* Unpublished PhD thesis, Rhodes University, Grahamstown.

Norrie, A. (2010). *Dialectic and difference: Dialectical critical realism and the grounds of justice.* New York: Routledge.

Nunez, I. (2013). Transcending the dualisms of activity theory. *Journal of Critical Realism, 12*(2), 141–165.

Tàbara, J. D., and Pahl-Wostl, C. (2007). Sustainability learning in natural resource use and management. *Ecology and Society, 12*(2), 3. www.ecologyandsociety.org/vol. 12/iss2/art3.

Sobh, R., and Perry, C. (2005). Research design and data analysis in realism research. *European Journal of Marketing, 40*(11/12), 1194–1209.

Virkkunen, J., and Newham, D. S. (2013). *The Change Laboratory: A tool for collaborative development of work and education.* Rotterdam: Sense Publishers.

Wals, A. J. (2007). Learning in a changing world and changing in a learning world: Reflexively fumbling towards sustainability. *Southern African Journal of Environmental Education, 24*, 34–62.

13 Exploring contradictions and absences in mobilizing 'learning as process' for sustainable agricultural practices

Tichaona Pesanayi

Introduction

Over 60 per cent of the staple the world eats is derived from rain-fed agriculture, a practice that covers 95 per cent of the agricultural area in Sub-Saharan Africa (FAOSTAT, 2005; Wani *et al.*, 2009). Nevertheless investment in its water potential has not been prioritized. Small-scale farming, while of varying importance in South Africa and Zimbabwe (the two countries referred to in this chapter), plays a critical role in mostly rural household food security and incomes in both countries. The quest for a more sustainable solution to agricultural water and appropriate seed systems is a key endeavour for small-scale farmers, but there is an interplay between different, ambivalent and sometimes contradictory information, knowledge and economic systems. Using agri-*cultural* water practices immersed in history and contemporary practices, this chapter explores the surfacing of contradictions and absences, framing possibilities for learning as process in collaborative developmental work in small-scale agriculture. In so doing, it opens up possibilities for formative intervention research and expansive learning to improve water supply, delivery and use challenges. The chapter shows how the emergence of contradictions and absences are related to generative mechanisms that have long-term effects on an object of activity, and how an understanding of these can be important for considering learning as a passage through the dialectic (Bhaskar, 1997; Nunez, 2013, 2014), strengthening especially the questioning phase in the expansive learning methodology used in cultural historical activity theory (Engeström, 1987).

Critical realism, agriculture and educational mediations

The current micro economies of both South Africa and Zimbabwe are characterized by large numbers of people living off the informal sector. Many of them live off the land but in a rapidly changing and variable climate which requires (re-)learning of new or forgotten skills at national and regional levels, and unlearning of cultural practices that deprive families of livelihood capabilities centred around water scarcity and increasing drought (SADC, 2012a, 2012b). According to reporting by the Intergovernmental Panel on Climate Change (IPCC), there is very high confidence that agricultural production and food security in Africa will be compromised (IPCC, 2012). South Africa is classified as a water-stressed and water insecure country, and relies on water imports from neighbouring Lesotho to ameliorate its freshwater requirements (Mashinini, 2010; Malzbender and Earle, undated). Climate change is threatening to worsen this water stress impacting on the agricultural sector from projected increases in temperatures and drought and variable rainfall within seasons (DEA, 2013). Ziervogel *et al.* (2014, p. 611) state: 'Using ... community-based adaptation planning, adaptation has become an iterative co-learning process, rooted in farmers' existing knowledge of what works in practice. Such "rooted" adaptation strategies have included ... soil and water conservation measures.'

In South Africa, more than 600,000 households engage in farming to produce the main source of food for the family. In addition, '*over a million households farm to supplement what they purchase*' (Watkinson and Makgetla, 2002, p. 2, my emphasis). Zimbabwe is also water insecure as demonstrated by repeated cycles of drought-related crop failure (Bird and Shepherd, 2003). The climate change projections for Zimbabwe paint an increasingly risk-prone picture for food security: 'Downscaled empirical models for Zimbabwe reflect similar expectations' for southern Africa of reduced precipitation coupled with high evapotranspiration in some seasons (Brown *et al.*, 2012, p. 4). The economy of Zimbabwe depends on agricultural production to a considerable extent (RBZ, 2005). According to the Reserve Bank Governor 'Agriculture is at the heart of the Zimbabwean economy, accounting for as much as 17% of GDP, about 27% of employment and 33% of total foreign exchange earnings' (RBZ, 2005, p. 6).

Educational processes can improve farmer, extension and educator agency by supporting learning from existing sustainability practices to improve biodiversity and food production (CBD Secretariat, 2012; Atchoarena and Delluc, 2002). But these need to be done well, i.e. they need to be responsive to changing conditions, extension and especially farmer needs. In previous educational research on investigating learning interactions in small-scale farming communities of practice, I found that it was important to understand the causal mechanisms that influence farmers' choices of cultivated food plants (Pesanayi, 2008, 2009), and the influence of persistent poverty on farmers' opportunities and freedoms. Farmers' choices are an expression of freedom that precedes or expands agentive action for sustainable development (Freire, 1998). Reflecting on persistent poverty from a critical realist perspective, Jones (2006) draws on Bhaskar's notion of 'presentism' – the overwhelming presence of the past's undesirable influences. However, the past and the present can be understood together. This need to connect with history and culture to provide a structural explanation to contradictions calls for a dialectical approach to research, which accounts for non-fulfilment of things such as aspirations (Bhaskar, 1993, 2008). 'Dialectics states that the way change takes place is through *conflict* and *opposition* … always looking for the contradictions within people or situations as the main guide to what is going on and what is likely to happen' (Rowan, 2000).

Dialectical critical realism stresses the ontological superiority of absence, where in this case the concept of dialectic means 'absenting of absence' (Bhaskar, 1993). This view of the dialectic goes beyond the Hegelian construction of the dialectic, and challenges postmodern relativist notions of truth. It suggests that absenting absences can be a process of dialectical negation of contradictions that reveal absences, as will be discussed later in this chapter through a historical analysis of water supply and use practice in Lloyd Village in the Eastern Cape, one of the sites of my study.

In my research I am concerned with cross-boundary learning in selected small-scale farming practices nested in and across activity systems. I have been working with Engeström's (1987, 2000; Engeström and Sannino, 2010) cultural historical activity theory and expansive learning approaches to surface tensions and contradictions. This methodology provides potential for enabling learning *as passage* through the dialectic. In this process, absences must be identified so that learning 'emerges as the need to absent what has been left out' (Nunez, 2014, p. xvii). A critical realist approach to activity theory 'sees a relationship between the learner and the world' (Nunez, 2014, p. 3) and involves learners as investigators who can potentially engage in processes of collective and self-emancipation. The dialectic in such a context must 'empower us to understand reality' (ibid.). Basic critical realism states that we have a stratified and differentiated world, and structures and associated generative mechanisms are found in the domain of the real, differentiated from the domain of the actual and empirical (Bhaskar, 1998).

Njihia (2011) argues that critical realism provides tools and concepts that can contribute to transformative emancipatory learning amongst African policy makers, and can generate knowledge that leads to institutionalization of change as it is taught in institutions of learning.

In this study, in combination with cultural historical activity theory (discussed below), dialectical critical realism provides for 'the deepening of ontology (that) enables us to expand the range of categories, e.g. to take in *absence* and negativity, including *contradiction, internal relations*, etc., which may be employed in our laminated explanations of open-systemic phenomena' (Bhaskar, *et al.*, 2010, p. 13, my emphasis).

Methodology and theoretical underpinnings

At a broad level, this research attempts to answer the question: *What learning processes, interactions and opportunities can scale up innovative sustainability practices or promote alternative practices as small-scale farmers cross (or avoid crossing) boundaries in water harvesting and seed saving?* The goal is to generate knowledge on how systems (e.g. extension, research, agricultural education and curriculum innovation and community learning practices) learn or can better learn from each other.

Studying cross-boundary learning in small-scale farming entails the study of complex individual, social and institutional learning and practice phenomena that may or may not be obvious. One way of studying these is to choose an object of inquiry that retains the systemic value of the social system while allowing for independent inquiry. The activity system approach offers such a unit of analysis, where for example the small-scale farming activity system or agricultural education water harvesting activity system is characterized by unique internal relations, tensions and tools at one level; and at another it networks or interacts with the agricultural extension and agri-business activity systems which are also unique. In critical realism, such ontological complexity is viewed as 'laminated', i.e. 'the meshing of explanatory mechanisms at several different levels of reality and possible orders of scale' (Bhaskar *et al.*, 2010, p. ix). The implications for methodology are summed up by the assertion in 'the fact that, in a laminated totality, natural and social determinants will often both be present means that interdisciplinary work must in general employ mixed methods' (Bhaskar *et al.*, 2010, p. 12).

In this study critical realism provides underlabouring for cultural-historical activity theory (CHAT) (Nunez, 2014), which provides a social methodology (in the form of developmental work research, discussed below; see also Figure 13.1) for advancing praxis-oriented learning and change in human activity. In CHAT an activity system is framed by the original Vygotskian mediation triad defined by an object, a subject and mediation artefacts (first generation activity theory), extended to include social relationships, i.e. community, division of labour and associated rules that determine activity. In an open system, activity systems often interact, giving the third generation activity system complex (Engeström, 2001) (see Figure 13.2).

To expand learning in response to contradictions and absences, this study uses a formative interventionist research design, called developmental work research (DWR) and active engagement with change processes (termed also change laboratories), derived from the CHAT developed by Yrjö Engeström, and based on the works of Vygotsky, Luria and Leont'ev (Vygotsky, 1978; Engeström, 1987, 2001). This research approach follows an expansive learning cycle involving a series of stages that are oriented towards expanding human activity via an expansive and interactive, dialogical learning process (outlined in Figure 13.1). The learning that is expanded is situated in activity systems, where people are collectively engaged with object-oriented activity, and the participants are intervening on a common object.

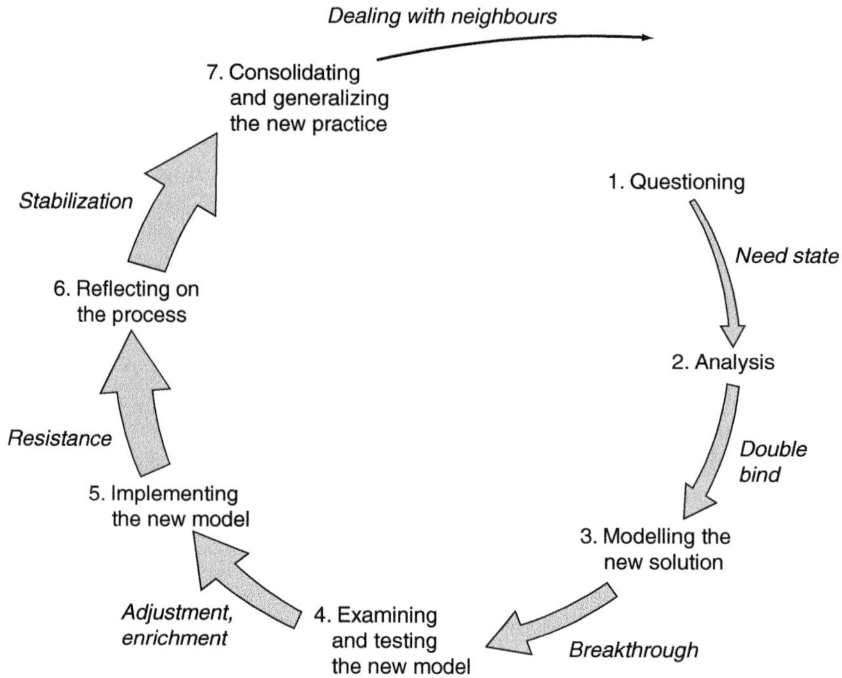

Figure 13.1 The expansive learning cycle used in DWR (source: adapted from Engeström, 2001, and as presented by Engeström, 2015, at a Change Laboratory Training Programme in Helsinki, Finland).

However, to engage with an expansive learning process as outlined in Figure 13.1 above via a formative intervention research process, there is need to understand the activity systems and their objects/shared objects. In this chapter I focus on understanding one boundary object, namely *water supply, delivery and use for productive small scale food growing*. A farming group may be interested in water supply, delivery and use. At the same time an agricultural college may be interested in the same object (as shown in Figure 13.2). Each would have their own interest in the object, but for transformative, boundary crossing praxis to occur, there would have to be a shared common point or engagement.

Such research is needed as it has been recognized that agricultural training is 'disconnected' from the real context of agricultural praxis in southern Africa, and especially so for the small-scale farmer, who is also increasingly at risk in the context of water scarcity, monocultural agricultural dominance and climate change (McIntyre *et al.*, 2009).

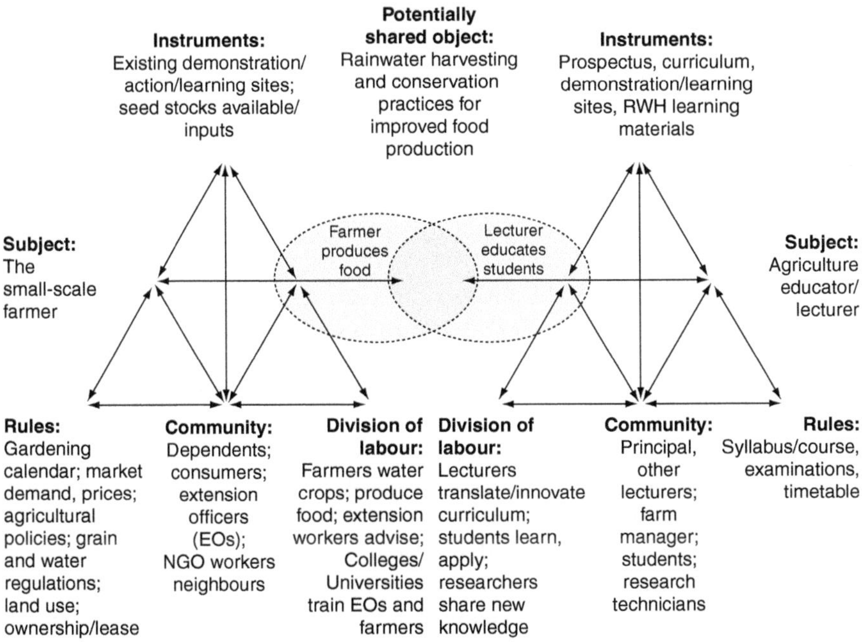

Figure 13.2 Third generation of activity showing small-scale farming activity system interacting with agricultural education activity system with potentially shared boundary object (source: adapted from Engeström, 2001, p. 136).

Historicity and 'following the object': the case of water supply, delivery and use for food gardening in the Lloyd Village Garden

A first phase of analysis involved identification of contradictions, which are deep-seated structural tensions shaped by generative mechanisms as explained by Engeström (1987). Identification of these contradictions requires careful historical analysis, combined with retroductive analysis to identify the generative mechanisms at work. This research process is significant for 'laying the foundations' for praxis oriented expansive learning, and is particularly important for boundary-crossing research and for beginning a process of learning as a 'passage through the dialectic' (Nunez, 2014).

Identifying the contradictions and tensions involves a process of questioning, which is significant in the further expansion of learning. Thus the first phase of the change process recommended by Engeström (2007, 2008) and colleagues (Warmington *et al.*, 2005; Virkkunen and Newnham, 2013) involves actors in an activity system or connected activity systems

(Engeström, 2007, 2008). Engaging in questioning of the activity helps to establish the 'need state' or 'learning as product' as explained by Bhaskar (1997; see also Nunez, 2014). This is also where primary contradictions are identified which provide the foundation for expanding learning (Engeström, 2001). In this regard, Engeström (1999) recommends giving attention to historicity, and suggests 'tracing the object': 'An expansive cycle is a re-orchestration of those voices, of the different viewpoints and approaches of the various participants. Historicity in this perspective means identifying the past cycles of the activity system' (Engeström, 1999, p. 35).

The case of 'tracing the object' in the Lloyd Garden case study (described in more detail below) demonstrates how when one 'follows the object' in cultural historical activity theory as recommended by Engeström *et al.* (2003), it is possible to see the emergence of new human activity over time. Critical realism helps to *explain this emergence* from the perspective of transformative social action, which is also a temporal perspective that accounts for the emergence of action in relation to structure via human agency (individual and collective in relation to four planar social being which includes intrapsychological and material planes) (Bhaskar, 2008). It also helps to identify absences for enabling further learning (Bhaskar, 1993; Nunez, 2014). As can be seen from the example below, emergence is also socio-cultural, socio-political and ethical in addition to being bio-physically linked, and temporal. Not all expansions of learning are transformative in an emancipatory sense, and critical realism helps us to decide 'which way is up' (Engeström, 1991) or how praxis may be considered transformative or not, and how one might develop an emancipatory perspective on praxis.

As recommended by Engeström in the citation above, 'historicity means identifying the past cycles of activity'. To do this, however, it is important to be clear on the object of activity which in this case is *water supply, delivery and use for productive food gardening* (referred to as Object A). This has been a key concern throughout the history of the Lloyd Village as will be seen below, but it has also taken different forms (Objects A1–4). In following the 'object' of water supply, delivery and use of water for productive food gardening in the context of the Lloyd Village Garden area in the Eastern Cape, I identified four object-historical distinctive cycles of expansive learning and change in human activity focusing on supply, delivery and use of water for food gardening. For ease of reference I call these Object A1 – traditional rainwater harvesting practices; Object A2 – pipe and pump supply; Object A3 – river to drum/tap to drum water supply practice; and Object A4 – emerging new practices for rainwater harvesting and conservation (RWH&C). Each of these is tied to forms of land use governance, market chain, and other social-historical circumstances as will be described below. Each of these are also shaped by the generative mechanisms that surround the historical social engineering projects of the colonial and apartheid states and which

Object historical layers and changes in water supply and use activity in Lloyd Village Garden

Pre-1970

Cultural disruptions and changes and gradual loss of traditional RWH&C practices (earlier RWH&C practices such as Gelesha = OBJECT A1)

Late 1970s – early 1990s

Homeland Act of 1971. Land re-allocated for communal production; pipe and pump water supply (with regular breakdowns) = OBJECT A2

Early 1990s – present

Abolishment of the Homeland Act of 1971. Land use restructured to allotment system; failure of pope and pump system; River to Drum / Tap to Drum system used (inadequate solution to breakdown of pipe and pump system) = OBJECT A3

Figure 13.3 Historical layers of the object of water supply, delivery and use in Lloyd Village garden (source: developed by author).

continue to hold the core contradiction shaping the activity in place, which appears to be strongly related to persistent inequality and capability deprivation that exists in spite of aspirations for change, and available resources and knowledge (see below; Westaway, 2012).

The importance of time and place in an analysis of the generative mechanisms that shape and connect political economy, institutional culture, capacity development and human emancipation is described by McGrath *et al.* (2004, pp. 5–6) who 'show that systems are not static.... The linking of the historical and contemporary ... bring(s) into clear focus that new meanings, policies and practices are based firmly in readings of the old' (see also Westaway, 2012). Bhaskar (1993) suggests that absenting of absences is a tensed-spatio-temporal affair.

Object A1. Pre 1970: historically constituted forms of water supply, delivery and use in community gardening contexts

Historical evidence from the Eastern Cape Province of South Africa shows that there has been a longstanding culture of practice focusing on rainwater harvesting and conservation. Denison and Wotshela (2009, p. 19) describe a practice of *Gelesha* as 'probably the oldest southern African recorded practice that entailed water harvesting and soil conservation'.

The Xhosa terminology *Gelesha Umhlaba* means 'hoeing or the tilling of the soil after a crop harvest'. In some areas it is described as *Ukuphethula Umhlaba* which means 'turning over the soil'. This practice was used to ensure that rain or dew could infiltrate the tilled soil so that it would provide moisture for the next planted crop. Denison and Wotshela go on to explain that: 'Historically Gelesha was done when a digging star (the Orion constellation) *Isilimela* appeared in the autumn and after the harvesting of the summer crops' (ibid., p. 19). This temporal approach via the appearance of a particular star constellation 'that signified the renewal of the soil hoeing season also marked the renewal of the year for sowing purposes' (ibid., p. 20). It coincided with the time of year (also reported as mid-winter, i.e. July) when cattle – usually oxen (for draught power) – were still strong enough due to food availability.

Gelesha-related practices were/are also part of the traditional farming culture in Zimbabwe where in some parts of the country it is known as *Kuvundukura*, literally meaning 'turning over the soil'. Other indigenous rainwater harvesting and conservation practices existing in this area include contouring which developed from the introduction of the ox-drawn plough. Other traditional water harvesting methods practised in the Eastern Cape of South Africa include homestead ponds and stone terracing (largely promoted by missionaries in the eighteenth century (Denison and Wotshela, 2009) and also practised in pre-colonial times in eastern and southern parts of Zimbabwe), among others.

This early practice showed that indigenous people used a form of retroductive or abductive analysis (judgement based on experience and observation of the elements and earth system cycles) to maximize the harvesting of precipitation (rainwater, dew or frost) for food production. It also shows the deeply intimate relationship between land, land use practices, agri-culture and the historically constituted forms of water supply, delivery and use in community food production practices. In the Lloyd Village Garden site in the Eastern Cape memory of this practice still existed, although this was only in the form of 'fragmented memory'.

As is well known, traditional forms of agriculture in the Eastern Cape in South Africa were severely disrupted by the various social engineering projects of the colonial and apartheid governments which were established to maintain white supremacy and to channel resource flows to those in power, issues which persist to this day (Westaway, 2012). Key amongst these social engineering policies, which have had a particular impact on the 'object of human activity' in focus in this paper (productive food gardening supported by successful supply, delivery and use of water) is the phenomenon of Bantu or Black Education which legislated for the limited development of the black person towards menial labour and associated vocations, and which also led to poor quality education for black people in South Africa (GoUSA, 1953; McGrath *et al.*, 2004). It reflects Antonio Gramsci's notion of the 'subaltern' who was subjugated by colonial policies, and its evolution into the new subaltern as contemporary reproduction of the inequity embedded in post-colonial policies and practices (Hoare and Smith, 1999) and opportunities for freedom. It also turned a longer-term tradition of agricultural practice amongst the amaXhosa into a practice that was seen to be a 'practice for the poor' and thus lacks appeal amongst modern-day youth.

The second major social engineering project of the apartheid state relates to ownership and distribution of farmland. In earlier colonial times in the Eastern Cape, productive farmland was concentrated into the hands of colonial masters (hence also the name of the village 'Lloyd Village'). The Bantu Homelands Citizenship Act of 1970 (Act No. 26 of 1970; successively renamed the Black States Citizenship Act, 1970 and later proclamation of the National States Citizenship Act, 1970) consolidated the apartheid state's policy of separatist development. This was to lead to legislated black de-naturalization, exclusion from participation as citizens in their own country, enforced ethnic nation-state development with meagre resources and hence a systemic institutionalized dual economy. Under this social engineering framework large tracts of land were reorganized and ownership transferred, restricting the Black population to about 13 per cent of the land. These two social engineering projects have had a profound and ongoing impact on the present capabilities of people who today occupy the area called 'Lloyd Village'.

Object A2. Late 1970s–early 1990s: pipe and pump water supply, delivery and use in a Bantustan structured communal gardening setting

Following the Engeström model of expansive learning actions it is possible to retrospectively analyse the emergence of the 'pipe and pump' water supply, delivery and use activity. The activity started soon after the proclamation of the 1971 Homeland Act when the land which was previously in the hands of the white famer family 'Lloyd' was transferred to the Ciskei 'homeland' government, who then allocated the land for communal land use (villagization) and food production. A decision was made to establish a communal garden on the site.

- **Questioning:** The key questions being deliberated soon after the land was allocated were related to how to establish a communal gardening system, and how to bring water to the garden. Some of the current questions were expressed as aspirations:

  > We would like to repair our pump ... but there are other problems. We don't have the funds at the moment, and our delivery pipe from the tank was damaged during road construction by the Municipality.
  >
  > (Garden Farmer, personal communication, 2 December 2014)

- **Analysis and modelling the new solution:** Historical records show that after some discussion the Ciskei government decided to install a pump and to pipe water to the garden from a nearby river. The supply of this water via the pipe and pump system thus enabled the communal gardening practice to become established.
- **Examining, testing and implementing the new model:** This farming model was then used for a number of years, but not without problems. Historical reports indicate that there were regular problems experienced with the pipe and pump approach to water supply, particularly related to ongoing maintenance of the pump to ensure an ongoing supply of the water for the food production activity.
- **Reflecting on the process:** Ongoing engagement with the activity surfaced a number of reflections which related to 1) ongoing problems with technical capacity to restore/maintain the pump for the 'pipe and pump' water supply system; 2) inequality in the shared benefits of the communal garden as some worked more in the garden than others, yet produce was shared. This problem was also linked to breakdowns in the marketing systems related to the produce.
- **Consolidating and generalizing the new process:** A broad view of the emergence of the pipe and pump approach to supplying water to the communal garden showed that the activity did not always work smoothly, given the problems reported on above. Thus, it is not surprising that, when given the chance, the community members involved in the farming practices in the Lloyd Village garden area chose to restructure the activity. This activity changed or a turning point in the activity occurred with the government change in 1994 after the first democratic elections in South Africa, when the Homelands Act of 1971 was repealed and land was then differently distributed.

Object A3. 1994–present: river-to-drum water supply, delivery and use in democratically structured allotment gardening setting

Following the Engeström model of expansive learning actions it is also possible to retrospectively analyse the emergence of the next phase of the object's history which can be termed 'river to drum' water supply, delivery and use activity. This cycle of activity emerged after the failure of the previous form of activity, as reported above.

- **Questioning:** Questioning in this phase was not disconnected to the reflections and dissatisfactions voiced in the earlier cycle of expansive learning and activity. However, the change in the Homeland Act created new questions which related to how to continue food production but within an allotment system, and also within the reality of the failing 'pump and pipe' approach to water provisioning

- **Analysis and modelling of the new solution:** Historical evidence shows that the people involved in the gardening practices within the allotment system in the Lloyd Village context turned to 'self driven' solutions to the water supply problems. They figured out that it would be possible to carry water from the river and temporarily store it in drums on the gardening site.
- **Examining, testing and implementing the new solution:** The river-to-drum approach has been used for a number of years, but with difficulty as the gardening community are mostly pensioners and struggle with the labour demands required by this practice. Also reported (but not officially) is the practice of running water from the local municipal tap (which is considered by the practitioners themselves to be a less than legal practice) into the drums.
- **Reflecting on the process:** There is some dissatisfaction with the current way of supplying, delivering and using water in the garden. Extension services have also failed to resolve the technological problems of maintaining the pump and pipe watering system, and economic conditions are such that the community members are not able to afford additional fuel for filling up the drums regularly. This leads to periodic crop failures. Additionally, the young people do not want to engage in agriculture because Bantu education has stigmatized rural agricultural practices and they prefer 'quick money' and jobs in the cities. Additionally, there appears to be little knowledge of, or engagement with alternative rainwater harvesting and conservation practices.

Object A4. Emerging interest in RWH&C practices in the allotment garden context

The three historical cycles of the object of water supply and use provided useful orientation to further expansion of the activity, as probed through empirical engagement in a new cycle of expansive learning, which, as yet remains incomplete.

- **Questioning:** At the time of the expansive learning intervention (2013/2014) a number of questions had therefore already arisen in the context. These were to do with the difficulties with ensuring a viable and sustainable supply of water to the food garden. In addition to these difficulties, an issue of questioning and making sense of their own apparently shifting and multiple identities seemed to be important in redefining the participating farmers' capabilities and agency. Sen (2004) describes this plural identity formation or emerging consciousness as an important emergent property in seeing beyond oneself and a group's capability (a limiting socially constructed categorization often reflected in tensions and contradictions) as being important in conceptualizing a wider choice of freedoms, possibilities and associations (negotiating internal contradictions); a journey through a dialectic. In the data generated for the study, the shift from incapacity-driven thinking associated with ageing and a history of dependence to a realization of an alternative self-driven path was expressed by participants in the following way: 'We are old, so how can we dig trenches?' (Garden Farmer personal communication, 1 February 2015); 'Maybe we could dig small dams in our garden?' (Garden Farmer, personal communication, 18 February 2015).

- **Analysis:** There was also some evidence of boundary crossing interactions taking place between the Lloyd Village garden farming community (the farming activity system) and the neighbouring activity systems (especially the training and extension activity system). Through these interactions, some analysis of the situation was taking place, and proposals were being put forward for implementing more sustainable Rainwater Harvesting and Conservation (RWH&C) alternatives. For example, some of the farmers had started tracing a diversion furrow stretching in an arc formation from the path leading to the garden gate down towards the fence on the southern side of the garden. These physical tracings were interpreted as manifestations of an aspiration to absent the absence of an adequate water supply and use system, and to implement the newly learnt but not yet fully understood water capture alternatives that were being shared by the local university and extension services, and our field research team.

Deepening insight into contradictions and identifying absences

As formative interventionist researcher, I was, at the start of the expansive learning process focusing mainly on Object A4 expansion above, as informed by mirror data that was based on the above historical analysis and the ethnographic observations. However, a key focus of the production of mirror data is also to mirror the contradictions so that these can be debated and deliberated as a potential starting point for expanding learning (Engeström, 1987). This requires the researcher to formulate an hypothesis or transfactual theory on what appears to be the central contradiction/s shaping the object of activity, and which influences the entire activity system. In doing this, I found critical realist analysis to be particularly helpful, as will be shown below. I formulated the following hypothesis on the contradictions shaping the central activity system and the object as follows:

- **Primary contradiction:** This appeared to pertain to the presence of persistent poverty, lack of realized alternatives and capability deprivation versus human rights-based policy and societal objectives for well-being and sustainability. This manifests in disjunctures in the actual use of the water technology (e.g. broken water pump) versus intended use (to facilitate easy food production); and disjunctures in actual agriculture and water policy implementation versus intended policy (municipality and the way the water allocation policies are functioning in practice).

- **Secondary contradiction:** A secondary contradiction was identified as being related to old ways of water supply, delivery and use (the object as historically emerged) versus the aspirations and livelihood needs of the pensioners who are the main actors in the Lloyd Village garden food production system (subjects who are old and are struggling with the labour demands) (object-subject contradiction).

Both of these contradictions stem from long-term structural inequalities and cultural disruptions which appear to be manifesting as generative mechanisms (see below). This has led to capability deprivation amongst the gardening community members, a lack of engagement in gardening practices amongst the youth, ongoing technological and extension support failures and persistent poverty that does not allow for resources to resolve the technological challenges (empirical).

Related absences are an absence of capability, appropriate and more sustainable technologies and technological skills, more benign and equitable structural and social-economic conditions, and the valuing of agricultural practices in the modern day, especially amongst youth, reflecting dynamics of Bhaskar's (2010) laminated system. These contradictions can be represented diagrammatically as shown in Figure 13.4 in relation to the activity system

As stated by Virkunnen and Newnham (2013, p. 11) contradictions can be viewed as obstacles encountered by participants in an activity due to

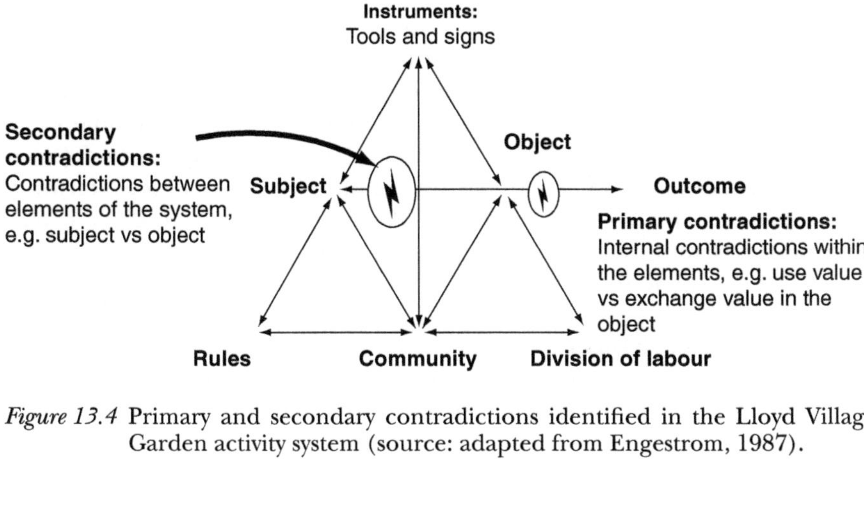

Figure 13.4 Primary and secondary contradictions identified in the Lloyd Village Garden activity system (source: adapted from Engestrom, 1987).

inconsistencies or ambivalences 'between the dominant logic of the activity and the logic of the new concept of the activity'. These, however have deeper roots and are shaped by underlying generative mechanisms, and as shown below, they manifest more widely than in just one activity system. The contradictions in the Lloyd Village garden area case identified are not the only ones identified in the study. I briefly share insights into some of the other contradictions that I have identified, after which I synthesize the chapter with some perspectives on the underlying mechanisms that shape the boundary crossing learning possibilities in this study.

Further and wider manifestations of the primary contradiction

As identified above in the Lloyd Village Garden case, the primary contradiction that I identified is: *Persistent poverty, lack of realised alternatives and capability deprivation versus human rights-based policy and societal objectives for transformation, well-being and sustainability*. This contradiction had wider manifestations than those found only in the Lloyd Village Garden case study. I highlight some of the related manifestations below that have emerged from some of the phase 1 data as these pertain to the agricultural education activity system. I frame them here as 'related contradictions' since they tend to exacerbate or contribute to the primary contradiction in the farming activity system, showing the related emergence of contradictions in a laminated system.

Wider contradiction 1: The agricultural college mission versus the capabilities of its graduates

Extension officers are unable to adequately advise farmers in climate constrained rain-fed farming conditions due to lack of or inadequate knowledge of rainwater harvesting options. Agricultural educators indicated that the curriculum does not prepare extension officers adequately. This was expressed by extension officers in the South African case study when they said that 'We teach for an ideal environment while farmers encounter often adverse and difficult conditions' (Extension officer, personal communication, 24 September 2014). This exacerbates the capability deprivation and lack of knowledge of alternatives found in the small-scale farming activity system. This is reflected in an evaluation of the South African Agricultural Curriculum which found issues of misalignment and inappropriateness of the curriculum with Agriculture Department Policy:

> despite policy changes to develop a single integrated extension service (serving both the commercial and emerging/small-holder farming sectors) most educational institutions offering agricultural programmes *still have commercial agriculture as their primary focus. Issues*

> *related to small-holder farming*, food security, *rural livelihoods and sustainable agriculture hardly feature* in the agricultural curricula.
>
> (DAFF, 2008, my emphasis)

Capability deprivation is not only a phenomenon related to South Africa, however, but similarly in the Zimbabwean case study, an agricultural educator observed that extension workers that she worked with in outreach were unable to advise farmers adequately on gulley-reclamation and rain-fed farming, due to inadequacies in the colleges' pre-service curriculum.

Wider contradiction 2: Isolated specializations in agricultural education versus demand for relational agency

Agricultural education has grown within a culture of separate development, which problematizes the relational agency required in creating new forms of human activity. This study has found that agricultural educators struggle to work out of the subject silo. This problem is illustrated by the difficulty found among agricultural college lecturers in Zimbabwe in rolling out the new environmental education (EE) curriculum across the traditional agricultural education disciplines: 'It was initially very difficult to mobilise other lecturers and even leaders to implement the new official EE curriculum.... However, after five years, there are still challenges to fully involve lecturers who are still locked in their disciplines' (Lecturer, personal communications, 13 February 2015). The EE curriculum was introduced in Zimbabwe in 2009 to respond to the problems of agricultural damage to ecosystems. This move is supported by an intensive extended study by the World Bank and UN which found that agro-ecological solutions to hunger are more effective than GM seed and other technicist solutions (McIntyre *et al.*, 2009). Progress of integrating environment and sustainability is slow in the classroom as agricultural educators are not yet committed to going beyond the familiar and the convenient, despite an increased international tendency in institutional change (ibid.). This was lamented by one lecturer who has been tasked with co-ordinating EE at an Agricultural College in Zimbabwe who said:

> Lecturers are still resisting visiting surrounding communities in our EE outreach activities ... yet I have found these activities useful in helping extension workers ... to deal with such problems as gulley reclamation and soil conservation ... which they were previously shying away from because they did not learn these practically when they were in college.
>
> (Lecturer, personal communication, 13 February 2015)

These two contradictions show the key role that *could potentially be played by agricultural educators* in relation to farmers in farming activity systems.

The data above shows that agricultural college educators need to develop the capability to work with fellow educators and with extension officers, and practising farmers in order to understand the meaning of their own work better. In this regard, Edwards (2007, p. 4) proposed cultivating relational agency as the 'capacity to align one's thoughts and actions with those of others to interpret aspects of one's world and to act on and respond to those interpretations'.

Wider contradiction 3: Elevation of techno-centric approaches versus simple and accessible solutions

The perceived need for advanced technology has been reified to the extent of excluding or undermining simple solutions typical of most effective RWH&C practices as also found in local indigenous knowledge practices. This has also happened at the expense of local and indigenous knowledge practices and innovations (Chambers, 1983). As shown in the historical review of the Lloyd Village garden case, earlier forms of RWH&C (e.g. *Gelesha*) were replaced by pump and pipe solutions that, when the technology was working, worked well, but failed the farmers at technology breakdowns. Increasingly, RWH&C is often associated with agricultural engineering solutions which are difficult to obtain in contexts where resources are scarce or not forthcoming due to other structural problems which hold persistent poverty in place. For example, when the pump and pipe technology failed in the Lloyd Village, garden community members were forced to use a river-to-drum solution which required elderly people to carry the water from the river to garden which put them under severe physical strain. A lecturer at an Agriculture College in Zimbabwe lamented the lack of financial resources to buy more advanced equipment as a drawback to effective teaching of RWH&C: 'We still use old technology such as dumpy levels to peg contours which make us move behind the rest of the advanced world' (Lecturer, personal communication, 12 February 2015). It was interesting to note that while memory existed of earlier culturally established RWH&C practices, these were not being used in the sites, due to the devaluing of cultural practices under colonialism and apartheid. Related to this is the lack of use of demonstration sites for learning in agricultural colleges, a problem identified in both research sites.

Wider contradiction 4: Curricula promoting intensive high-input conventional (mono-cultural) farming versus mixed cropping and organic farming systems

The agricultural curricula that are being used in both the South African and Zimbabwean case studies are based on conventional green revolution agriculture which is 'high input' regarding agro-chemicals, and externally supplied seed. High input poses a risk to small-scale farmers who usually have very low capital; hence there is a contradiction in the curriculum design with respect to the intended final 'recipients' of the knowledge in the current activity systems. Yet there is an alternative – sustainable farming practices incorporating locally adapted farmer innovations – which is absent from the curriculum, which would otherwise reduce farmers' risk.

A laminated system analysis of the generative mechanisms and underlying structures relevant to small-scale farming

As can be seen from the above, the primary and secondary contradictions identified in the Lloyd Village garden case stem from underlying causes and generative mechanisms that are also visible in the other activity settings (the colleges) and these can be seen within the laminated system. In the case of this study it would seem that there are a number of structural forces at work that are (a) holding persistent poverty in place, despite progressive policies to the contrary, (b) continuing capability deprivation in the form of lack of knowledge of alternatives, and lack of access to technology competences for maintaining technological equipment, (c) alienating youth from what used to be taken-for-granted valuable practices of food production, (d) marginalizing use of indigenous knowledge and organic/more sustainable farming methods, (e) maintaining academic and disciplinary silos that fragment college-farmer relationships.

Using retrodictive analysis, and the laminated system framework used by Bhaskar (2010) (as adapted by Price (2010) and referred to using [square brackets] in the section below), it is possible to suggest that these are related to wider global forces that continue to exacerbate inequalities and that devalue organic farming, and promote mono-cultural thinking and disembedded educational curricula and programmes [geo-historical trajectories and global trends]. These trends have been widely reported in studies and assessments of sub-Saharan African and Asian systems by both official (government) and civil society research (McInytre *et al.*, 2009), and show up at case-based micro levels of praxis over long periods of time as has been shown by the Lloyd Village garden area case outlined above, and the extracts of data from the Zimbabwe and South African agricultural education and training settings.

The culture of governmentality in post-apartheid South Africa established policies meant to transform the agriculture sector, remove hegemony of institutionalized inequity and assist the disadvantaged through social grants [South African society, including structural characteristics of society]. However, the latter are believed to have a negative side of creating dependency and the shunning of productive work, especially by young people (DAFF, 2008). Despite their relatively small amounts these grants to some extent absent the desire for gainful work through self-validating reduction of the already stigmatized occupations with and in farming, including gardening [human interactions at the small scale].

The culture of working of Government Extension Services in this area is to promote commodities such as field crops, poultry, sheep for wool, piggery, cattle for red meat … maize, sheep, etc. (personal communication, Senior Extension Officer, Department of Rural Development and Agrarian Reform, South Africa, 24 September 2014). While such approaches help farmers to access some markets, they also promote monocultural designs for crops and livestock, privileging income and business (dialectical materialism) at the expense of soil integrity, integrated pest management (biodiversity) [human-material interactions]. However, the local municipal authority has shown some good leadership in promoting open pollinated, locally adapted seed varieties linked to a market chain in its smallholder agriculture programme which is implemented by its economic development agency.

There is also need to give attention to the personal and social psychological aspects of valuing of agricultural practices, which have been devalued as noted above [conscious-unconscious motivation]. This is especially also important amongst the youth, who, as described by the pensioners in the case study, are no longer interested in agriculture as they do not see it to be a valued social activity.

To transform this situation, there is a need to engage with expansive learning processes in ways that are not naïve and that do not fall prey to localized contextualism or naïve constructivism. Thus careful analysis of contradictions, underlaboured by critical realist retrodictive analysis within a laminated system perspective shows potential for identifying absences at different levels of the laminated system, avoiding such a problem in CHAT-oriented formative intervention research.

Next steps in the expansive learning process: mediation of agricultural curriculum innovation for adaptive transformative praxis

The birth pains of transformative practice for mainstreaming rainwater harvesting and seed saving in agricultural education appear to be stimulated in the theatre of realizing and acknowledging one's limited knowledge and trying out new knowledge practices. Mediating agents and tools are needed. In its ongoing engagements in the expansive learning process, this study is finding that praxis is emancipatory and emergent. Mobilizing praxis that is located in a clearly understood and critical realist historically informed analysis, in response to absences, can potentially offer a 'pedagogy of hope' for those educators aspiring to deliver quality agricultural learning experiences. This is demonstrated by the following agentive reflection by an agricultural college lecturer in the Eastern Cape of South Africa:

> Techniques like (tied) ridging, terracing, contour-planting … are given attention but more on their functions not the design. They only

> are talked about as a way of managing the available water, not as a tool of capturing rain water and to reduce runoff.... Rain water harvesting must therefore be addressed in depth as a technique to back up the dams and the rivers when there is not enough rain.
>
> (Lecturer, Course Assignment 1, September 2014)

The expansive learning cycle and boundary-crossing laboratory engagements are potentially powerful approaches mobilizing agency and praxis across institutions as has been shown by a number of southern African researchers (Mukute, 2009; Masara, 2010; Lindley, 2014; Kachilonda, 2015). The study is showing that the usually perceived 'powerless' farmer can in fact be seen to be very powerful when they possess or exercise powerful knowledge practices in such innovations as RWH&C that extension workers and agricultural educators usually do not have. The fear that this reversal of power generates was expressed by an extension officer in South Africa as follows: 'we will also participate in the trainings because farmers must not know more than us' (Extension officer, personal communications, August 2014). In the Zimbabwean case study, an innovative farmer indicated that government institutions were curtailed to spread innovations by the political order, while another indicated that there is need for more agency among extension workers.

Conclusion

Small-scale farmers in remote marginal villages in Zimbabwe and South Africa are experiencing water-related problems with productive food production, but are also demonstrating water harvesting and seed saving practices as real possibilities for transforming adverse farming conditions and livelihoods. Deeper engagement with these activity systems can provide new knowledge and direction for agricultural education institutions, which can facilitate improved boundary crossing between college-and-farmer. This is necessary as current modes of formal agricultural training hardly equip extension officers with the capability to help farmers in these regions as was shown in the Lloyd Village garden case study, but also more widely in the study data. Paradoxically, this can be quite disempowering for extension officers who find themselves unable to offer meaningful advice and services to farmers.

In the case study sites in focus in this study, Agricultural Education Colleges and Faculties have started taking up agency to integrate such solutions as they learn together in networks with other agricultural professionals and small-scale farming practitioners. Some of these farmers' knowledge practices provide potential 'germ cells' for modelling new solutions across boundaries (Engeström and Sannino, 2010). Unfortunately such inter-sectoral fora are usually not available to facilitate boundary crossing, and seem to need external mediation to set up. The expansive learning approach, underlaboured by critical realism, provides tools for supporting the establishment or strengthening of such learning networks, to potentially foster transformative learning and praxis. The study concludes that transformative agentive action in agricultural education needs to mobilize praxis in order to contribute to the public good, and that the knowledge to do so is available in non-traditional sources, but needs to be mediated through transgressive and boundary-crossing processes that are not naïvely constituted, but which take careful account of the generative mechanisms that structure and potentially constrain transformative praxis. Warmington (2005, as cited by Nunez, 2014) argues that the study of activity systems 'should be the study of contradictions located within labour activity of production and reproduction' (p. 119). In the case of community learning in southern Africa, these are associated with the continuities of master-slave relations emerging from colonialism and apartheid, as is clear from this case study, which have become manifest in the forms and approaches to curriculum and extension training in agricultural colleges, with continued impacts on the marginalized, exacerbating forms of marginalization, especially also marginalization of farmers' knowledge and practices. It is here where boundary crossing processes need to be conceived with new foundations that are not reproductive of the past patterns of inequality and marginalization.

Acknowledgements

This study forms part of a PhD study currently being undertaken in the Environmental Learning Research Centre at Rhodes University by the author (Pesanayi, in process). The author would like to thank the South African and Zimbabwean farmers and water harvesting innovators who made enormous contributions to the generation of data that fed into the chapter. The South Africa Netherlands Partnership for Academic Development (SANPAD) made it possible to generate mirror data in both South Africa and Zimbabwe, while the Water Research Commission has enabled cross-boundary expansive learning of rainwater harvesting in the South African case study.

References

Atchoarena, D., and Delluc, A. (Eds). (2002). *Revisiting technical and vocational education in sub-Saharan Africa: An update on trends, innovations and challenges.* Paris: IIEP-UNESCO.

Bhaskar, R. (1993). *Dialectic: The pulse of freedom.* London: Verso.

Bhaskar, R. (1997). On the ontological status of ideas. *Journal for the Theory of Social Behavior, 27*(2/3), 135–147.

Bhaskar, R. (1998). *The possibility of naturalism.* London: Routledge.

Bhaskar, R. (2008). *A realist theory of science.* London: Routledge.

Bhaskar, R. (2010). Contexts of interdisciplinarity: Interdisciplinarity and climate change. In R. Bhaskar, C. Frank, K. G. Høyer, P. Naess, and J. Parker (Eds), *Interdisciplinarity and climate change: Transforming knowledge and practice for our global future* (pp. 1–24). London: Routledge.

Bhaskar, R., Frank, C., Hoyer, K. G., Naess, P., and Parker, J. (Eds) (2010). *Interdisciplinarity and climate change transforming knowledge and practice for our global future.* London and New York: Routledge.

Bird, K., and Shepherd, A. (2003). *Chronic poverty in semi-arid Zimbabwe.* CPRC Working Paper No. 18, June 2003.

Brown, D., Chanakira, R., Chatiza, K., Dhliwayo, M., Dodman, D., Masiiwa, M., Muchadenyika, D., Mugabe, P., and Zvigadza, S. (2012). *Climate change impacts, vulnerability and adaptation in Zimbabwe.* IIED Climate Change Working Paper No. 3, October 2012.

CBD (Convention on Biological Diversity) Secretariat. (2012). Outcome Document on Education and Biodiversity Conservation. XIth Conference of the Parties to the Convention on Biological Diversity (COP 11), 13–14 October 2012. Hyderabad: CBD/CEE.

Chambers, R. (1983). *Rural development: Putting the last first.* New York: Longman.

DAFF (Department of Agriculture, Forestry and Fisheries, Government of South Africa (GoSA)). (2008). *Evaluation of agricultural education and training curricula in South Africa.* Pretoria.

DEA (Department of Environmental Affairs). (2013). *Long-Term Adaptation Scenarios flagship research programme (LTAS) for South Africa: Climate trends and scenarios for South Africa.* Pretoria.

Denison, J., and Wotshela, L. (2009). *Indigenous water harvesting and conservation practices: Historical context, cases and implications.* WRC report no. TT 392/09, April 2009.

Edwards, A. (2007). Relational agency in professional practice: A CHAT analysis. *Actio: An International Journal of Human Activity Theory, 1*(1), 1–17.

Engeström, Y. (1987). *Learning by expanding: An activity-theoretical approach to developmental research.* Orienta-Konsultit, Helsinki, Finland.

Engeström, Y. (1991). Activity theory and individual and social transformation. *Multidisciplinary Newsletter for Activity Theory, (7)*8, 14–15.

Engeström Y. (1999). Expansive visibilization of work: An activity-theoretical perspective. *Computer-Supported Cooperative Work, 8*, 63–93.

Engeström, Y. (2000). Comment on Bhaskar *et al.* activity theory and social construction of knowledge: A story of four umpires. *Organization*, 7(2), 301–310.

Engeström, Y. (2001). Expansive learning at work: Toward an activity theoretical reconceptualization. *Journal of Education and Work, 14*(1): 133–156.

Engeström, Y. (2007). Enriching the theory of expansive learning: Lessons from journeys toward coconfiguration. *Mind, Culture and Activity, 14*(1–2), 23–39.

Engeström, Y. (2008). *From teams to knots: Activity-theoretical studies of collaboration and learning at work.* Cambridge: Cambridge University Press.

Engeström, Y., and Sannino, A. (2010). Studies of expansive learning: Foundations, findings and future challenges. Educational Research Review. doi:10.1016/j.edurv.200212.002.

Engeström, Y., Engeström, R., and Kerosuo, H. (2003). The discursive construction of collaborative care. *Applied Linguistics, 24*(3), 286–315.

FAOSTAT (Food and Agriculture Organization Corporate Statistical Database).

(2005). *Database*. Rome: Food and Agriculture Organization. Retrieved 8 December 2013, from http://faostat.fao.org.

Freire, P. (1998). *Pedagogy of freedom: Ehics, democracy and the civic courage*. Lanham: Rowman & Littlefield.

GoUSA (Government of the Union of South Africa). (1953). *Bantu Education Act: Act No. 47 of 1953*. Parliament of South Africa.

Hoare, Q., and Smith, G. N. (1999). *Selections from the prison notebooks of Antonio Gramsci*. London: ElecBook.

IPCC (Intergovernmental Panel on Climate Change). (2012). Summary for Policymakers. In C. B. Field, V. Barros, T. F. Stocker, D. Qin, D. J. Dokken, K. L. Ebi, M. D. Mastrandrea, K. J. Mach, G.-K. Plattner, S. K. Allen, M. Tignor, and P. M. Midgley (Eds), *Managing the Risks of Extreme Events and Disasters to Advance Climate Change Adaptation: A Special Report of Working Groups I and II of the Intergovernmental Panel on Climate Change* (pp. 1–19). Cambridge, UK and New York: Cambridge University Press.

Jones, B. G. (2006). *Explaining global poverty: A critical realist approach*. New York: Routledge.

Kachilonda, D. (2015). *Investigating and expanding learning in co-management of fisheries resources to inform extension training*. Unpublished PhD thesis, Rhodes University, Grahamstown.

Lindley, D. (2014). *Can expansive (social) learning process strengthen organizational learning for improving wetland management in a plantation forestry company, and if so how? A case study of Mondi*. Unpublished PhD thesis, Rhodes University, Grahamstown.

Malzbender, D., and Earle, A. (undated). *Water resources of the SADC: Demands, dependencies and governance responses*. The Institute for Global Dialogue (IGD) and the Open Society Initiative for Southern Africa (OSISA). Retrieved 25 February 2015, from www.acwr.co.za/pdf_files/IGD_Water%20Resources.pdf.

Masara, C. (2010). *Learning commercial beekeeping: Two cases of social learning in Southern African community natural resources management contexts*. Unpublished master's thesis, Rhodes University, Grahamstown.

Mashinini, V. (2010). *The Lesotho Highlands Water Project and Sustainable Livelihoods Policy Implications for SADC*. AISA POLICY Brief No. 22, June 2010. Retrieved 25 February 2015, from www.ai.org.za/wp-content/uploads/downloads/2011/11/No-22.-The-Lesotho-Highlands-Water-Project-and-Sustainable-Livelihoods.pdf.

McGrath, S., Badroodien, A., Kraak, A., and Unwin, L. (Eds). (2004). *Shifting understandings of skills in South Africa*. Cape Town: HSRC.

McIntyre, B. D., Herren, H. R., Wakhungu, J., and Watson, R. T. (Eds). (2009). *Agriculture at a crossroads: International Assessment of Agricultural Knowledge, Science and Technology for Development (IAASTD) Global Report*. UNDP, FAO, UNEP, UNESCO, The World Bank, WHO, GEF. Washington: Island Press.

Mukute, M. (2009). *Expanding and exploring learning in sustainable agriculture activity systems*. Unpublished doctoral thesis, Rhodes University, Department of Education, Grahamstown.

Njihia, J. M. (2011). Critical realism and its prospects for African development research and policy. *Thought and Practice: A Journal of the Philosophical Association of Kenya, New Series, 3*(1), 61–85.

Nunez, I. (2013). Transcending the dualisms of activity theory. *Journal of Critical Realism, 12*(2), 141–165.

Nunez, I. (2014). *Critical Realist Activity Theory. An engagement with critical realism and cultural-historical activity theory.* London: Routledge.

Pesanayi, V. T. (2008). *Investigating learning interactions influencing farmers' choices of cultivated food plants: A case of Nyanga and Marange communities of practice, Manicaland Province, in Zimbabwe.* Unpublished MEd (Environmental Education) thesis, Rhodes University, Grahamstown.

Pesanayi, T. (2009). Sigtuna Think Piece 6: A case of exploring learning interactions in rural farming communities of practice in Manicaland, Zimbabwe. *Southern African Journal of Environmental Education, 26.*

Price, L. (2010). *Using an articulated, laminated totality to give an interdisciplinary explanation of the oppression of women in South Africa.* Paper presented at the Critical Realism Seminar, Institute of Education, University of London, 26 April 2010.

RBZ (Reserve Bank of Zimbabwe). (2005). *Drought mitigation and resuscitation policies for the agricultural sector in Zimbabwe.* Supplement 2 of 3 to the 1st quarter monetary policy statement delivered by the Governor, May 2005.

Rowan, J. (2000). Dialectical thinking and humanistic psychology. *Practical Philosophy,* July, 19–22. Retrieved 21 January 2015, from www.society-for-philosophy-in-practice.org/journal/pdf/3-2%2020%20Rowan%20-%20Humanistic%20 Psychology.pdf.

SADC (Southern African Development Community). (2012a). *Climate Change Adaptation Strategy in SADC: A strategy for the water sector.* Gaborone: SADC Secretariat.

SADC (Southern African Development Community). (2012b). *Regional Climate Change Programme Document.* Gaborone: SADC Secretariat.

Sen, A. (2004). *Rationality and freedom.* Louiseville: Belknap Press.

Virkkunen, J., and Newnham, D. S. (2013). *The change laboratory: a tool for collaborative development of work and education.* Rotterdam: Sense Publishers.

Vygotsky, L. (1978). *Mind in society: The development of higher psychological processes.* Cambridge: Harvard University Press.

Wani, S. P., Rockström, J., and Oweis, T. (Eds). (2009). *Rainfed agriculture: Unlocking the potential.* Oxfordshire: CABI.

Warmington, P., Daniels, H., Edwards, A., Brown, S., Leadbetter, J., Martin, D, Middleton, D., Parsons, S., and Popova, A. (2005). *Surfacing contradictions: Intervention workshops as change mechanisms in professional learning.* British Education Research Association Annual Conference, University of Glamorgan, September 2005.

Watkinson, E., and Makgetla, N. (2002). *South Africa's food security crisis.* National Labour and Economic Development Institute (NALEDI).

Westaway, A. (2012). Rural poverty in the Eastern Cape Province: Legacy of apartheid or consequence of contemporary segregationism? *Development Southern Africa, 29*(1), 115–125.

Ziervogel, G., New, M., van Garderen, E. A., Midgley, G., Taylor, A., Hamann, R., Stuart-Hill, S., Myers, J., and Warburton, M. (2014). Climate change impacts and adaptation in South Africa. *WIREs Clim Change, 5,* 605–620. doi:10.1002/wcc.295.

14 Exploring critical realist insights into transformative environmental learning processes in contexts of social-ecological risk

Ingrid Joan Schudel

Introduction

This chapter uses the concept of ecological footprinting as an indicator of social-ecological risk to reflect on the ways that critical realism offers insights into a transformative intent that dominates many perceptions of environmental learning (Ferreira, 2013). For example, O'Donoghue suggests a transformative approach to environmental learning, which entails 'practice-based deliberations that might allow the better mediation of choices that are more reality congruent and socially responsible' (2007, p. 153). This transformative intent is also evident in the work of Fien (1993, p. 16) who describes education *for* the environment as a 'counter-hegemonic process' and as having 'an overt agenda of values education and social change'. While Jickling and Spork (1998) have reservations about the determinism implied in the preposition 'for' sustainable development, they acknowledge the importance of the counter-hegemonic or socially critical dimension of environmental learning and its political and socially critical role in society.

With an understanding that transformational learning requires a nuanced understanding of change, this chapter has two main aims. First, it explores the product/process, spatio-temporal, intransitive/transitive, laminated, emergent and value-oriented nature of change by drawing on the ideas of original critical realism (Archer *et al.*, 1998) and dialectical critical realism (Bhaskar, 1993; elaborated by Hartwig, 2007 and Norrie, 2010). Second, these aspects of change are discussed in relation to a transformational intent in environmental learning characterized by open-ended, ontologically grounded and action-oriented approaches to learning in this broad field.

Measuring social-ecological risk: the concept of footprinting

The concept of ecological footprinting, as used here, provides the context of the ensuing discussions. An ecological footprint is an indicator of the sustainability of the lifestyle and activities of a person (country or other group). It is determined through calculating the square area of land or sea needed to supply the resources consumed and to absorb the waste produced for this particular individual or group. Conventionally this measurement is compared with our 'earthshare' – that is the area of land available to each person if the world's resources were to be shared equally between all the people of the world (Chambers *et al.*, 2000). Ideally, a country will achieve a low footprint but at the same time will meet its people's developmental needs (Malik, 2013, p. 94). When 151 countries were assessed in terms of both their ecological footprint and their 2012 Human Development Index (HDI),[1] only four countries occupied the space which represented the 'ideal' balance between development and living within ecological limits (Malik, 2013).

Such footprinting exercises highlight inequalities between individuals and countries in relation to this concept of earthshare, thus challenging a 'discourse of progress [which] threatens to narrow democracy by thwarting attempts to achieve social justice and develop a critical citizenry that is reflexive in a rapidly growing consumerist society' (Le Grange, 2013, p. 131). The importance of this concern for equality and social justice is particularly pertinent in sub-Saharan Africa, as evident in a recent Human Development Report which points to sub-Saharan Africa as having the lowest regional HDI at 0.475 and the lowest Inequality-adjusted Human Development Index (IHDI)[2] at 0.309 (Malik, 2013, p. 155). Malik explains that: 'Low HDI countries suffer most because they tend to have greater inequality in more dimensions. Low HDI countries lose a third of HDI value to inequality, whereas very high HDI countries lose only 11%' (ibid., p. 29).

From within this social-ecological context, this chapter reflects on a critical realist understanding of the nature of change and on the ways that such an understanding offers insights into contemporary approaches to environmental learning.

The nature of change and implications for environmental learning

Change has transitive and intransitive dimensions

When working with the concept of ecological footprinting it is important to recognize it as simply one way of describing human-environmental relationships. It is an indicator of human impact on the biophysical world. Using Bhaskar's (1993) distinction between a transitive reality and an intransitive reality helps to highlight the significance of this. Bhaskar highlights the necessity of understanding 'the truth of or reason for *things* and phenomena, *not propositions*, as genuinely ontological, and in this sense as objective in the intransitive dimension' (1993, p. 217). This enables a description of a reality outside of the knowing subject. For example, planet Earth, at a particular time, has a particular biocapacity. Certain human activities make use of and use the Earth's resources to absorb their waste either within the Earth's biocapacity or not. This happens in the intransitive dimension of reality regardless of whether we measure or assess human use of resources or not. The intransitive dimension is very specific at a particular moment in time and space and so has significance and power in the world. It is important to note that the intransitive dimension is not fixed or unchangeable. The Earth's biocapacity can change as badly managed ecosystems can degrade and lose biocapacity; or degraded systems, with careful management, can become more ecologically sustainable and increase biocapacity in the process.

The transitive dimension of reality is more fluid and is also subjective. Ecological footprinting is an attempt to develop an indicator of the amount of resources we use and waste that we produce in relation to the Earth's capacity to provide these resources and to absorb the waste we produce. This positions it in the realm of the transitive because it is an attempt by humans to know and describe this relationship. The ecological footprint is based on scientific and mathematical extrapolations from 'best available' data; it is dependent on available resources to gather and interpret that data and is even open to manipulation from those with a particular purpose in mind. Also, alternative ways of describing ecological systems can paint different 'pictures' or present different 'angles' of a situation.

This then raises the question of how to deal with different interpretations of reality in the transitive dimension of reality. Approaches to learning such as mirroring – described by Wals as 'learning by mirroring one's own ideas, views, values and perspectives with those of others' (Wals, 2009, p. 385) – imply that there will be different interpretations of reality amongst people. Wals proposes that the mirroring process enables people to 'recognize, evaluate and, when needed, potentially transcend or break with existing social norms, group thinking and personal biases' (Wals, 2007a, p. 39).

This mirroring process suggests that there is a 'need to transcend' and in this he implies a transformative intention for environmental education. Such an intention would not accept a relativist 'anything goes' outcome for environmental deliberations. Price highlights the problem of relativism in her explanation that this 'leads to the inability to act, since how can we act if we cannot choose between better or worse?' (Price, 2007, p. 107). For example, the concept of sustainability is open to relativist interpretations leading to an 'ambiguity that, somewhat paradoxically, provides the focus for discourse about a variety of issues within the contemporary problematique, while undeniably also providing fuel for those who would procrastinate in their resolve to seek sustainability' (Bawden, 2004, p. 21).

Lotz-Sisitka (2008, p. 142) suggested that by probing 'deeper, more complex and reality congruent understandings' of sustainability issues, we can construct visions and interpretations of sustainability in particular contexts that do not leave us with a sense of 'no direction' cultivated by the relativism of strongly constructivist notions of learning. Instead we can conceive of an approach to learning in which views of reality can be constructed in the minds of learners, but remains open to critique and needs to stand up to tests of truth, ontological grounding and normative judgements (ibid.).

So what possibilities are offered by critical realism in order to enable people to transcend and transform? In the educational context, Shipway (2001) argues that 'the curriculum must do more than just balance multiple local interests; it needs to have a defensible ontological link to the deeper strata of reality that underpins the "local", such as the concept of social justice' (p. 139). This argument suggests an ontologically grounded approach to education in which both the intransitive and transitive dimensions of reality are acknowledged. This is seen in Price's suggestion that 'neither the world nor the mind is the ultimate authority. Truth is co-constructed by the mind and the world, where the world is made up of networks or relationships of interlinked, mutually constructing, humans and non-humans' (2007, p. 39).

Thus a relational approach to environmental learning, including the exploration of the relationships between the transitive and the intransitive, provides a potential 'grounding', an iterative truth-seeking process between what is and what is perceived to be. Furthermore, such an approach enables 'the exercise of constructing models of alternative ways of living on the basis of some assumed set of resources' (Bhaskar, 1993, p. 395). This is what Bhaskar terms 'concrete utopianism' which functions to 'pinpoint the real, but non-actualised possibilities inherent in a situation, thus inspiring grounded hope to inform emancipatory praxis' (Bhaskar, 1998a, p. 215). Put more simply, Bhaskar describes concrete utopianism as playing a 'role of creative fantasy ... that yields at once hope *and* possibility' (1993, p. 209, my emphasis). Bhaskar elaborates on the meaning of 'alternative ways of living' by specifying that concrete utopianism 'involves thinking about how a situation of the world could be otherwise, with a change in the use of a given set of resources or with a different way of acting subject to certain constraints' (Bhaskar, 2010, p. 22). When considering options for changing our ecological footprint this would mean considering possibilities we have at our disposal for making individual and collective changes and what needs and possibilities there are for influencing change at a structural level.

Change depends on complex laminations

Bhaskar (1998b) describes reality as 'stratified' in terms of the existence of events which are independent of experiences of these events, and mechanisms and structures which exist independently of events and actions of humans that interact with them. Although these mechanisms and structures have underlying causal powers, these powers are not always exercised or triggered. This is elaborated by Archer's (1998a, p. 190) description of a realist theory of science which entails a 'quest for non-observable generative mechanisms whose powers may exist unexercised or be exercised unrealized, that is with variable outcomes due to the variety of intervening contingencies which cannot be subject to laboratory closure'. A further complexity of the nature of reality is Bhaskar's explanation that mechanisms interact in complex open systems as a 'conjunctive multiplicity or laminated totality constituted by several ontologically distinct but interacting mechanisms' (Bhaskar, 2010, p. 14).

Applying this concept to understanding an ecological footprint, consider an alternative indicator to the more commonly applied gross domestic product (GDP) for measuring development and well-being. The ecological footprint brings in another lamination relevant to development questions. Beyond purely an economic view of the world, it considers an ecological lamination and combined with the HDI, the ecological footprint ensures consideration of a further lamination – that is the socio-economic aspect of human activities. When probing our footprint further, the laminations, for example, of our energy footprint might be made up of biophysical laminations which would require understanding of the accessibility of coal, hydro, wind or solar power on a national scale; political laminations that influence technological and scientific research into alternative systems; economic laminations (for example, on a personal level), including personal financial security for small-scale, short-term 'investments' such as buying an energy efficient light bulb or long-term investments such as solar power; physiological and psychological laminations which would require understanding of needs and wants that influence electricity needs; and more.

Focusing on Bhaskar's inclusion of 'interactions' referred to above, a view of these laminations as static parts making up a particular whole (or totality) would still not, however, provide a complete picture of change. This would require looking at this laminated system from a processual perspective. It is the interactions between these laminations that help to give us a more complete picture of change so that a change interest needs to be addressed both at the level of understanding the laminations as well as the relationships between laminations. For example, consider a factory that raises concern regarding the risk of pollution it poses to a community. First, at a physiological level, are the goods it produces of benefit to the health of plants, animals or people and what are the physiological effects of the manufacturing process on these same beings? At the psychological level, what are the benefits (or counter-benefits) of the manufacturing processes and produced goods for animals and people? At the biophysical level, what are the effects of the industry on ecological systems and earth systems and how do these effects influence the capacity of the biophysical environment to further support the industry and to support physiological and psychological well-being at other levels? At the economic level, what employment opportunities and trade benefits are created by the industry and how do these contribute to the well-being (or not) of people employed in the industry? At the political level, what legislation and associated implementation strategies are in place in order to govern industrial practices and how are these targeted at the different levels in the laminated system?

The importance of examining the relationships between different laminations has also been highlighted in the field of environmental education. For example, Hattingh (2004), in reflecting on the concept of sustainability, draws a distinction between a view of sustainability seen as underpinned by three 'pillars', namely the financial, social and environmental; and sustainability seen as three embedded spheres of the economic within the socio-political within the ecological. The latter model of embedded spheres creates a complex image of interlocking values and inseparably intertwined logics (ibid.; see also Parker, 2010) so that any change in the system needs to be considered in terms of how this would affect relations within the system. The former 'three pillars' model creates the impression of three opposing pillars which need to be balanced and weighed against one another in an impossible prioritizing exercise which invariably lead to arguments of false necessity, trade-offs and subsequent calls for change compromises that are unlikely to have the power to change the status quo.

This propensity for compromise can be further examined through Bhaskar's (1993) notion of a 'TINA compromise' – a notion he coined in critiquing Thatcherite economic reforms in Britain in the 1980s. Norrie (2010) explains how the notion can be used to critique arguments from the point of 'false necessity'. A TINA compromise refers to a false statement that 'There Is No Alternative' and it is through identifying and critiquing TINA compromises that alternative ways of responding to societal 'ills' can be identified. For example, consider the 'polluter pays principle' which has been legislated in some countries as a strategy for regulating industry's environmental impact. A common argument attached to any call for increased manufacturing costs in industry is the tendency to threaten cut-backs and redundancies to those employed by the industry, thus creating an example of impossible prioritizing or a TINA compromise in the tension between 'environment' and 'development'. However, the relationship presented between the ecological footprint and HDI of the four countries which represented the 'ideal' balance between development and living within ecological limits, referred to above (Malik, 2013), is an illustration that such a compromise is not essential. The existence of such a balance illustrates that it is possible to address needs related to both of these equally important aspects of life on earth. In this example, the use of the notion of a TINA compromise illustrates its use as a tool to think about the 'feasibility of desirable alternatives, asking counterfactual questions, conducting thought experiments and scrutinising critical standpoints' (Sayer, 2000, p. 178).

Change as product/process complexity

The previous section highlighted the processual nature of laminated systems. So what does that mean for the nature of change? What kinds of action in society might constitute societal change? Consider the act of replacing an energy guzzling bulb with an energy efficient bulb to reduce your energy footprint. To make this decision you may have a change in mind/understanding from 'I haven't compared bulbs before' to 'This bulb is more expensive, but it lasts longer', and/or a change of heart from 'There is no point, the action is too small to have any impact on our global carbon footprint' to 'This is a significant symbolic action', and/or a change of values in the form of 'I don't really care about my energy footprint as the world has more pressing concerns' to 'I feel it is important to take personal responsibility for my footprint'. These examples of change present a particular view of change as one static 'thing' or 'product' that can be replaced with another.

An educational programme which set out with a *pre-determined* vision of particular outputs would be operating with a goal of changing one idea or way of being to another. Huckle (1986 in Gough, 2013) describes such an approach to environmental education as an evangelical mission of changing hearts, minds and values. Gough describes earlier approaches to environmental education, for example, in the 1990s in the United States where behaviour change was seen as the 'acquisition of responsible environmental behaviour' (2013, p. 17). An emphasis on the acquisition of particular behaviours can be seen in the first international summit focusing on environmental education, where common ground for environmental education was sought in Tbilisi in 1997. At this summit, one of the goals for environmental education was formulated as 'to create new patterns of behaviour of individuals, groups and society as a whole towards the environment' (UNEP, 1978, p. 24).

However, the notion of 'behaviour change is regarded by many as too technical or technicist an approach to change' (Ferreira, 2013, p. 63). The problem with such an approach is that there is an assumption that change is as simple as 'out-with-the-old' and 'in-with-the-new'. Yet the 'old' is not a static 'thing' that can be discarded because integral to this 'thing' is also the world and the history from which it emerged. Bhaskar's (1993) dialectical critical realism, and Norrie's elaboration on this as 'structured being and becoming' (2010, p. 192) present a processual aspect of change that gives insight into change as emergent in this way.

Norrie explains how 'being involves becoming, for product is always in process, and becoming is always attached to being, for process is always in relation to product' (2010, p. 47). Furthermore, he explains that 'an example of process-in-product is the way in which our geo-historical location influences what we are, while an example of product-in-process is the way in which any entity has effects in the world' (p. 233). So, a person is a product-in-process because whether they change or do not change to an energy efficient light bulb depends on various technological processes which determine, for example, the quality and efficiency of their 'chosen' bulb, various policy processes influencing trade and industry, and social and political processes that include or exclude them from an economic bracket which can afford more expensive light bulbs. If they do enact a decision to change a bulb, they have an effect in the world, however miniscule, as less electricity will be needed in order to power the light bulb. However, the effect they have on the world is still influenced by other products such as the type of electricity available from the national electricity grid (into which the light bulb is plugged) or the efficiency of its production.

This product/process complexity can be captured as 'Recursive embedding [which] means that one can also talk of a process-in-product-in-process to denote the constant interaction between entities and their environment' (Norrie, 2010, p. 233). Therefore certain processes enable or constrain the choices or non-choices and actions or non-actions of human agents who then have varying degrees of power to transform the processes from which their actions emerged. Changing a light bulb is a simple behaviour change that is not without value or meaning. However, understanding the educational process means working through and understanding the processes that influenced and enabled this change to happen; or if the light bulb was not changed, then understanding the processes that constrained this same action. It also means working through the power of the action or non-action to transform or reproduce: (a) products, such as the amount of electricity used, the way electricity is generated, the alternatives available for consumption, the policies that drive processes and the institutions that drive processes; and (b) processes, such as the political and socio-economic co-operative and competitive relations at play.

In the field of environmental education, reactions against a technicist interest in behaviour change have also led to reflections on the difference between product and process phenomena in society. For example, Wals declared that 'moving towards sustainability as a social learning process is more interesting than sustainability as an expert predetermined transferable product' (2007b, p. 39). However, what Bhaskar's product-process complexity suggests is a warning not to lose sight of product-process relations while still heeding Wals' concern with behaviourist notions of 'pre-determination' and 'transferability'. This signifies a call for approaches to environmental education and learning that are both open-process and open-ended but not pre-determined.

A further argument for the need for open-process/open-ended approaches to environmental education and learning is that a particular behaviour is not simply the choice of one individual or even one group of people, but is enabled (and constrained) by a dynamic and changing society and culture. Such insights can help environmental educators understand the limitations of educational interventions which 'strive for one coherent knowledge system; are underpinned by assumptions of universality, foundation, homogeneity, monotony and clarity; and consider unknowns, or conflicting knowings, as "not-yet-resolved-but-in-principle-resolvable" imperfections' (Bauman, 1993, in Gough, 2013, p. 18). Lotz-Sisitka (2011) calls for working with environmental knowledge 'in open-ended, and innovative ways with learners, who need to develop a deep understanding of the dynamic nature of this knowledge' (p. 28). A further comment regarding the dynamic nature of change is posed by Wals who states that 'we need to think in terms of challenges to be taken on in the full realisation that as soon as we appear to have met the challenge, things will have changed and the horizon will have shifted once again' (Wals, 2007b, p. 37).

Change as spatio-temporal

Having argued that change is product and process in nature, it is useful to explore what tools critical realism can offer in order to provide a different angle on the product/process relationship. One of these tools is the notion of spatio-temporality. Whatever totality we attempt to describe is always 'becoming' and always has the potential to change in different places, times and contexts.

Bhaskar (1993) argues that the temporality of change processes needs consideration both in terms of the ongoing presence of historical events as well as potential change into the future. He explains that

> the future is paradigmatically shaped possibility of becoming (a possibility which may be closer or more distant from us, more or less about, and more or less likely to be actualized), the presence of the future is a perfectly kosher locution, but … is always mediated by the presence of the past.
>
> (Bhaskar, 1993, p. 142)

Significant here in Bhaskar's reflection on tense is the emphasis on 'presence'. Both past and future are present in the sense that they have the potential to manifest even though their powers may or may not be exercised. In order to explore this idea, four modes for thinking about (spatio-)temporality offered by Bhaskar are discussed below in relation to the food component of an ecological footprint.

First, Bhaskar invites reflection on existential constitution which Norrie elaborates through explaining that 'any entity contains within it, as part of its being, the process that constituted it. Since process is spatio-temporal "becoming and begoing", we are existentially constituted by what is now apparently absent because [it is in the] past' (Norrie, 2010, p. 32). For example, consider Africa's history of colonization, which is now existentially constituted in the present in the form of a neo-colonialism. Neo-colonialism is enabled by globalization whereby rich countries use paper wealth to acquire the real wealth (the remaining accumulations of energy and resources) of developing and resource-rich countries (Kissinger and Rees, 2010). In addition to the temporality evident in this example, Kissinger and Rees illustrate the spatial nature of this problem with their explanation of how this process of neo-colonialism enables consuming populations to support themselves on the output of distant ecosystems and low-cost labour. Ecological footprinting is a tool that

> explicitly connects people's consumption, product by product, to several types of appropriated ecosystem area and reveals the gap (positive or negative) between local demand and supply of biocapacity. It can therefore be used to highlight the study population's trade-based dependence on distant ecosystems.
>
> (Kissinger and Rees, 2010, p. 2619)

Such thinking enables reflection on 'geographical displacement of sources and sinks' (Martinez-Alier, 2002), in other words, to think about which countries are the sources of carbon pollution and which countries serve as important carbon sinks, or which countries reap the benefits of ecological exploitation and which countries experience the environmental and often associated social (in)justice repercussions.

A second lens for viewing the presence of the past and presence of the future is co-presence. Bhaskar explains that 'This paradigm includes the "intersecting" combination of rhythmically differentially sedimented structures on a single episode' (1993, p. 141). For example, as illustrated by viewing an ecological footprint in relation to a country's HDI, all countries that contribute to the global ecological footprint are co-presently influenced by different development needs and agendas and different imperatives in terms of preserving the ecological integrity of their ecosystems. So, globally, within a country or even within a small community there may be 'disjoint, and possibly contradictory, rhythmics' (ibid., p. 141) which influence a footprint and the capacity for changing it.

Third, Bhaskar invites us to think about 'lagged, delayed potential or realized efficacy carried across a (spatio-)temporal level-specific void' (1993, p. 141). This idea is particularly important in understanding ecological footprinting. Footprinting uses the notion of 'ecological overshoot' to understand the point at which the Earth's resources are being utilized beyond the ability of ecological systems to regenerate. Acknowledging the existence of delayed efficacy enables an understanding that we can still continue to operate beyond the Earth's capacity up until a point in the *future* when the Earth becomes too degraded to recover its essential ecological functionings. A spatial element of such delayed efficacy is that greenhouse gas pollution is a problem shared by the whole planet and is more likely to be felt *in the future* by people and countries most vulnerable in terms of their low resilience to climatic change, a point that has clearly been made in the most recent International Panel of Climate Change reporting (IPCC, 2014).

Fourth, Bhaskar invites us to reflect on agentive perspectivality. Bhaskar explains that from the perspective of the agent, 'wherever she goes, "metaphorically" there has always been someone else before. She is always – from birth to death – living in a pre-constituted world. She is always living in the past' (1993, pp. 141–142). Thus, globally, nationally or at the level of a small community, a footprint is differentially enabled by different 'sedimenting structures' (ibid., p. 141) which influence their capacity and choice in terms of the consumption part of their footprint – such as availability of goods, different technologies for energy and goods production, and waste management structures and technology.

The above discussion illustrates how these four modes of spatio-temporality operate in a time/space continuum. Combined with the notion of change as process/product relationship, and the laminated nature of totality as discussed above, this can be represented as in Figure 14.1[3] as a time/space continuum of change where both past and future have significance as represented by the time continuum. In Figure 14.1, the dotted space continuum lines represent change in different (geographical, social and cultural) spaces over time. Superimposed on the space/time continuum is an emergent representation of product and process with each leading from the next, but representing different moments in time. The parallel lines represent the concurrent and interacting laminations of totality that affect change.

Consideration of space-time context is a well-known concern in environmental and environmental education fields; for example, Janse van Rensburg and Mhoney (2000) acknowledge that 'both environmental and educational issues are context-specific in time and space, and seldom amenable to universal solutions' (p. 47). Cognisance of the product-process, spatio-temporal and laminated complexities of change presented in this chapter can help address a concern in the field of environmental education and learning of reliance on 'simple' logic and over-simplification of, and hasty judgements about, environmental issues (Smyth, 1995).

Figure 14.1 with its emphasis on present and past and changing geographic locations (introduced through the notion of spatio-temporality), can provide environmental educators with tools to inform open-ended educational processes which can accommodate the dynamism and specificity of each situation. Figure 14.1 can be used to help environmental

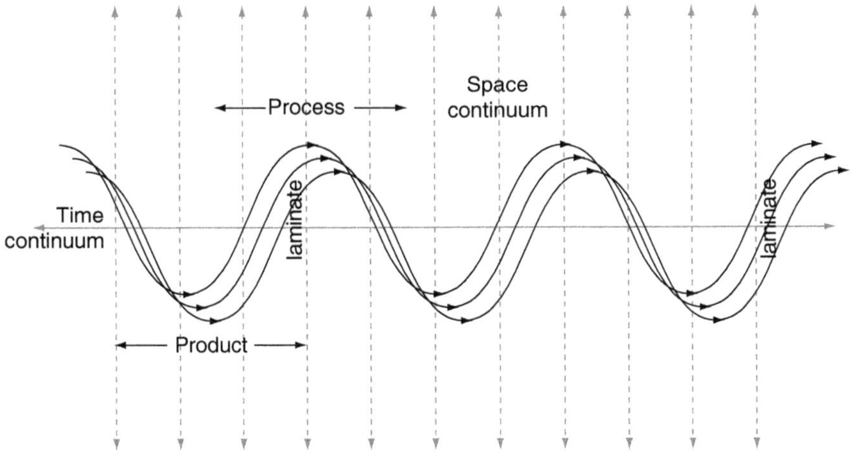

Figure 14.1 Time/space continuum of change in product/process laminated totality.

educators to understand, for example, past influences on (non-)choice and (non-)actions and what future choices and actions might emerge from this past in different geographical, cultural and socio-economic contexts. This understanding of spatio-temporality also highlights Wals's (2007a) concern regarding the non-transferability of actions from one context to another as well as the problematic of predetermining and imposing a change action which has not emerged from a particular spatio-temporal context. This point is also raised by Stables and Scott (2002, p. 54) who argue that they are opposed to an approach 'that assumes we should be aiming for a single and uncontested set of understandings and for complete consensus concerning future action'.

Change requires agency

Previous sections have touched on the question of choice and actions. Considering the relationship between individuals and society provides the tools to think about the relationship between choice and action and to understand how society influences people's actions and how their actions influence society in return.

Providing insight into the agential aspect of change is Bhaskar's transformational model of social activity in which he illustrates that social structures and human agency are mutually interdependent. This is through the explanation that 'social structure is a necessary condition for, and medium of, intentional agency, which is in turn a necessary condition for the reproduction or transformation of social forms' (Bhaskar, 1993, p. 154). Seeing the agency/structure relationship in this way enables an understanding that social structure will influence the choices that people make regarding environmental actions, although these choices might be compliant/acquiescent or resistant in relation to the structures from which they emerge and might be consciously or unconsciously expressed.

Bhaskar presents a view of societal structures as not only able to impose limits on, but also *offering possibilities* for agency. In this emergent relationship 'society is both the ever-present condition for human action, and as such both enabling and constraining' (1993, p. 469). For example an urban household might be enabled by government financial subsidization to install a solar geyser while a rural family might be constrained by both a lack of financial power (despite subsidization) and/or a lack of technical support in their remote communities to do the same.

In an elaboration of his 'transformational model of social activity' Bhaskar (1993) gives further insight into agency, highlighting that agency can represent various permutations of passive (for example, not to buy consumer goods that have a particularly high environmental footprint) or active (for example, to consciously buy fair-trade consumer products), attentive (to research life cycles and actively read labels on consumer products) or inattentive (to buy goods without paying attention to the life cycle of the product and its ingredients), and direct (to make and/or sell one's own alternative products) or indirect (to participate in the development of policy that legislates the boundaries of agential action).

Agency can also be examined from the point of view of whether it results in a change in the status quo or not, that is whether it has the capacity to reproduce or transform societal structure (Archer, 1998b, p. 375). For example, one might transform societal structure by changing legislation at a national level, or mobilizing indigenous practices at a local level; or simply reproduce structures by consuming and responding to consumerism in a way that keeps the power in the hands of the dominant economic forces.

Change involves power relations

The section above highlights the importance of understanding that structures and mechanisms can enable and constrain change. A further aspect of understanding the nature of reality involves the role that power relations play in such enablement and constraint. Deepening insight into change processes requires a greater or lesser 'stretching of the moral imagination' (Bhaskar, 1993, p. 265), for example stretching consideration in one's choices from one's immediate needs or health concerns to deep, global and structural problems such as child labour, environmental pollution, and cycles of deepening poverty and inequality. This enables an illumination of certain relations in society that are characterized by 'exploitation, domination, subjugation, oppression, repression and control' which Bhaskar terms 'Power2' relations (1993, pp. 175–176).

Bhaskar's notion of Power2 relations can help illustrate how there is not always a linear relationship between choice and action. Bhaskar's notion of Power2 refers to the 'transfactually efficacious capacity to get one's way against either (i) the overt wishes and/or (ii) the real interests of others (grounded in their concrete singularities)' (1993, p. 153) or 'negative characteristics such as domination, subjugation, exploitation and control that can be identified in given social structures' (Hartwig, 2007, p. 372). Power2 might take the form of:

- Agent A making B do something that she would otherwise not do. This was illustrated in the discussion above of the neo-colonial exploitation of the natural resources of developing world countries as an entry into the global market economy.
- The construction of compliance by the exclusion of known alternatives, thus narrowing choice, as in the case of a country that produces electricity for its national grid using predominantly coal and nuclear power while failing to explore alternative energy sources.
- Socialization such that the 'worldview' of individuals precludes an understanding of real interests. For example, in a country where inequality is high, promises of employment in unsustainable industries precludes exploration of environmental and personal health (ibid., p. 372).

In response to the oppressive nature of such Power2 relations, Bhaskar highlights the possibility for Power1 relations, by which he refers to:

> The general causal powers of human agency whose characteristics entail the possibility of human emancipation, such as our capacity to investigate, communicate, plan, construct moral and ethical systems, feel and care for others, and come to agreement based on judgmentally rational argument directed at practices that transform our lived circumstances.
>
> (Hartwig, 2007, p. 372)

Knowledge of the power relations described above provides one tool for developing a 'judgementally rational argument' for change or to guide the direction of change. It is the emancipatory interest in critical realism that calls for a move from 'unneeded, unwanted, and oppressive to needed, wanted and empowering sources of determination' (Bhaskar, 1989, p. 6). Bhaskar argues for this emancipatory interest by explaining that any ill will imposes oppressive constraints and inequities. He then reasons that 'inasmuch as they are (i) *unwanted,* (ii) *unnecessary* and (iii) *remediable* or removable, they should be transformatively negated, i.e. absented' (1993, pp. 259–260).

For example, consider a debate often faced by communities experiencing unemployment when promised the possibility of a new industry in the area.

Couched in an argument regarding the right to development, an executive summary presented to South African press in September 2012 argued the national Department of Mineral Resources' decision to allow the exploration of opportunities for hydraulic fracturing ('fracking') in South Africa's semi-arid hinterland known as the Karoo. The summary outlined a potential 'R1-trillion boon to the economy' from the proposed fracking and highlighted economic benefits for the 'severely deprived communities in the Karoo' (Mail & Guardian, 2012, no pagination). If educational processes can help learners to question the 'real reasons' for promoting the project – for example, is the promise of economic boon going to primarily enrich unemployed locals or those in power in the petroleum industries? Is the use of natural gas going to solve the problem of climate change or only delay the problem of a limited supply of non-renewable energy resources?

Bhaskar highlights a further important tool for supporting transformative environmental learning that is 'eliminating the prejudices, errors, unsupported claims and philosophical false trails which have covered or disguised reality for us; and he talks of using this reclaimed reality as the only basis for emancipatory social practice' (Corson (1991) in Shipway, 2001). Focusing on Power2 and Power1 relations can be a useful tool for developing judgementally rational arguments which enable a critique of power relations which are unneeded, unwanted and oppressive. This focus on power relations can provide environmental educators with tools for supporting 'a shift or switch to a new way of being and seeing' (Wals, 2009, p. 385).

So what does this interest in transformation mean for the way in which environmental education processes are mediated? Environmental educators Mogensen and Mayer (2005: 19) argue that 'actions must first and foremost be seen in relation to their educational and/or epistemological value – not in the first hand in relation to any possible societal and material consequences of the activity'. This argument is also evident in the words of Jickling who argued that:

> For many practitioners, educational achievement should enable individuals to act intelligently and with some measure of independent thinking. People will not think and act intelligently if they have been trained, conditioned, coerced, or otherwise manipulated to behave in a certain way.
>
> (Jickling, 1997, p. 95)

These arguments are not saying that learners should not be involved in environmental action, but they are questioning the purpose and motivation behind environmental action. In other words environmental learning should focus on 'human development rather than human behaviour' (Wals and Jickling, 2009, p. 78). The reflections on the relationship between structure and agency provided by Bhaskar above can be useful in helping students to understand how structure influences choice (or non-choice) and the possibilities (or not) for action so that even if they are not necessarily able to change the world directly they can at least develop a critical understanding of how change can be enabled or constrained. They can also develop critical thinking tools in order to understand questions such as what the difference would be between passive and active change or direct or indirect change in particular contexts. A further tool would be identifying and probing cases of 'inattentive' change for their implications for socio-ecological health and well-being, and the unequal ecological footprint of humanity.

Concluding summary

This chapter briefly presented ecological footprinting as a particular tool of environmentalists for exploring human-environment relationships. Some benefits of this tool are that it highlights the ecological limits within which humans must operate and elucidates inequalities and unsustainable practices. Combined with the challenges posed by the human development index and inequality co-efficient, ecological footprinting is a useful indicator for raising questions about sustainability and development. The chapter explored various critical realist tools and their potential to give us further insight into these sustainability/development questions and the possibilities for change. Change was explored in terms of its nature as transitive and intransitive and thus requiring change in ways of doing as well as change in ways of knowing. Change was also presented as involving interactions within and between different laminations in an emergent product/process relationship on a varied spatio-temporal landscape, thus requiring consideration of context-specific relational explorations of change possibility. In addition, change was explored as dependent on agency which can be both enabled and constrained by structure and which can manifest in terms of individual and structural change processes. Finally, the dependence of change on taking cognisance and responding to different power relations was discussed with emphasis on developing judgementally rational arguments for a different future.

These views of change were discussed in relation to contemporary thinking in the field of environmental education, highlighting a consistency with open-process, grounded and transformative approaches to environmental learning.

These tools for viewing reality and change questions can provide environmental educators with a 'systemic and reflexive way of thinking and acting with the realisation that our world is one of continuous change and ever-present uncertainty' (Wals, 2007b, p. 37), thus calling for an approach to environmental learning that cannot be pre-determined but which is responsive to changing need and to the unexpected. This means a move away from a narrow focus on communicating environmental 'messages' and raising awareness about environmental problems to processes of 'deliberative co-engagement in the challenges and risks of the day' (O'Donoghue, 2007, p. 148). While consciousness of the ecological limitations of a footprint is important, no generic or universal way of being or changing can be imposed on a person, organization or country. In educational contexts this means exploring possibilities for finding context-specific and morally defensible ways of living within the Earth's ecological limits.

Notes

1 An HDI is a composite measure of 'information on life expectancy, schooling and income' (Klugman, 2010, p. 3).
2 An IHDI 'examines the average level of human development and its distribution along the dimensions of life expectancy, educational attainment and command over the resources needed for a decent living. Where there is no inequality, the IHDI equals the HDI' (Malik, 2013, p. 29).
3 Figure 14.1 was inspired by Archer's (1998b) insistence on the importance of temporality in her representation of morphogenesis and morphostasis as well as Bhaskar's (1993) representation of the relationship between time and space in his diagram of four-planar social being.

References

Archer, M. S. (1998a). Introduction: Realism in the social sciences. In M. Archer, R. Bhaskar, A. Collier, T. Lawson, and A. Norrie (Eds), *Critical realism: Essential readings* (pp. 189–205). London: Routledge.

Archer, M. S. (1998b). Realism and morphogenesis. In M. S. Archer, R. Bhaskar, A. Collier, T. Lawson, and A. Norrie (Eds), *Critical realism: Essential readings* (pp. 356–381). London: Routledge.

Archer, M. S., Bhaskar, R., Collier, A., Lawson, T., and Norrie, A. (1998). *Critical realism: Essential readings.* London: Routledge.

Bawden, R. (2004). Sustainability as emergence: The need for engaged discourse. In P. B. Corcoran and A. Wals (Eds), *Higher education and the challenge of sustainability: Problematics, promise, and practice* (pp. 31–32). Dordrecht, Boston: Kluwer Academic Publishers.

Bhaskar, R. (1989). *Reclaiming reality.* London: Verso.

Bhaskar, R. (1993). *Dialectic: The pulse of freedom.* London: Verso.

Bhaskar, R. (1998a). *Possibility of naturalism* (3rd edn). London: Routledge.

Bhaskar, R. (1998b). Philosophy and scientific realism. In M. Archer, R. Bhaskar, A. Collier, T. Lawson, and A. Norrie (Eds), *Critical realism: Essential readings* (pp. 189–205). London: Routledge.

Bhaskar, R. (2010). Contexts of interdisciplinarity: Interdisciplinarity and climate change. In R. Bhaskar, C. Frank, K. G. Hoyer, P. Naess, and J. Parker (Eds), *Interdisciplinarity and climate change: Transforming knowledge and practice for our global future* (pp. 1–24). London: Routledge.

Chambers, N., Simmons, C., and Wackernagel, M. (2000). *Sharing nature's interest: Ecological footprints as an indicator of sustainability.* London: Earthscan.

Ferreira, J. (2013). Transformation, empowerment, and the governance of environmental conduct. In R. B. Stevenson, M. Brody, J. Dillon and A. E. J. Wals

(Eds), *International handbook of research on environmental education* (pp. 63–68). London: American Educational Research Association/Routledge.

Fien, J. (1993). *Education for the environment: Critical curriculum theorising and environmental education.* Geelong, Victoria: Deakin University Press.

Gough, A. (2013). The emergence of environmental education research: A 'history' of the field. In R. B. Stevenson, M. Brody, J. Dillon, and A. E. J. Wals (Eds), *International handbook of research on environmental education* (pp. 13–22). London: American Educational Research Association/Routledge.

Hartwig, M. (2007). *Dictionary of critical realism.* London: Routledge.

Hattingh, J. (2004). Speaking of sustainable development and values ... A response to Alistair Chadwick's viewpoint: Responding to destructive interpersonal interactions: A way forward for school-based environmental educators. *Southern African Journal of Environmental Education, 21,* 157–165.

IPCC (Intergovernmental Panel on Climate Change). (2014). *Climate Change 2014: Impacts, adaptation, and vulnerability. Contribution of working group II to the fifth assessment report of the Intergovernmental Panel on Climate Change.* (C. B. Field, V. R. Barros, D. J. Dokken, K. J. Mach, M. D. Mastrandrea, T. E. Bilir,... L. L. White, Eds). Cambridge: Cambridge University Press.

Janse van Rensburg, E., and Mhoney, K. (2000). The spiral model of Learning for Sustainability: An exploration of process-oriented teacher development in environmental education. In E. Janse van Rensburg and H. Lotz-Sisitka (Eds), *Monograph: Learning for Sustainability: An environmental education professional development case study informing education policy and practice* (pp. 41–68). Johannesburg: Learning for Sustainability Project.

Jickling, B. (1997). If environmental education is to make sense for teachers, we had better rethink how we define it! *Canadian Journal of Environmental Education, 2*(Spring), 86–103.

Jickling, B., and Spork, H. (1998). Education for the environment: A critique. *Environmental Education Research, 4*(3), 309–328.

Kissinger, M., and Rees, W. E. (2010). An interregional ecological approach for modelling sustainability in a globalizing world: Reviewing existing approaches and emerging directions. *Ecological Modelling, 221*(21), 2615–2623.

Klugman, J. (Ed.). (2010). *Human development report 2010: The real wealth of nations: pathways to human development.* United Nations Development Programme.

Le Grange, L. (2013). The politics of needs and sustainablity education. In R. B. Stevenson, M. Brody, J. Dillon, and A. E. J. Wals (Eds), *International handbook of research on environmental education* (pp. 126–132). London: American Educational Research Association/Routledge.

Lotz-Sisitka, H. (2008). Utopianism and educational processes in the United Nations Decade of Education for Sustainable Development. *Canadian Journal of Environmental Education, 13*(1), 134–152.

Lotz-Sisitka, H. (2011). Teacher professional development with an education for sustainable development focus in South Africa: Development of a network, curriculum framework and resources for teacher education. *Southern African Journal of Environmental Education, 28,* 30–71.

Mail & Guardian. (2012). *Fracking spin-offs trump concerns.* Retrieved 24 September 2012, from http://mg.co.za/article/2012-09-14-fracking-spin-offs-trump-concerns.

Malik, K. (2013). *The rise of the South: Human progress in a diverse world.* Human Development Report 13. New York: United Nations Development Programme.

Martinez-Alier, J. (2002). *The environmentalism of the poor: A study of ecological conflicts and valuation.* Cheltenham, UK: Edward Elgar.

Mogensen, F., and Mayer, M. (2005). *Eco-schools: Trends and divergences. A comparative study on eco-school development processes in 13 countries.* Vienna: Austrian Federal Ministry of Education, Science and Culture.

Norrie, A. (2010). *Dialectic and difference.* London: Routledge.

O'Donoghue, R. (2007). Environment and sustainability education in a changing South Africa: A critical historical analysis of outline schemes for defining and guiding learning interactions. *Southern African Journal of Environmental Education, 24,* 141–157.

Parker, J. (2010). Towards a dialectics of knowledge and care in the global system. In R. Bhaskar, C. Frank, K. G. Hoyer, P. Naess, and J. Parker (Eds), *Interdisciplinarity and climate change: Transforming knowledge and practice for our global future* (pp. 205–226). London: Routledge.

Price, L. (2007). *A transdisciplinary explanatory critique of environmental education: Business and industry.* Unpublished PhD thesis, Rhodes University, Grahamstown.

Sayer, A. (2000). *Realism and social science.* London: Sage.

Shipway, B. (2001). *A critical realist perspective of education.* London: Routledge.

Smyth, J. (1995). Environment and education: A view of a changing scene. *Environmental Education Research, 1*(1), 3–20.

Stables, A., and Scott, W. (2002). The quest for holism in education for sustainable development. *Environmental Education Research, 8*(1), 53–60.

UNEP (United Nations Environmental Programme). (1978). *Final report of intergovernmental conference on environmental education, Tbilisi 1977.* Paris.

Wals, A. (Ed.). (2007a). *Social learning towards a sustainable world: Principles, perspectives, and praxis.* Wageningen: Wageningen Academic Publishers.

Wals, A. (2007b). Learning in a changing world and changing in a learning world: Reflexively fumbling towards sustainability. *Southern African Journal of Environmental Education, 24,* 35–45.

Wals, A. (2009). Mirroring, gestaltswitching and transformative social learning. *International Journal of Sustainability in Higher Education, 11*(4), 380–390.

Wals, A., and Jickling, B. (2009). A framework for young people's participation in sustainability. In P. B. Corcoran and P. Osano (Eds), *Young people, education, and sustainable development* (pp. 77–84). Wageningen: Wageningen Academic Publishers.

15 Emergent properties and position-practice system of university educators in the mainstreaming of Education for Sustainable Development

Adesuwa Vanessa Agbedahin

Introduction

This chapter examines higher education responses to environmental and sustainability issues in Africa, specifically in the southern African region. It highlights the roles and attempts of universities in mainstreaming these issues into their core functions in making their contributions to the objectives of the United Nations Decade of Education for Sustainable Development (UNDESD 2005–2014) which sought to expand the influence and scope of education for sustainable development into education systems around the world (UNESCO, 2014). The chapter draws on Archer's morphogenetic approach and Bhaskar's position-practice system as frameworks to explore and analyse the above. It focuses on the International Training Programme (ITP) on Education for Sustainable Development (ESD) in Higher Education, initiated by the United Nations Environment Programme (UNEP) in phase two of Mainstreaming Environment and Sustainability in African (MESA) University Partnership programme as a broader case context, and the University of Botswana as a specific case. The thrust of this chapter is to show that the position-practice system of university educators before, during and after their participation in change-oriented professional development training programmes on ESD has significant influence on the mainstreaming of ESD into universities and beyond. The case study has revealed that in addition to individual and corporate changes in positions and practices influencing ESD mainstreaming positively, there is evidence of national and regional social-ecological influence.

Sustainability issues in southern Africa

Africa is the second largest continent in the world by area and population after Asia, and is characterized with diversity, beauty, culture and natural resources; all these assets make it a strategic region for global development opportunities (UNEP, 2008). Amongst the world economies, Africa holds a marginal position and is the region with the lowest human development index (0.502) (UNDP, 2014). Of the 42 countries classified as having a low development index, 31 of these are located in sub-Saharan Africa (ibid.). According to the most recent global climate change report produced by the Intergovernmental Panel on Climate Change, Africa is highly vulnerable to the impact of climate change, although it contributes the least of the greenhouse gases in the atmosphere (IPCC, 2014a, 2014b). Other threats to the natural environment in Africa include over-exploitation of resources, deforestation, poor water quality, poverty, food insecurity, drought, and deterioration of marine and coastal ecosystems (Lotz-Sisitka, 2011). The contemporary social-ecological condition that is being experienced is characterized by increased environmental degradation and risk, which therefore provide new ontological conditions that shape human and non-human vulnerability, as well as potential prospects and pathways for social-ecological change (Lotz-Sisitka and Price, 2012).

Fifteen countries make up the Southern African Development Community (SADC): Angola, Botswana, Democratic Republic of the Congo, Lesotho, Madagascar, Malawi, Mauritius, Mozambique, Namibia, Seychelles, South Africa, Swaziland, Tanzania, Zambia and Zimbabwe. The aforementioned environment and sustainability issues are considered critical to livelihoods and well-being in the southern African region, which has been identified as one of the world's regions most at risk to the intersecting implications and consequences of social-ecological risks (Pesanayi, 2009; Lotz-Sisitka and Price, 2012; IPCC, 2014b). These contemporary conditions require new approaches and provide opportunities for all education sectors, especially Higher Education Institutions, to rise to the challenge, as they are considered as potential change agents and drivers of development in terms of the global knowledge economy, development knowledge production, sustainable development, social change, community engagement and practice (Kotecha *et al.*, 2012).

Sustainable development and Higher Education

The concept of sustainable development (SD) was first introduced in the World Conservation Strategy and later by the United Nations World Commission on Environment and Development (WCED) in its 1987 Brundtland report, *Our Common Future.* The United Nations (UN) conference on Environment and Development in 1992 (Rio Summit) with its Declaration of action plan known as Agenda 21 and the World Summit on Sustainable Development (WSSD) are two examples of major conventions that have ushered in better understanding and operationalization of SD. SD has three dimensions, namely the ecological, societal and economic spheres which in critical realist terms can be extended and viewed as a laminated system with emergent properties as described by Bhaskar (2010) (physical, ecological, personal, social, cultural). To explicate the most commonly referred to spheres of SD (namely social, ecological and economic) and their significance, meaning and interplay, several models have been put forward; some are considered 'strong', while others are considered 'weak' (Lotz-Sisitka *et al.*, 2006; Neumayer, 2013). An example of a strong model of SD is that presented by Hattingh (2004), which recognizes that the social and economic are embedded in and emergent from (but not reducible to) the ecological, and that well-being of the social and economic sphere is dependent on the state of the ecological system. Neumayer (2013, pp. 1–2) suggests that the model of 'weak' sustainability emerges from neoclassical economics which refutes the value of natural capital in economic thinking, while 'strong' sustainability suggests that natural capital is 'non-substitutable'. He suggests that the latter is still 'difficult to define' in terms of its implications which is a direct result of the dominance of neo-classical and neo-liberal economics.

Several international declarations, conferences and initiatives have highlighted the role of Higher Education in actualizing the principles of SD. The aforementioned 15 member SADC states are all part of the Southern African Regional Universities Association (SARUA), established in 2005 (SARUA, n.d). It currently assists over 60 universities with the vision of revitalizing and developing leadership and institutions of Higher Education in the Southern African region, thus enabling the regional Higher Education sector to meaningfully respond to the developmental challenges facing the region (ibid.). Socio-ecological issues and risks are developmental concerns affecting the region and the Higher Education Institutions (HEIs) herein have the responsibility of overcoming these challenges in their core functions: teaching, research and community engagement through mainstreaming Education for Sustainable Development and responding to key issues such as climate change (SARUA, 2014). SARUA has recently published (2014) a large mapping study outlining regional university responses to climate change across 12 SADC countries. In framing a knowledge co-production strategy for regional co-operation amongst universities in the SADC region based on this mapping study, SARUA (2014, p. 31) states that:

> In addition to the usual roles of universities as expressed in the 'normal' mandates of universities, i.e. to conduct research, teaching and community outreach, the climate change and CCD/SD research trajectory 'also expects universities to take up stronger innovation and leadership roles…' and to '…contribute to … solutions and actual changes in practice'.

Mainstreaming environment and sustainability in African universities

The mainstreaming of environment and sustainability in African universities (particularly Southern Africa) has a short history but its beginning was heralded by strong individual and institutional commitment (UNEP, 2008). For example, the university alliance entitled Mainstreaming Environment and Sustainability in Africa (MESA) was formed in 2004 and the Global Environment Partnership in Environment and Sustainability (GUPES) was formed in 2010 and the International Training Programme (ITP) on ESD in Higher Education was initiated in 2008. These are contextual examples of partnerships constituted with the aim of assisting and providing ample opportunities for universities and their educators to mainstream environment and sustainability concerns into the core functions. Many African countries and specifically countries of the southern African community were party to them.

The ITP is a change-oriented professional development programme on ESD important in that professional development is necessary for the enhancement of the capacities of Higher Education professionals in order for them to be equipped for the challenging task of enhancing SD. The aim of professionalism is to 'improve practice and enhance accountability by creating means for ensuring that practitioners will be competent and committed' (Darling-Hammond, 2009, p. 49). Among the available approaches to change, the professional approach (unlike the bureaucratic, market or democratic approach to change) 'seeks to invest in knowledgeable practitioners who can make sound decisions about how to shape education' (Darling-Hammond, 2009, p. 46). This approach is necessary and has been employed in the field of environmental education (EE) and ESD for many years (Lotz-Sisitka *et al.*, 2015). Training of all employees is the most important aspect in the process of any organizational change, and only in this way can awareness of any issue be developed in any organization (Lynton and Pareek, 2000).

Investing in professional development of educators not only increases the expertise and quality of educators but also affects students' achievement rate as well as institutional improvement (Darling-Hammond, 2009). There is an urgent need to increase human capacity and skills to improve development opportunities and respond to sustainability risks in Africa (Lotz-Sisitka, 2011). This need is evident in a recent UNEP foresight publication stating that the transformation of human capabilities to meet environmental challenges is the second most rated emerging issue in the world in the twenty-first century (UNEP, 2012). Lotz-Sisitka argued that SD is, in itself, a change-oriented learning process since no society (or sector of society) has yet been able to figure out how to develop sustainability (SAQA, n.d). Change-oriented professional development training on ESD, such as the ITP, takes cognizance of the changing societal context, the emerging critical state of the environmental and the sustainable development issues which are significant in the twenty-first century (Lotz-Sisitka and Hlengwa, 2012; Lotz-Sisitka *et al.*, 2015). It is also aware of the impact that unsustainable generational practices have on disciplines, the knowledge structure and pedagogical process. Furthermore, it anticipates how universities are going to be able to withstand consequent challenges (ibid.). The ITP on ESD in Higher Education therefore is a timely and important change-oriented learning approach.

However, after university educators have participated in the ESD mediational process included in the professional development training programmes on ESD, there is a need to know how their practices can be critically investigated, understood, interpreted and documented. In order to achieve the above, I present case studies of two university educators from the University of Botswana. The data was collected through: semi-structured, one-on-one interviews; and a questionnaire, as well as document review. The data was analysed using the concepts and notion of position-practice system and emergent properties, with specific reference to and use of Archer's morphogenetic social realist theory of change framework (Bhaskar, 1979, 1989, 1998; Archer, 1995, 1998, 2010); these are explained in the next section. An analysis of the positions and practices of the educators before, during and after the ITP, and the resultant emergent properties are presented below.

Relevant critical realist perspectives

Position practice system

Social things are activity dependent and the existence of social structure is only possible in and through the activities of human agents (Bhaskar, 1989; Archer, 1998). The operation of universities is a social thing and the institutions themselves are social structures. The activities that occur in universities are only made possible by the operations and activities of human agents, therefore the existence of the university structure is dependent on human agents comprising academic and non-academic staff (Agbedahin, 2015). The PhD research from which this chapter emanates focuses only on academic staff, particularly those that have participated in the change-oriented professional development international training programme on ESD. These academic staff occupy positions and participate in certain expected activities in accordance with their professional roles and obligations. There is an assumption that there is a link between the positions of university educators, their practices and effective/ineffective pursuance of ESD mainstreaming in universities. This chapter explores this assumption in the case study. From a critical realist vantage point, this assumption can also be viewed as a transfactual theory. As such, the case study below will be exploring this theory, or testing it out, by applying it to the situation. I do this both to better understand the case studies, but also to check whether or not the theory stands the test of the facts of the case study. In this way, the case study research can also be seen as an example of applied science, using the RRREIC model, that uses resolution (understanding what is), re-description (describing using theories), retrodiction (analysing to explain why things are the way they are), elimination (eliminating alternative accounts that don't adequately explain what is going on), identification of alternatives (using induction and deduction), and correction (suggesting alternatives for improved practices) (Bhaskar, 2010).

In critical and social realism, society is considered as an open system which remains open simply because society is 'peopled', i.e. agent directed (Archer, 1998, p. 190). Nothing happens 'behind their backs, [hence] everything that happens, happens in and through their actions' (Bhaskar, 1998, pp. 39–40). People in a society or institutions have properties and powers including self and social reflection and reflexivity which they exhibit when confronted with challenges in their social context in order to reproduce, recreate or transform their context (Archer, 1995, 1998; Bhaskar, 1979, 1989, 1998). Bhaskar advised and Archer concurred that it is very important to distinguish categorically between people and society, as well as between human actions and changes in social structure; this is because 'the properties possessed by social forms may be very different from those possessed by the individual upon whose activity they depend' (Bhaskar, 1989, p. 41; Archer, 1998, p. 200). Archer further pointed out that 'the distinction between positions and practices is crucial and it is by maintaining it and working on its implications that what is *sui generis* to society can be extracted' (Archer, 1998, p. 200). She clearly stated that positions should not be conflated with practices. Therefore considering the educators in universities who are the subject of the case studies presented in this chapter, their 'positions must predate the practices they engender: although activity is necessarily ceaseless for society to be, it is discontinuous in nature because changes in society's structure [e.g. positions created by structure] then condition practices in distinctively different ways' (Archer, 1998, p. 201).

In order to understand the continuous reproduction or transformation of social structures that exist and are exercised by virtue of human agency, Bhaskar postulated that we need a system of mediating concepts, i.e. a system of concepts describing the 'point of contact' between human agency and social structure (1998, p. 40). He explained that such a point of contact which links action to structure must *both* endure and be immediately occupied by individuals (ibid.). Bhaskar called the mediating system the 'position-practice system' (Bhaskar, 1998, p. 41). This mediating system involves 'the positions (places, functions, rules, tasks, duties, rights, etc.) occupied (filled, assumed, enacted etc.) by individuals, and of the practices (activities etc.) in which, in virtue of their occupancy of these positions (and vice versa), they engage' (Bhaskar, 1998, pp. 40–41; Archer, 1998, p. 200). Bhaskar added that positions and practices, though seemingly distinct and unrelated, can only be explored relationally, i.e. one in relation to the other. Therefore, this chapter takes cognizance of the position-practice mediating system which is essential to the understanding of the social reality of the case university lecturers, their institution, the ITP, ESD and SD learning and change processes.

Emergent properties and powers

Generally, when there is emergence, Bhaskar explained that 'new beings (entities, structures, totalities, concepts) are generated out of pre-existing material from which they could have been neither induced nor deduced' (2010, p. 45). He referred to its occurrence and existence as a 'quantum leap, or nodal line ... the materialized imagination' (ibid.). Indeed the conceptualization and planning of ESD mainstreaming by university educators and their institutions remains an imagination until the point at which it is materialized at the level of actual implementation and change in practice. According to Archer, emergent properties are 'those entities which come into being through social combination ... they exist by virtue of interrelations' (Archer, 1998, p. 192). Structure and agency possess distinct generative emergent properties although they are interdependent and interrelated; hence they are not considered to be equally established and are therefore analytically separable (Archer, 1995). A variety of emergent properties exist in society, including structural, cultural and personal (or agential) emergent properties, which are relatively continuous, independent and irreducible to the other (ibid., p. 175). Structural emergent properties (SEP) depend primarily on the availability of human and physical material resources, i.e. internal relations that exist in the structure itself (ibid.). Structural Emergent Properties (SEP) and Cultural Emergent Properties (CEP) could, in this context, be institutional change and learning outcomes such as community engagement, campus management, teaching programmes, policy innovation, or new pedagogy etc. Examples of Personal Emergent Properties (PEP) include cognitive gains, motivation, reflexivity, ethics, self-consciousness, resolution, self-monitoring ability, valued being and doings of research participants in relation to their positions and practices when mainstreaming ESD.

In order to secure the reality of relational concepts like positions and practices, the causal efficacy has to be demonstrated, that is, a causal criterion needed to establish reality must be employed (Archer, 1995; Bhaskar, 1998). The mediation done by the ITP (what I call the macro mediating system) and subsequent participants (what I call the micro mediating system) are the causal criteria employed by this study to investigate positions and practices and emergent properties and powers in the case studies presented. There are non-observable generative mechanisms which possess powers that may exist even if they are exercised unknowingly or not exercised at all (Archer, 1998; Bhaskar, 1998). The presence of such powers has 'variable outcomes' which occur as a result of various 'intervening contingencies' existing in an open system (Archer, 1998, p. 190) like universities. Examples of generative mechanisms (powers) with variable outcomes and intervening contingencies (known and unknown, exercised unknowingly or not exercised at all, considered or unconsidered) will be the broader socio-ecological, global, regional, national systemic realities. This is exclusively captured in the laminated system analysis of the phenomenon being investigated (Agbedahin, 2015).

The change-oriented intervention of the ITP through the university educators and the implementation of their ESD change projects is the intervening contingency this chapter seeks to explicate and the emergent properties and powers being the variable outcomes of the position-practice system of university educators (exercised generative mechanisms). The activities of generative mechanisms are 'continuous and invariant' and this characteristic is as a result of their powers and properties that are relative; this is referred to as 'transfactuality of mechanisms' (Archer, 1998, p. 195). Archer (1998, p. 201) also put forward that the ways that pattern social interaction (e.g. social conditions and positions) are incomprehensible without back reference to the conditional influences (which are thus prior) of the position (e.g. influence of professional development training programmes), the resources associated with it, the interests vested in it, none of which can be captured by a seamless web of 'practices'.

> A position has to exist prior to its occupancy and even if the same people become the incumbents of newly elaborated positions, the new set of internal relations into which they are then embroiled exert a *sui generis* conditional influence upon them – which is causally detectable precisely through their changed practices as agents and the elaboration of agency itself.
>
> (Archer, 1998, p. 202)

The morphogenetic framework

The principles of morphogenesis originated from Buckley's system theory (Zeuner, 1999). Archer (2010, p. 228) defined morphogenesis as 'the complex interchanges that produce change in a system's given form, structure or state (morphostasis being the reverse), but it has an end-product, [which is] structural elaboration'. The society has a distinct morphogenetic nature which characterizes its ability to change its form or shape; hence social things are not unchallengeable (Archer, 1998). The morphogenetic cycle therefore provides 'social realism with a *method* of explaining social structure over time in terms of the interplay between structure and agency – which can be used to generate practical social theories in particular domains' (Archer, 1998, p. 202).

The morphogenetic approach has the potential to help analyse generative mechanisms that originate from structures and cultures as well as from people involved in the structural and cultural milieu – in the form of emergent properties (Archer, 1995). There is an assertion within the morphogenetic perspective that structure and action operate over different time periods and this claim is based on two simple propositions that (1) structure necessarily pre-dates the actions which transform it; and (2) that structural elaboration necessarily post-dates those actions (Archer, 1995, 1998). Bhaskar ascertained in a similar vein that 'social structures always pre-exist human agency, but are reproduced or transformed only in virtue of it and in the course of on-going social activity' (2010, p. 9).

Morphogenesis consists of a three-part cycle, namely (1) structural conditioning, (2) social (cultural) interaction and (3) structural elaboration (Archer, 1995, 1998, 2010) (see Figure 15.1).

Figure 15.2 presents an exploration of the case study context and application of the position-practice system and emergent properties to the morphogenetic cycle. A detailed explanation follows.

Structural conditioning

T^1

T^2 Socio-cultural interaction T^3

Structural elaboration (morphogenesis)

Structural reproduction (morphostasis) T^4

Figure 15.1 Archer's basic morphogenetic/static cycle with its three phases (source: Archer, 1998, p. 375).

T^1 Structural conditioning (national, institution context, ESD problem/origin of change project)

Agential conditioning (micro mediating system: position-practice system)

T^2 Socio-cultural interaction (learning and change processes) T^3

ITP (macro mediating system); Micro mediating system (position-practice system)

Structural/cultural elaboration or reproduction (SEP, CEP)

Agential elaboration or reproduction (position-practice system) (PEP) T^4

Figure 15.2 Applying the morphogenetic cycle and position-practice system to the case study (source: adapted from Archer, 1998, p. 375).

The case study data: Botswana and the University of Botswana

T1: Structural conditioning – Botswana and the University of Botswana (UB)

The University of Botswana, the only university in Botswana, was established in July 1982 by an Act of Parliament, stemming from the then University of Botswana, Lesotho and Swaziland (University of Botswana, 2013). The University is closely involved in the national development process of Botswana (ibid.) through a curriculum that responds to contextual knowledge discourses informed by national policies (Ketlhoilwe and Jeremiah, 2010). Some of the national policy documents that support environmental education, SD and ESD include the *Revised National Policy on Education, National Development Plans,* the *Environment Education Guidelines, National Environmental Education Strategy and Action Plans 1 and 2* (ibid.). It is a large and rapidly expanding university with over 21,000 students (Participant 1[1]). At present there are seven established faculties: Business, Education, Engineering and Technology, Humanities, Science, Social Sciences, Faculty of Health Sciences and the School of Graduate Studies; in general, departments in the same Faculty work closely together in offering degree, diploma and certificate programmes (University of Botswana, 2013). The Faculty of Education, the site of these case studies, has about seven departments and specializes in training teachers (Participant 1).

T1: Problematic institutional sustainability issue and risk that the university educators sought to address before participation in the ITP

Research that develops and generates ample knowledge on SD is promoted at the University of Botswana and the outcome of the 2005 and 2008 audit of the university programmes indicated that almost all faculty programmes were involved in sustainable development related activities (Ketlhoilwe and Jeremiah, 2010). However, by 2007 and before the staff participation in the ITP in 2008, ESD was not integrated into all Education faculty curricula (the faculty of the educators/participants in this research) and students were not abreast of sustainability issues and risks. Although the staff occupying higher positions were somewhat conversant with ESD, 'some HODs [Heads of department] were not aware of what was going on in some units as far as sustainability performance is concerned' (ibid.) and most junior staff were totally unaware. Regarding the formal curriculum 'there were limited environment and sustainability components/topics across the UB faculties particularly in the Faculty of Education courses' (Ketlhoilwe and Jeremiah, 2010). The following data excerpts shed more light:

> When I started in 2008 … all these courses were not there; the only course on sustainable development was in Environmental Science Department and there was another one on sustainable development in the Faculty of Social Science under Developmental Studies, but in terms of Education for Sustainable Development (ESD), it wasn't much. It [ESD] is lacking and then looking at the capacity and the competences … it is very difficult to introduce that [ESD] across the university curriculum.
>
> (Participant 1)

> The implementation of UN Decade on Education for Sustainable Development (UNDESD) was minimal or none existent. There is no strategy to guide UNDESD implementation. The findings [of the audit we did] showed that the Heads of Departments understood the concept of ESD, which is reflected in the Mission and Vision of UB and respective departments,… most of the academic staff members have not internalised them.
>
> (Participant 2)

T1: Agential conditioning – position and practice

Before the ITP mediating system, both Participants 1 and 2 were educators at the University of Botswana, Gaborone, Faculty of Education, Department of Social Science Education, Environmental and Sustainability Education unit. Participant 1 was the head of the unit and Participant 2 was a senior lecturer. Their practices before the ITP mediation were quite similar and regular including lecturing, supervising, researching, developing courses and course contents, training teachers to teach social sciences. The following data excerpt puts this into perspective: 'To teach undergraduate, graduate and post graduate students, do and supervise research, do community service and any other relevant job related to the mandate of the university and my profession' (Participant 2). Before the ITP in 2008, the other participant seemed to lack skill and capability to initiate ESD into course contents and programmes but displayed the zeal to bring about change in UB in this regard: 'To influence change within the university … I was actually preparing for a master's programme … so when the (ITP) invitation came, I said "this is an opportunity for me to actually undertake this change project"' (Participant 1).

T2–T3: Social-cultural interaction (ITP mediation – learning and change process)

During the ITP mediating system in early 2008, these two change agents decided to use their positions and practices by mainstreaming environment and sustainability issues. They revised existing courses in their unit and introduced a new master's and PhD programme. They planned to influence curriculum and policy change in order to effectively mainstream environment and sustainability issues, thereby contributing to the UNDESD. They attended the ITP and collaborated with their institution and colleagues. During the programme, their ESD related activities included a change project development which was influenced by the realities in their institutional and organizational structure as well as their agential conditionings. During the ITP intervention, a Unit-Based Sustainability Assessment Tool (USAT) was used to conduct an audit of selected departments in the Faculty of Education and other Faculties. The participant explained that:

> The USAT tool came.... I actually administered it ... that helped me to actually confirm that there are some gaps as far as Sustainable Development and Education for Sustainable Development is concerned. So that was how I actually undertook that change project.
>
> (Participant 1)

The change project implementation process started. A seminar which involved the UB academic and administrative staff was organized to present the change project. One of the participants went to Sweden as part of the ITP international exchange session. There was further participation, deliberation and collaboration with stakeholders at the faculty and in the university at large. The ITP regional co-ordinators visited the participants' institution and had a meeting with stakeholders; this move was very significant in authenticating and highlighting the efficacy of the project. This interview excerpt explains:

> I think it is multifaceted the way we are working together.... I organized a seminar ... I only invited 50 people, I invited all heads of departments, deans of faculties and then I got some people from outside ... there were a lot of supports from the university administrators particularly the vice chancellor [who] was happy ... actually promised me support.... I had the opportunity to go to Sweden ... [the ITP co-ordinators] visit to my institution helped in influencing the institutions' management to support me even more.
>
> (Participant 1)

It was decided to focus on one course, the master's course, and not all the university courses while developing the concept of mainstreaming. Departmental and local stakeholder consultation was held on the design of the MEd programme. A report was produced. The framework for the development of the MEd in Environment and Sustainability Education course design was developed during the ITP regional programme session. This interview excerpt sheds more light: 'I started small; and in starting small, I started with my courses that I was teaching to actually infuse and integrate the elements of environment and sustainability' (Participant 1).

T4: Structural, cultural and agential elaboration (PEP, SEP and CEP)

Structural emergent properties (SEPs)

After the ITP mediation by the end of 2008, participants' activities included the submission of the course proposal to the university curriculum board, and further department level consultations being held. The MEd course was developed and this change project has influenced and is influencing the practices of other university educators. The participants explained:

> So that is how we started growing, and then we went back to the Master's programme and then we had a Master's programme in environment and sustainability education.... So from there to my department. A good number of lecturers actually accepted and they revised their courses and then integrated some elements of environment and sustainability.
>
> (Participant 1)

> Yes, our change project is ongoing. New programmes are ongoing. MESA has been accepted and University of Botswana is a focal point for MESA in the Southern Africa region.
>
> (Participant 2)

Apart from the evidence of curriculum innovation, including the development of an MEd programme on environment and sustainability, there was the commencement of a PhD programme. The research agenda and focus of both participants and their department have changed to imbibe the values and principles of SD. The following data supports this:

> Our change project influenced curriculum innovation in terms of courses revision, review, and development of new courses to integrate sustainability issues and concerns. This strengthened our courses and updated our curriculum. Our change project outcome was the ongoing Masters programme and a PhD programme on environmental and sustainability education. These programmes are now being implemented. Our change project has also influenced our research agenda and focus.
>
> (Participant 2)

The experience and impact of the first ITP participant and co-participant in the Faculty of Education led to the participation of more ITP participants, and has made the actualization of the principles and strategies of ESD and SD more achievable at the University of Botswana. More courses were revised and developed in other departments of the Faculty of Education. The following interview excerpt sheds more light:

> My colleagues had also attended the ITP course and their projects are also curriculum innovation oriented. I had two more [colleagues] attending ITP after me, so my *head of department* was the first to attend the ITP, then when he came back, he did the same process, he developed two new courses in the languages, language and sustainable development, the other one is literature and sustainable development, so he developed that he had a co-participant, so now in the same department we are four.... The other thing is my deputy dean's project.... So as it went up the hierarchy, the administrative hierarchy, more and more people were getting interested; that is why we have ... new courses to enrich the programme.
>
> (Participant 1)

Participants' colleagues have been influenced positively and the institutional impact of the ITP change projects has gone beyond the confines of the University of Botswana to the country at large (Ketlhoilwe and Jeremiah, 2010). For example, the two course participants are now running national training programmes for teachers involving all the national teacher education colleges using the model and content of the ITP. There is also great future potential to include ESD in more sections of the University of Botswana (ibid.). Due to the participants' involvement in the ITP, the broader MESA activities and the substantial elaboration of ESD related practices at both individual and institutional levels, a SADC MESA Chair now exists in the Education Faculty which has a focus on curriculum and policy change. These participants' change project is already part of the SADC REEP/MESA Network activities and UB has one of the three MESA Chairs in Southern Africa. Hence, the pre-ITP positions and practices of the university educators have metamorphosed and made way for elaborated positions and practices. This interview excerpt highlights this:

> Most of our colleagues are infusing ESD, more applications for PhD are flowing in, and some faculty members are developing interest in researching on sustainable development. In the country we have an annual ESD change projects course.... Environment and sustainability concerns would be mainstreamed in curriculum, policies and practices at institutional, regional and organisational levels.
>
> (Participant 2)

Personal emergent properties (PEPs)

Some of the PEPs that are notable in the participants include *motivation, sustained interest* and *enthusiasm* in ESD related issues and concerns. As Ketlhoilwe and Jeremiah (2010, p. 7) argued, the '[ITP change] project has *stimulated interest* in education for sustainable development among some staff members and students'. Other PEPs are: *enhanced active participation* in the project of mainstreaming ESD and courses/programme development in ESD; *better understanding* of ESD issues across countries; and *appreciation* of a variety of change projects from other countries and institutions:

> Thorough knowledge about ESD, sustained interest in ESD related issues, active participation in MESA and courses/programme development in ESD.
>
> (Participant 2)

> [My trip to] Sweden has increased my enthusiasm about ESD. It has also helped me understand more about ESD issues across the countries. [During the South African component of the course:] An appreciation of a variety of change projects from other countries and institutions. This encouraged me and my co-participants to work harder and reconceptualise our change project looking at local contexts.
>
> (Participant 1)

Participants have also *gained and exhibited the capability to publish* ESD related papers. They have developed *increased insight, and acumen to address and overcome institutional challenges* and frustrations encountered in the process of implementing the change project. In addition, participants' *reflexive capabilities seem to have been enhanced to work from wider to narrower forms of action,* and expectations have been and are being actualized as shown in the interview excerpts below:

> Our change project has also influenced our research agenda and focus. We managed to research and publish papers on environment and sustainability.
>
> (Participant 2)

> The initial change project was just to influence change within the university, I was hoping that using the feedback I got from them, my kind of insight, new insight I got from ITP, people were going to buy in the idea of integrating environment and sustainability right in their courses as well as in their programme; majority of them were going to do that but it was an overwhelming thing; because I still have the

> energy and the zeal I had then I thought I was going to do it, but with the challenges it wasn't possible, but starting small I saw that it was actually happening just a little bit. But as far as my work was is concerned I felt like I was kind of overworking myself, but at the same time, with the interest I felt like well let those who can take it take it, those who cannot take it not take it, but overtime perhaps they will eventually take it. And to some extent some people actually bought the idea.... So I will say my expectations to some extent are being fulfilled slowly slowly moving from the bigger ... to a narrower or smaller and then going back to a much bigger approach to change project.
>
> (Participant 1)

Cultural emergent properties (CEPs)

Despite the different positions and changed practices of the case ITP participants and their other colleagues who later participated in the ITP, there have been enabling CEPs such as *consultation and collaboration in terms of contemporary sustainability issues*. The participants have *developed a culture of learning from and sharing with one another in academic forums and meetings*, even beyond their immediate department and faculty. The challenge of implementing change in social structure by agents was also experienced by participants, but as time went by, they *overcame some of the institutional limitations*, as outlined below:

> So within this group of now six people when we have seminars we invite other people, we invite each other; so when I have seminar or workshop even outside university, I actually tap from this expertise, yes so they go and resource, wherever we have workshop, wherever I cannot reach I approach them they also attend ... so we are working very well together.... How people in term of position got interest and then doing it, having their change project, implementing change project, and mentioning it several times in departmental meetings, mentioning it in the faculty meetings.... That actually brought people together.
>
> (Participant 1)

Some constraining CEPs include the reality that 'academic staff members are closed in little boxes called "areas of expertise and disciplines"' (Ketlhoilwe and Jeremiah, 2010, p. 8). They seem to be informed more by their training and research interests, and although they do research on a regular basis, they are not open to being involved in research that examines their practices (ibid.). Some academics are not prepared to be trained in a 'strange area' such as environment and sustainability issues (ibid.), this therefore limits transdisciplinary learning opportunities, transformation and self-reflexivity of their own practices. The following data supports this:

> Well I will say it [implementing the change project] was difficult. People claiming their own areas, findings, environment and sustainability issues … to me environment education was difficult, but after seeing other people like those who attended the ITP also having their seminar and workshops … people actually started wondering what is this actually coming … it was not easy but it is doable yeah.
>
> (Participant 1)

> People claim that they are within their territory, they don't want anything that will be a burden, people don't want to do anything extra apart from what they are trained for.… Except there is extra money, or some reward, some kind of reward. It is difficult that is the thing … extending my territory I have to … more and more; and so forth and people lacking the idea of how really this [ESD] could be part of what I am doing, it is part of what I am doing … integrating environment and sustainability issues … research, people do not see those opportunities.
>
> (Participant 1)

Conclusion

This chapter has briefly touched on some prevalent environment and sustainability issues and risks in Africa and particularly in the southern African region. The roles of higher education institutions in mainstreaming these environmental and sustainability concerns in African universities have been highlighted. The need for professionalism was presented and an example of a change-oriented professional development training programme on ESD was used as a broad case; university educators and their university were used as specific cases. As indicated in the introduction of the chapter, there is an assumption that there is a link between the positions of university educators, their practices and effective/ineffective pursuance of ESD mainstreaming in universities. This chapter explored this assumption in the case studies. I was able to apply this transfactual theory, assumption to the situation of the University of Botswana, and to test it out using the case data.

I did this by understanding the situation in the university (resolution) and the positions and practices of the participants before they joined the ITP as these pertained to ESD. I was then able to re-describe what was occurring using empirical insights shared from the participants, combined with Archers' morphogenic theory in which I was able to use retrodiction to explain how emergent properties (SEPs, PEPs and CEPs) interacted to lead to changes in the situation, in the university programme being discussed, and more widely in the university and even beyond the university. It is therefore possible to say that the ITP was able to effect changes in the positions and practices of the participants on the ITP, but that this was brought about via a range of interacting emergent properties, not all of which were directly related to the course, but rather to the interacting relations between the course and the participants' context. This has provided some insight into how such professional development programmes can work in open systems.

The main insight gained from this study is that the positions and practices of university educators, before, during and after their participation in ESD training have the potential for more than personal and institutional influence; in addition there is potential for a national and regional commendable impact as was shown by the actualized potential in the UB case. This case can provide a framework for guiding further analysis of other cases, from which it will then be possible to see if the assumption guiding the programme (or the transfactual theory) is consistent or diverse across a wider range of contexts and cases; especially over a longer period of time.

Acknowledgements

This chapter is based on PhD research that is being conducted by the author in the Environmental Learning Research Centre at Rhodes University (Agbedahin, 2015). The study has been partially supported by the Environmental Learning Research Centre's Murray and Roberts and Gold Fields Scholarship Trust Fund.

Note

1 Full details of research participants can be found in the PhD thesis (Agbedahin, 2015) on which this chapter is based.

References

Agbedahin, A. V. (2015). *A morphogenic and laminated system explanation of how position-practice system and professional development training programme shape education for sustainable development in African universities*. Unpublished draft PhD thesis, Rhodes University, Grahamstown.

Archer, M. S. (1995). *Realist social theory: The morphogenetic approach*. New York: Cambridge University Press.

Archer, M. S. (1998). Realism and morphogenesis. In M. S. Archer, R. Bhaskar, A. Collier, T. Lawson, and A. Norrie (Eds), *Critical realism: Essential readings* (pp. 356–381). London: Routledge.

Archer, M. S. (2010). Morphogenesis versus structuration: On combining structure and action. *The British Journal of Sociology*, *61*(s1), 225–252.

Bhaskar, R. (1979, 1989, 1998). *The possibility of naturalism*. (1st edn) Sussex: The Harvester Press; (2nd edn) Hemel Hempstead: The Harvester Press; (3rd edn) New York: Routledge.

Bhaskar, R. (2008). *Dialectic: The pulse of freedom*. New York: Routledge.

Bhaskar, R. (2010). Contexts of interdisciplinarity: Interdisciplinarity and climate change. In R. Bhaskar, C. Frank, K. G. Hoyer, P. Næss, and P. Jenneth (Eds),

Interdisciplinary and climate change: Transforming knowledge and practice for our global future (pp. 1–23). Abingdon: Routledge.

Darling-Hammond, L. (2009). Teaching and the change wars: The professionalism hypothesis. In A. Hargreaves and M. Fullan (Eds), *Change wars* (pp. 45–68). Bloomington: Solution Tree.

Hattingh, J. (2004). Speaking of sustainable development and values: A response to Alistair Chadwick's viewpoint 'Responding to destructive interpersonal interactions: A way forward for school-based environmental education'. *Southern African Journal of Environmental Education, 21*, 157–165.

IPCC (Intergovernmental Panel on Climate Change). (2014a). *Climate Change 2014: Impacts, adaptation, and vulnerability. Part A: Global and sectoral aspects. Contribution of working group II to the fifth assessment report of the Intergovernmental Panel on Climate Change.* (C. B. Field, V. R. Barros, D. J. Dokken, K. J. Mach, M. D. Mastrandrea, T. E. Bilir,... L. L. White, Eds) Cambridge: Cambridge University Press.

IPCC (Intergovernmental Panel on Climate Change). (2014b). *Climate Change 2014: Impacts, adaptation, and vulnerability. Part B: Regional aspects. Contribution of working group II to the fifth assessment report of the Intergovernmental Panel on Climate Change.* (C. B. Field, V. R. Barros, D. J. Dokken, K. J. Mach, M. D. Mastrandrea, T. E. Bilir,... L. L. White, Eds) Cambridge: Cambridge University Press.

Ketlhoilwe, M. J., and Jeremiah, K. (2010). Mainstreaming environment and sustainability issues in institutions of higher education: The case of the University of Botswana. *International Journal of Scientific Research in Education, 3*(1), 1–9.

Kotecha, P., Wilson-Strydom, M., and Fongwa, S. N. (2012). *A profile of higher education in Southern Africa – Volume 1: A regional perspective.* Johannesburg: Southern African Regional Universities Association.

Lotz-Sisitka, H. (2011). The 'Event' of the modern sustainable development and universities in Africa. In Global University Network for Innovation (GUNi), *Higher education in the world 4: Higher education's commitment to sustainability: from understanding to action (GUNi series on the commitment of universities)* (pp. 41–68). London: Palgrave Macmillan.

Lotz-Sisitka, H. B., and Hlengwa, A. (2012). *Seeding change: The socio-ecological condition and professional development of university teachers.* Paper presented at the International Association of Critical Realism Conference, Rhodes University, Grahamstown, July 2012.

Lotz-Sisitka, H., and Price, L. (2012). *Critical realism, environmental learning and societal change.* Unpublished book proposal for the Sandisa Imbewu Project, Grahamstown.

Lotz-Sisitka, H., Agbedahin, A. V., Hlengwa, A., and Neeser, M. (2015). 'Seeding Change': Developing a change-oriented model for professional learning and ESD in higher education institutions in Africa. In press.

Lotz-Sisitka, H., Gumede, M., Olvitt, L., and Pesanayi, T. (2006). *History and context of ESD in Southern Africa: Supporting participation in the UN decade of education for sustainable development.* Howick: SADC REEP.

Lynton, R., and Pareek, U. (2000). *Training for organizational transformation – Part 1: For policy makers and change managers.* New Delhi: Sage.

Neumayer, E. (2013). *Weak versus strong sustainability: Exploring the limits of two opposing paradigms.* Cheltenham, UK: Edward Elgar.

Pesanayi, T. (2009). A case of exploring learning interactions in rural farming communities of practice in Manicaland, Zimbabwe. *South African Journal of Environmental Education, 26*, 64–72.

SAQA (South African Qualifications Authority). (n.d.). *SAQA's research model and agenda*. Pretoria: Compiled and distributed by the Directorate: Strategic Support, SAQA.

SARUA (Southern African Regional Universities Association). (2014). *SARUA Climate Change Counts mapping study: Knowledge co-production framework (Volume 1)*. Johannesburg: SARUA.

SARUA (Southern African Regional Universities Association). (n.d.). *Vision of SARUA*. Retrieved 30 December 2013, from www.sarua.org/?q=content/vision.

UNDP (United Nations Development Programme). (2014). Human Development Report. *Sustaining Human Progress: Reducing Vulnerabilities and Building Resilience*. New York. Retrieved May 2015, from www.undp.org/content/undp/en/home/librarypage/hdr/2014-human-development-report.html.

UNEP (United Nations Environment Programme). (2008). *Maintreaming Environment and Sustainability in African Universities Partnership (2004–2008 Report)*. Nairobi: UNEP.

UNEP (United Nations Environment Programme). (2012). *21 issues for the 21st century: Results of the UNEP foresight process on emerging environmental issues*. February 2012. Nairobi: UNEP.

UNESCO (United Nations Educational, Scientific and Cultural Organization). (2014). *Shaping the future we want: UN Decade of Education for Sustainable Development (2005–2014). Final report*. Paris: UNESCO.

United Nations. (1992). *Agenda 21*. United Nations Conference on Environment and Development. Rio de Janerio: United Nations. 3–14 June 1992. Retrieved 10 June 2013, from http://sustainabledevelopment.un.org/content/documents/Agenda21.pdf.

University of Botswana. (2013). *University of Botswana*. www.ub.bw.

Zeuner, L. (1999). Margaret Archer on structural and cultural morphgenesis. *Acta Sociologica, 42*, 79–86.

16 Steel Valley and the absence of environmental justice in the new South Africa

Critical realism's kinship with environmental justice

Victor Munnik

Introduction

This chapter explores how dialectical critical realism (DCR) can strengthen Environmental Justice (EJ) activism.[1] It argues that DCR can help EJ challenge scientific reductionism in immediate confrontations (in this example about pollution), as well as challenge current hegemonic thinking that obstructs us in envisioning and working towards a society that is based on environmental justice. Critical realism enables us to rethink and reword our realities. It brings ethical practice or activism into the centre of knowledge production. It anchors knowledge in our practical interaction with the world. At its heart is a desire for freedom, empowered by the ability to think change.

The argument in this chapter is based on a review of a case of community resistance to pollution: 'Steel Valley' in South Africa's industrial heartland, the Vaal Triangle, which was the subject of my recent PhD study (Munnik, 2012). The chapter proceeds by showing that historically the EJ movement and how it developed in South Africa in the early 1990s works with similar concerns to critical realism: an ethic of freedom and solidarity, an holistic approach, looking for the best feasible explanation, and actively working for change. It explains the centrality of these concerns in critical realism. It then introduces the empirical material of the case study – the pollution struggle in Steel Valley (1952–2012) – in three phases with very different historical contexts: under apartheid; in the context of a new, democratic South Africa with a constitutional right to a healthy environment; and the current phase in which EJ is used as a conscious discursive and organizing principle.

It addresses two audiences: researchers working within the broad environmental justice (EJ) tradition, to show that critical realist concepts can enrich and enable the analysis of an EJ case study, and researchers who work within the critical realist tradition, in order to show how critical realist concepts[2] can be applied in environmental justice. To achieve this, we[3] will:

1 use the critical realist approach of the best possible, or most reasonable explanation to confront the reductionist strategies of 'prove beyond doubt that it is us', which has been so useful to the polluter in our case study;
2 work with the concept of negativity, absence and ills (and their absenting), giving a number of examples from the Steel Valley struggle. The absence of environmental justice – in the form of environmental injustices – is crucial to EJ thinking and its history;
3 explore the possibilities of creating coherence between disparate logics and levels of analysis, via the concept of open totalities and the application of the seven scale analysis of the laminated system. This helps us understand how there can be contradictory logics at work at different levels of causality, for example a global neo-liberal economic system enabling, but also challenged by, a global system of information exchange like the internet, which is in turn challenged by activist interpretations of lived realities at a local level;
4 see how placing the notion of the irrepressible drive to freedom at the centre of DCR thinking leads to a detailed consideration of ethical agency, and binds this system together, in a dynamic or open totality (open totality allows us to think inclusively but retain the dynamic ideas of change, and of subtotalities or parts that are split off, isolated or repressed);
5 consider the implications of a concept of an always already value-saturated social science, which holds important implications for the relationship between supportive EJ academic work and activism. The sheer potential of this relationship challenges DCR to find ways of dealing with its plethora of new concepts ('jargon'), in order to enable its use in combination with a public sociology and citizen science, which are arguably jargon-free approaches underlying EJ intellectual practice.

Crucial questions in the Environmental Justice tradition

Environmental Justice (EJ) first appeared as a critique of environmental racism in the United States, when 'communities of color' discovered that they were targeted to be victims of pollution. From an immediate resistance via community activism to the racist distribution of pollution, this tradition moved to reflection on the causes that were producing the pollution: that is the capitalist economy, dominated by the idea of permanent growth (Pellow and Brulle, 2005; Bakan, 2004; Kovel, 2002; Henderson, 1996), described as a 'treadmill of production' (Schnaiberg, 1980) which,

> continues to create ecological problems through a self-reinforcing mechanism of ever more production and consumption. The logic of

> the treadmill of production is an ever-growing need for capital investment in order to generate goods for sale on the market. From the environment, it requires growing inputs of energy and material.... In practice the state has often acted to accelerate the treadmill in the hope of avoiding political conflict.... The ecological result of this process is that the use of natural resources continues to increase, regardless of the consequences on the sustainability of the ecosystem. The social result is that inequalities increase and working-class populations receive less and less material benefit for their labour. Thus, both ecological disorganisation and race and class inequalities are inherent by-products of the social order.
>
> (Pellow and Brulle, 2005, p. 4)

The EJ movement incorporates Beck's description of risk positions in an advanced industrial society: 'like wealth, risk adheres to the class pattern, only inversely; wealth accumulates at the top, risks at the bottom' (Beck (1986) cited in Pellow and Brulle, 2005, p. 5). Where the poor are also racially defined, this creates a pattern of environmental racism. Such analyses involved a recasting of environmentalism away from its preoccupation with the conservation of biodiversity. Pellow and Brulle (2005, p. 3), editors of a collection of essays reflecting on EJ tactics, declare that:

> The EJ movement has sought to redefine environmentalism as much more integrated with the social needs of human populations, and, in contrast with the more eco-centric environmental movement, its fundamental goals include challenging the capitalist growth economy as well.

Since its early days in the 1980s, environmental justice has grown to include a wide range of issues, including transport, housing and health (Sze and London, 2008) but it remains important to remember its origin in the imposition of pollution on poor and black communities, the issue that we take up in the Steel Valley case study.

Environmental Justice in South Africa

South African activists enthusiastically adopted the EJ framework in the early 1990s, at the time of, and as part of, the opening up of politics in the public sphere. In 1992, an Environmental Justice Network Forum (EJNF) was founded. The EJNF saw Environmental Justice as an extension of the anti-apartheid struggle, just as black communities and activists in the United States of America (USA) built their EJ movement on the basis of the civil rights struggle. In another parallel with the USA experience, the EJ movement grew in South Africa as an alternative to the narrowly focused and elite conservation movement, not as an outgrowth of it. EJ activists positioned themselves in opposition to many forms of conservation as they supported black South Africans in their efforts to claim back land that they had lost to conservation (Cock and Koch, 1991). Apartheid – and also colonial history – presented clear forms of environmental injustice – resettlement on the worst land, second class services, if any – and the almost automatic location of waste dumps near black communities mirrored the USA experience (Ramphele, 1991; Cock and Koch, 1991; Cock and Munnik, 2006; Hallowes and Munnik, 2006). The mines themselves, and the migrant labour system that enabled it, were immediately seen as environmental injustices.[4]

The early 1990s (up to 1996) in South Africa was a constitution-making phase. It included the National Environmental Policy Process (CONNEPP) which formulated South Africa's progressive National Environmental Management Act (RSA, 1998), a framework law that laid down the principles for further legislation, and also formulated Section 24 of the new Constitution (RSA, 1996), known as the 'Environmental Justice right'.

The South African EJ movement focused on organizing fenceline communities bearing the brunt of pollution on the fencelines of waste dumps, mines and polluting industries. It confronted many challenges, including (1) integrating environment and social justice concerns in a transition, where it was confronted with ongoing racial division driven by the momentum of a colonial and apartheid past, (2) the hegemony or dominance in the new South Africa (post-1994 democracy) of the neo-liberal growth model, which pitted environment versus social rights via growth discourse. Particularly the trade unions saw environment concerns as a Northern concept, and a threat to jobs in a segmented society where class and race largely coincided, and (3) the colonization of ecological thinking by environmental management dominated by a framework of ecological modernization which turned nature into a storehouse of resources to be both shepherded and used by those controlling the economy (Luke, 1999).

This created a profound challenge to envision a new, ecological society, and immediately deal with pollution, which counted amongst a host of other (partly related) challenges including access to water, energy, housing, health and education services, and an HIV/AIDS epidemic. Underlying all were extreme inequality and poverty. In confronting this complex mix of realities and the many confusing ideas about them, I argue here that EJ can benefit from the long-run philosophical/activist project of Critical Realism.

Dialectical critical realism deals with the same concerns

The shared concerns between DCR and EJ can be seen from a consideration of CR history. Critical Realism emerged in a context, around 1968, where change was in the air – the same change that enabled the emergence of the environmental justice movement:

> [at] the end of the post-war boom, more than a whiff of revolt and even revolution and the rebirth of a free Marxist current in the new eclecticism of a still malaised social science.... Relativity theory, quantum mechanics, the liberation of the colonies, the threat of a nuclear holocaust and looming ecological crisis rendered conventional assumptions obsolete. The time was ripe for ontology; and as the seventies made way for the eighties and the events of 1989, for a new account of change, especially in the context of the collapse of communism, the poverty of most materialist dialectical philosophy and the monstrous inequities of the strife-torn, crisis-ridden chaotic new world order that Bush, Benetton and Hayek were in the process of ushering in.
>
> (Bhaskar, 1993, p. 280)

Critical realism was initiated with a concern with problems of modernization and underdevelopment (Bhaskar, 2002). But in dealing with these problems, critical realism's core philosopher Roy Bhaskar found that he was being obstructed by a philosophical hegemony that said that there could be no argument or discussion about the world, about how the world was in itself, how change comes about, and can be talked about (Norrie, 2010).

So Bhaskar first 'set up the philosophical respectability of talking about the world in itself' (2002, p. 124). He attacked the positivist idea of 'constant conjunctions of events' as the basis of science, investigated the preconditions of science and concluded that 'reality is stratified, structured, and also that the surface of reality was differentiated' (ibid.).

The engagement with positivist science was a first step in a historical trajectory for critical realism, resulting in a 'menu' from which those who wish to work with it can choose specific phases or tools created in these phases. The second phase engaged social science to overcome the many classical binaries in it: 'structure/agency, individual/collective (or whole), meaning/behaviour; reason/cause, mind/body, fact/value and theory/practice' (Bhaskar, 2013, p. 8). A third, dialectical phase (see below) gave attention to absence, metacritique and ethics, which was then followed by a 'spiritual turn', which although consonant with environmental justice thinking about nature, spirituality and the role of civil society (Meta-Reality, Bhaskar 2002) is beyond the scope of this chapter (see, however, Price's Chapter 18 in this volume). In our application to the Steel Valley case study, we focus on Dialectical Critical Realism, because of its ability to think about change, open totalities and freedom at the same time, and coherently.

The creation and legitimation of pollution, Steel Valley, South Africa

It is in the nature of pollution – an externalization of costs onto unwilling or unknowing third parties – that the creation and legitimation of pollution run together. Three phases of the case study are presented in the following description. The first phase starts with the first evidence of pollution in 1961, and runs to the dawn of a new democracy in South Africa in 1990. The second phase is from resistance inspired by new democratic rights, to the defeat and dispersal of the Steel Valley community in 2000 to 2002. From 2002 a third phase is taken up: the introduction of the concepts of EJ in a growing and internationalizing movement that grew around a core organization from the second phase – the Steel Valley Crisis Committee. In the first two phases, the polluter was an apartheid parastatal, Iscor Ltd. (hereafter referred to as Iscor), which during the third phase became a subsidiary of the biggest steelmaker in the world, ArcelorMittal.

Steel Valley pollution struggle under apartheid

The Vanderbijlpark steel factory was built as part of a South African industrialization strategy after the Second World War, on the Vaal coalfield, which had been the site of early industrialization since the last quarter of the nineteenth century (Mendelsohn, 1991; Trapido 1986). Evidence for the pollution was first recorded in 1961, but it had most likely been occurring since 1952, when the factory started operations. In 1956 the first South African water pollution legislation was passed, but regulation remained minimal (Fuggle and Rabie, 1992, 1983).

Up to the early 1990s, the attempts by the white residents among the around 3,000 residents on the smallholdings of Steel Valley to resist the pollution had very little impact, even though they were participants in the 'whites only democracy' of the time.[5] Their greatest 'successes' were that complaining individual smallholders received alternative water supplies or were bought out, but in return had to sign confidentiality agreements preventing them from discussing the pollution.

This first phase showed serious limits to activism, in a very narrow public sphere as referred to by Habermas (1996), limited by the oppressive politics of apartheid. Apartheid inherited from its predecessor segregation in support of strong industrialism, driven by the needs and influence of the Minerals Energy Complex (MEC) (Fine and Rustomjee, 1996). The MEC was – and still is – the dominant accumulation strategy that was crucial in creating South Africa as a political entity. This also created and sustained the system of migrant labour and a physical reality which fits closely the description of a treadmill of production (Schnaiberg, 1980) with its massive and often toxic legacies of waste from both mining and heavy industry. The MEC played a crucial role in South African history, reaching back into colonial times. These historical trajectories that determine the present as 'the past in the present', will be reflected as part of a totality below, in our discussion of the seven scales of a laminated reality.

Democratic rights, and the right to a healthy environment

In 1996, the new Constitution gave all South Africans the right to live in an environment not harmful to their health and well-being, as part of a new post-apartheid dispensation, reflected in Section 24 of the Constitution (RSA, 1996). Residents of the Steel Valley community – now racially mixed but still deeply marked by racial and class segmentation – responded by using their control over the newly elected local government[6] and its resources to challenge the pollution.

Residents involved the polluter, Iscor, in a pollution discussion forum, which was later upgraded to a more official forum by the Department of Water Affairs. Consultants carried out a cost-benefit analysis, comparing the buy-out and remediation options. It found that the buy-out option was preferable. Iscor immediately (in 2000) implemented the buy-out, as the only buyer in the market (it was well known by most if not all other potential buyers that the smallholdings were polluted[7]), at prices that many smallholders were unhappy with. The houses were bulldozed and the area turned into an artificial wilderness serving today as a buffer zone for the steel factory. Efforts by residents and the local government to have the area declared a disaster zone, to resettle the community as a whole in another area, to receive compensation for losses due to the pollution, and to establish a trust fund to deal with future health problems were all unsuccessful. Through two court cases, affected residents challenged Iscor's strategies. The first was folded into the buy-out, after a settlement out of court, and the second stopped when funds and energy ran out. The community was largely dispersed from Steel Valley. Government, in its role as regulator, chose the route of ecological modernization, trading justice aspects such as compensation, and an acknowledgement of transgressions, for a promise of cleaner production in future.

Residents were left with two burning questions: (1) How was it possible that such obvious pollution could be successfully denied in court, using scientific evidence? This completely defied common sense! And (2) how was it possible that the polluter could escape all possibility for compensation for damage, especially since the new South African Constitution promised its citizens the right to live in an environment 'not harmful to their health and well-being' (RSA, 1996)? These questions formed the basis of my PhD thesis (Munnik, 2013), which analysed the play of discursive power in various discursive arenas that led to these very material outcomes. Global discourse/practices including reductive science, scientific-commercial confidentiality practices, environmental management, sustainable development, legally limited liability of companies, the growth hegemony and its everyday discourses enabled the continuance of pollution and its legitimation. Clearly, these discourses and the way their discursive power was deployed, were marked by oppressive practices, absences and ills. These will be discussed below.

Environmental Justice phase

In a third phase of resistance to the pollution, the surviving organization of the scattered residents, the Steel Valley Crisis Committee (SVCC) was exposed to and connected with national and worldwide environmental justice networks, teamed up with other communities in the Vaal, and later formed an international watchdog network monitoring the activities of the steel factory's globalized owner, ArcelorMittal. In this phase, residents and activists resisted closure and the reification of the outcome of history.

In 2002, the SVCC took part in the World Summit on Sustainable Development (WSSD) in Johannesburg, connecting to international networks. In 2004, the SVCC core called together organizations representing those affected by pollution in the Vaal area and formed the Vaal Environmental Justice Alliance (VEJA). For the first time Boipatong on the other (Eastern) side of the factory was included, as well as Samancor workers suffering ill health from working in a nearby manganese factory. Leadership of the resistance passed from white to black citizens.

The third phase was quite different from the first. The concept of EJ was introduced and became a central organizing principle. It was active on discursive and institutional levels, it enabled engagement with power and social relations, and the imaginary (Harvey, 1996). The EJ analysis of three mechanisms that create environmental injustice became working principles for political engagement. The pollution was now clearly understood as an environmental injustice: the *externalization* of production costs (the steel factory was saving money by neglecting current pollution and not compensating for past damages), which it achieved by *excluding* residents and activists from decision making – for example withholding information, such as refusing to release the Master Plan which detailed the extent of the pollution and the company's plans to attend to it – and *enclosure*. The company's financial and legal power resulted from previous enclosures of resources through colonization and under apartheid, making it by far the biggest and most powerful employer and political influence in the area.

International solidarity became important as technical and political information was exchanged within what was now a global movement watching and opposing ArcelorMittal in many countries. Locally, the resistance was now much more strongly rooted in social movements also dealing with local services, housing and health issues. To this analysis, DCR can add a practical understanding of open totality thinking, and ethical agency. The rest of this chapter now explores specific DCR concepts that have the potential to strengthen EJ analysis.

Confronting reductionism

To EJ activists it is crucially important to be able to argue and demonstrate that pollution and its effects are real. Basic critical realism identifies three levels in our engagement with reality. The 'real' contains generative structures and mechanisms, that is, the necessities in nature that 'make things happen'. These mechanisms may not always result in actual events.[8] Moreover, an event may actually take place, but not be witnessed. Radiation, for example, cannot be seen, heard or tasted, but it is real in that it has effects. Exposure can, for example, lead to cancer developing in an exposed person 15 years later. For critical realists, there is obviously a real world (how else would we have knowledge of it?), and that world contains necessity, including the powers and liabilities of objects to behave in certain ways. This necessity allows, for example, monitoring of pollution, and establishing sequences of effects of exposure, e.g. heavy metals in the body. Such events may be hidden from perception (Nixon, 2011). For example, in the Steel Valley case heavy metals like cadmium would be taken up by the body, covered in proteins and then excreted. While this incurs damage to the kidneys, the effects are only visible if a post-mortem biopsy is undertaken.[9] The effects of heavy metal pollution may thus be actual, but not empirically available. So the critical realist style of thinking allows us to respond to science that would reduce everything to what can be immediately perceived, made worse by legal strategies (also in the Steel Valley case) that reduce knowledge to what can be 'proven'.

Reality is stratified. Life emerges out of chemistry, and while chemistry is necessary for life, life cannot be reduced to it. Pollution evidence is often reduced to chemistry alone, ignoring emergence, even on a simple level such as synergistic effects between toxicants.

How does this enable us to confront reductionism? Reductionism in the natural and social sciences constrains our thinking by first imaging the world as made up of discrete entities, and then insisting that these entities be linked in unilinear causal chains – that is, if it does not banish causality by only recognizing correlation in events. In contrast, critical realism allows us to think about relational entities – entities that are whole or partial aggregates of other entities, that flow through each other, change into and constitute each other in a fluid manner – and to identify causal mechanisms in order to arrive at a best explanation for events rather than, as in the case of the Steel Valley court cases, an absence of explanations and therefore responsibility.

Absences in environmental justice

Absence and change

EJ thinking is continuously at risk of being constrained, silenced, blinded and bound by the slipstream of conventional 'scientific' thinking. The DCR concern with liberating human possibilities is a good starting point to develop an anti-hegemonic or freedom oriented politics of knowledge. Alan Norrie (2010, p. 22) argues that 'At its heart, Bhaskar's dialectic …

involves a vision of a world in which dominant social and economic power suppresses human possibilities, and the resulting lack is then repressed in its philosophy.'

What is repressed on the level of thinking and representation is, more often than not, a real contradiction, conflict or repression in society. Absence is thus 'a concept that denotes both a geo-historical and a philosophical problem' (Norrie, 2010, p. 27) and the point of dialectical critical realism is to deal with this problem (absent it, or get rid of it), in this case, by confronting the power structures that hold in place the conditions of possibility for environmental injustice. An analysis of absence points to the causes of environmental injustice, and how to address them, which involve real change in social structures. Absence is intrinsic to change (ibid.).

Bhaskar argues forcefully that the category of absence is far more important than that which is currently present, when he says that, 'by the time we are through, I would like the reader to see the positive as a tiny, but important, ripple on the surface of the sea of negativity' (Bhaskar, 1993, p. 5).

In the dialectical critical realist view:

> A purely positive account of reality cannot ultimately consider how the world might change in fundamental ways, because it cannot grasp that radical futures containing what is absent in the present or past might emerge into being in the future. It therefore legitimates the way things are.
>
> (Norrie, 2010, p. 44)

How do we understand and use absence?

Absence or negativity appears in many forms and processes. Bhaskar presents a diversity of negativities that we can think with, for example 'to deny, reject, contradict, oppose, exclude, marginalize, denigrate, erase, separate, split, sunder, cancel, annul, destroy, criticize and condemn' (Bhaskar, 1993, p. 8). And 'to absent, to constrain, to condemn, to injure, to refine, to qualify..., to limit..., to cancel, to suspend, to undo, to erase,... to nihilate, to annihilate; and more technically to mediate, transform, contradict, marginalize, sunder and void' (ibid.). Negativity 'is also present in ... specifically dialectical concepts concerning negativity: the hiatus, the margin, the void, the hidden, the empty, the anterior, the exterior, the excluded, the omitted, the forgotten and the feared' (Bhaskar (1993), cited in Norrie, 2010, p. 23). Absence can be seen as non-being (something that did not happen, e.g. environmental justice or discussions about it are absent under apartheid) or as partial non-being (the Constitutional right to a healthy environment in South Africa is hollow), or the process of becoming (something comes into being that was not there before, and what was, is now absent: that is the process of change).

Absences are real because they have consequences. Most EJ activists would be familiar with such absences that have consequences.

The absence of environmental justice is the rallying point in the EJ movement. The negative of injustice and the ideal of environmental justice represent a constant switching of perspectives from a present to a desired state. In the Steel Valley case, the gap between the painfully lived reality of residents, the evidence they discovered during their participation in the forums of decades of pollution (Munnik, 2012), the promises and shows of state intervention (via the forums) that came to nothing, the hollowness of the environmental right in the Constitution, the lack of political will in the state to act on the side of its citizens, or the lack of power against a powerful corporation to implement new policies and legislation – these are so many absences that constitute the case study of the Steel Valley fight against pollution. They all present constraints on dealing with the pollution with all the resources both as local, affected citizens, and as a new democratic state, a constitutional democracy. In EJ terms, these moves constituted the successful exclusion of not only the community but also the state from (effective) decision making about the pollution. Exclusion – one of the three mechanisms of imposing environmental injustice – is an achieved absence that is maintained at many levels in a representative democracy – despite claims and appearances to the contrary (Jessop, 1990; Munnik, 2012).

Absence as a cause

In the Steel Valley case, a fundamental absence was the denial of Iscor and later ArcelorMittal, of the reality of the pollution of Steel Valley, as introduced in an earlier section of this chapter. A number of national and international activists and academics were attracted to the Steel Valley case by its very unreasonableness. It attracted support from the Friends of Steel Valley, the WSSD tour, Heinrich Boell Foundation and other funders, and the ongoing interest of the funders that brought networks and resources (financial and intellectual), in the form of solidarity to this struggle. The denial of the existence of pollution struck most observers as unreasonable, even fascinating; nevertheless, it was exactly these contradictions and injustices that attracted the attention of activists and funders.

Another absence was also interesting (and seemed to have far-reaching implications): the outcomes in the Steel Valley case contradicted the new Environmental Right in Section 24 of the SA Constitution (RSA, 1996). What did it mean? Was the Constitution powerless and the government not serious about environmental issues?

The Steel Valley case seemed to show a strong use of discursive power, able to override common sense and local and wider knowledge of the 'obvious' pollution. It was not only an absence, but a strong absenting of the knowledge needed by fenceline communities to gain environmental justice. This absence invites comparison with other industrial polluter strategies of 'deceit and denial', and legitimating practices, studied in *agnotology*, the study of how ignorance is created (Proctor and Schiebinger, 2008), with examples from the tobacco and lead industries, climate change denial by the same institutions, and the social structures and political agendas behind them. It raises the question of how general this approach is among polluters, and specifically how much of the case study of this phenomenon (this 'determinate absence') is more broadly applicable to the Minerals Energy Complex in South Africa, extractive industries in general, mobile transnational corporations, and how this knowledge could be useful to fenceline communities, activists, regulators and other decision makers.

The creation of silence and erosion of evidence in Steel Valley relied on active interventions. In response to unfavourable media coverage in 2001 Iscor imposed a gagging court order on 16 applicants in a court case against it (residents of Steel Valley going to court for a stop to the pollution and for compensation). Acting on legal advice the children of the applicants formed an organization, the Steel Valley Crisis Committee, which was able to speak out, and ironically grew to become the interface of Steel Valley with the national and international environmental justice movements, and formed the core of the Vaal Environmental Justice Alliance and eventually the Global Alliance against Arcelor Mittal. The silencing of the 16 applicants was a cause of eventual eloquence, growth in the EJ movement locally, nationally and internationally.

'You can't prove that it's us'

An important aspect of the absence of justice in Steel Valley was the constant absenting of information. Iscor was able to reduce a rich complexity of knowledge, that clearly pointed to Iscor as the source of the pollution, to a very small positivity: the official, reduced, scientific, legally arguable forensic evidence that remained after various 'erosive forces' had washed over and through it. This 'positivity' or remainder was not able to deal with the pollution: understand and explain it, clean it up, and provide current and future compensation. Its success was to reduce not only the understanding, but also the human possibilities of dealing with it. The small 'positivity' of reduced, forensically valid, scientific knowledge, guarded by privileges, commercial and power politics based can be contrasted to the common sense insight that Iscor and later Mittal caused the pollution and was responsible for it. This reductionism in order to avoid responsibility is a common tactic among corporate polluters, and confronting it is crucial for EJ activists. In the next sections, three absences are discussed: (1) dissimulation in argument; (2) censorship; (3) the hollowness of the South African environmental right in the Constitution (RSA, 1996).

Drowning in their own 'bacteriological contamination'

One particular example from the Steel Valley struggle shows a deceptive use of the authority of science, in the form of partial truth or dissimulation. Iscor consistently argued that the pollution issue in Steel Valley was 'more complex' than a matter of their industrial pollution. There was also 'bacteriological contamination' from the pit latrines of Steel Valley residents. This, Iscor maintained, was the result of a combination of a high water table and deficient French drains. In a newsletter that Iscor used to communicate with residents about the buy-out process (mentioned above), it also attempted to establish certain strategic 'truths' about the pollution, relying on the findings of the cost-benefit analysis. The newsletter, *InfoLink Feb.1999*, first argued that:

> Bacteriological contamination (e-coli) is generally the result of poor sewage systems and cannot be related to Iscor or any of Iscor's processes. The sewage from the Works does not flow into the canal but into the municipal sewage system.

So, if Iscor's sewage is going into the municipal sewage system, the poor sewage systems at fault can only belong to the Steel Valley residents. Next, the newsletter used the idea of 'bacteriological contamination' to displace the industrial pollution in the following explanation of why Steel Valley was not habitable, and residents needed to move:

- Organic, inorganic and bacteriological contamination is present in the ground water in a large area of the community.
- The polluted ground water and soil, without remediation, makes the area unsuitable for human habitation.
- Most of the smallholdings are unsuitable for habitation because of e-coli (bacteriological) contamination that affects 95 per cent of the area.
- Most of the area is not ideally suitable to sanitation based on drainage. The existing systems will have to be upgraded from time to time.
- The area is not suitable for planting and sowing and stock farming alone is not economical because of the size of the smallholdings.
- Infrastructure in the form of water and sewage is not available. The report says that the local authorities do not regard it as a priority to provide such services.
- The area does not have good foundation soils for building. The soil can affect the structural stability of buildings. (Iscor, *InfoLink Feb.1999*)

Iscor's industrial pollution is completely absent from this list! Moreover, it was simply not true that the bacteriological contamination took place independently of Iscor's actions. Over nearly 50 years of operation Iscor used huge quantities of water, most of which were released to the West, on the surface via the Rietspruit canal, the Rietkuilspruit, overland seepage and via the groundwater. It was Iscor's activities that had raised the water table, which then connected the residents' pit latrines underground, and led to the 'bacteriological contamination'. According to water expert, Carin Bosman, the slagheap itself raised the water table because of its sheer weight (C. Bosman, ex-official Water Quality Department of Water Affairs, currently director CBSS Consultants, Pretoria, personal communication, 22 April 2009). Moreover, since the 1950s, the effluent in the canal was available to Steel Valley residents for irrigation. The victims of the pollution were being blamed for their own pollution – bad sanitation, allowing their own excrement to enter their own drinking water – while Iscor's pollution was minimized. The 'bacteriological contamination' was not only useful to confuse, it had material outcomes. It became an important factor in deciding who Iscor would supply water to, and allowed Iscor to charge the municipality for the water it supplied to the smallholders it had polluted. When Iscor finally took over and fenced in Steel Valley, they were able to include public space like roads and verges in their enclosure because of this water supply debt from the council.

The confusing absence and presence of the Iscor Master Plan

The steel factory management's denial of the reality of the pollution reached an absurd point in the Master Plan (Iscor, 2003), in which the pollution they claimed did not exist, was described in fine detail by a team of consultants contracted to them. The Master Plan – an absence in the sense that it was secret, and could not be accessed, quoted, or used in evidence – was nevertheless 'present elsewhere' and used as a real planning tool in official administrative negotiations with the Department of Water Affairs. When a 2003 water use licence application required public participation, the Master Plan could be read in a library, but not copied or taken away: it was non-absent on a temporary and spatially limited basis! The Master Plan continues to beguile by its absence. VEJA (Vaal Environmental Justice Alliance) has taken the steelmaker to court to gain access to the Master Plan (VEJA won the case, but at the time of writing, ArcelorMittal was taking it on appeal). In the meantime, the Master Plan has been put into the public domain by an ex-Iscor employee and whistleblower. It confirms the pollution.

The arguments in the 2013 court case, used by the steelmaker's legal team, justified denying access to the information by arguing that the activists had no right to monitor the steel factory's pollution. This makes the connection between thought and practice that critical realism argues for. The Master Plan is an instance of the 'denial of social change by an elite'

(Norrie, 2010), through a politics of knowledge which has a material effect of allowing the continued pollution of the community in the present, in the future through health effects and unremediated groundwater pollution, and in the past through lack of compensation.

The hollowness of the environmental right in the SA Constitution

As a result of their experiences, many SV residents concluded that the environmental right in the SA Constitution was hollow. The lack of capacity or political will to enforce the constitutional right to a healthy environment can also be understood as an absence. On one level, the environmental right can be seen as an aspirational right, so that its 'hollowness' can be filled out over time by exactly the type of confrontation that took place in Steel Valley. But the hollowness is also argued to consist in 'lack of capacity' to enforce, and 'lack of political will'. The first refers to inadequate staff (numbers or skills), or a system not strong enough to monitor and enforce. In Steel Valley the pollution was obvious and efforts to enforce were made by more junior officers, only to be discouraged by more senior people who decided on a strategy of modernization: effectively trading future compliance for absolution of past pollution – thus sealing the historical externalization of pollution onto the fenceline communities (Munnik, 2012). In other South African contexts, politicians with interests in mining and connections to the ANC elite are not prosecuted when contravening, for example, water use licenses (see www.benchmarks.org.za/publications/mines_operating_without_water_use_licenses.pdf). This is an example of 'lack of political will' in which political networks have the effect of suspending the need for compliance with environmental legislation. Dealing with such an absence will need a separate strategy, on another – national – level. It is a challenge for EJ activists to think all these issues together, while sorting them into the appropriate levels and categories for analysis and action. For this task, the third moment of Bhaskar's dialectic, the art of constellation in an open totality, provides specific insights and tools.

Open totalities: coherence, emergence, levels of analysis and strategy

DCR provides categories that can hold together differentiated, contradictory entities and thoughts that are, however, somehow related: the diffracted realities of existence in the world. Figures of thought such as emergence, inside/outside, duals and the grounds from which they arise, perspective switch, and the presence of the past in the present and the future can be constellated together in open totalities. Thinking open totalities can bring a large variety of experiences, knowledges and strategies into a coherent form. One such constellation, Bhaskar's seven

Table 16.1 A laminated system analysis of pollution in Steel Valley

The global level (vii)	Climate change, globalization and a global steel industry, a global EJ movement and international activist solidarity, a global neoliberal framework dominating the world economy, sustainable development discourse, environmental management, Agenda 21; EJ reframing of contexts and meanings.
Level of whole cultures and traditions (vi)	Colonialism in Southern Africa, segregation, extractivism, creation of waste legacies and pollution in regional (southern African) Minerals Energy Complex.
Macro level orientated to understanding the functioning of whole societies or their regions (v)	Industrial growth and apartheid spatial ordering in South Africa since 1950s. South Africa transition to post-apartheid state, politics of hegemony. Constitutional Right to a healthy environment (RSA, 1996).
	The tax dependent national state (Yudelman, 1984). South African history of regulation of water quality, air quality, waste in industry and mining. South African public opinion. History of South African EJ movement, within South African civil society. National organization of fenceline communities.
Meso level concerned with sub-national institutions and functional roles (iv)	Vaal Triangle pollution history. Iscor and ArcelorMittal histories of externalization of pollution, exclusion of communities from decision making and enclosure of resources.
	Citizens' resistance. Forums involving polluter and national department of water. Buy-out. Citizens' monitoring of water quality. VEJA and its relationships with social movements in the Vaal.
Micro level studied, for example, by ethnomethodologists and others (iii)	Social relations. Steel Valley community, fenceline communities, activism. Mobilization of Steel Valley community, negotiations around individual farms, relationships with neighbours.
Individual or biographical level (ii)	The body in the environment, health effects of using polluted water sources. Meaning making in personal trauma narratives making sense of pollution.
Sub-individual psychological level (i)	Personal experiences of pollution, participation, resistance and protest. Identification with life style, neighbours, environmental activism. Understanding of EJ and activism.

Source: Developed by the author.

planar lamination (Bhaskar, 2010), allows us to arrange these issues in a laminated system for the Steel Valley pollution case study (Price, 2014) so that is possible to see how and where (at what level, with what questions) different knowledges can be applied to a necessarily stratified social reality, as outlined in Table 16.1.

From this multi-level laminated system analysis it can be seen that activism needs to engage on levels from the individual to the global, that there are complications and opponents, but also allies and resources at each level, and potentially emergent relations that could strengthen EJ. To really choose among these, it is necessary to animate this constellation with a conception of ethics that provides a basis for activism.

Ethics, activism and social science

Dialectical critical realism is ethical. It is driven by a desire for freedom and solidarity – something that clearly resonates with EJ. Bhaskar sees this desire as inherent in a human being, as part of our species being, because we can move from an understanding of our own needs and desires to an understanding for a universal human need for freedom. Real as this drive 'to universal human emancipation in a society in which the free flourishing of each is the condition of the free flourishing of all' (Bhaskar, 1993, p. 98) is, it is only realized in history subject to the constraints of historical-material conditions, the actually existing world in which human beings exercise agency.

The drive to freedom works on what is absent by evaluating absences, such as that of justice, as ills, within a bigger conception of a system/praxis/discourse constraining and denying human flourishing, and absenting them. This is not a formal argument only, Bhaskar insists that it is possible to reach a conclusion, using knowledge of the real world, about how to evaluate ills. EJ activism is an instance of the underlying ability and drive of people to achieve a just society, under conditions that are not yet in existence. Such a society, although we cannot in an open totality know what it will be like, is likely to see,

> 'massive and global redistribution of resources, tending to a core equality in virtue of our common humanity' (Bhaskar, PDF: 300). 'There would be a vast expansion of rights and of participation in democracy' (Norrie, 2010: 153).... Social solidarity is the only context in which human agency can survive, develop and flourish ... while a domain of personal choice ... subject to the recognition of others' rights (Norrie, ibid.), will exist within 'a normative order based on care, solidarity and trust' (Bhaskar, PE: 150). Such a society would be free of master-slave relationships, and would be respectful towards nature and the wellbeing of other species.
>
> (Norrie, 2010, p. 153)

Again, this is in keeping with the grounding values of Environmental Justice. According to Hallowes and Butler (2002, p. 51), for example,

> Environmental justice obtains where relations between people, within and between groups of people, and their environments are fair and equal, allowing all to define and achieve their aspirations without imposing unfair, excessive or irreparable burdens or externalities on others or their environments, now and in the future.

Solidarity, which is crucial to EJ organizing on many levels, is derived by Bhaskar in an interesting exploration of what is implied when humans give sincere advice to each other (Norrie, 2010, pp. 133–135). Simply put, giving advice animates expectations that the advice is truthful, that it is fitting for the other person in her specific situation, that the advice giver identifies with the person against the constraints (including oppression) that she is facing. Beyond the helpfulness of the advice, there is an implication that the advice giver may or should practically help the person to deal with the situation. Finally this can be universalized to helping all people deal with all oppression.

This universalizing tendency is crucial to activism. It can be applied to the development (growth in activist consciousness) from a 'not in my backyard' to a 'not in anyone's backyard' logic. This is exactly what has driven EJ to contest not only the distribution of environmental 'bads', but also their production in the first place. One of the unintended results of environmental activism is that pollution externalities move from a stronger, better resourced community able to object and defend itself, to other, weaker communities (Sze and London (2008) give examples of this). Superficial and locally isolated resistance merely relocates pollution. The answer to this must be something like the activism in Steel Valley which grew from an initial base of individual smallholders defending their livelihoods (and settling individually with the polluter), to activists who work with other communities neighbouring on the polluting factories of the global giant they face in Vanderbijlpark.

A challenge to all activists is how to live and exercise agency within a system that they are trying to change, or as Norrie puts it, in the face of an 'elite interest denying change' (Norrie, 2010, p. 221) both on the discursive and practical levels. What the options are for ethical action – and so for EJ activism – need to be judged contingently according to the situation. It speaks directly to the divisive effects of debates that confuse tactics and principles (Munnik and Wilson, 2003). Three situations are discussed:

1 Acting morally in an immoral world means that individuals may 'do the right thing' even when this does not change the mainstream activities. While such action may seem futile, it is a critique in practice and keeps alive the possibility of future freedoms. Such actions may also

provide the starting points for resistance and transformation and thus not be futile.

2 Exercises in concretely utopian ethical thinking and practice provide the home ground for much environmental activism because they project alternative societies and what they may look like.

3 There is also the danger that accounts of freedom can be co-opted into oppressive power relationships, for example 'legitimation through participation'. It warns us to identify and use opportunities in such processes, and not to be used. While occupying spaces in the mainstream can be a powerful strategy in the right circumstances, it constantly moves close to or sometimes over the edge of co-optation (Harvey, 1996).

Critical realism enables us to think through these contradictory situations with an underlying consistency, derived from (1) an understanding of the actual geo-historical situation, with all its specificities and contingencies, and (2) based on a fundamental understanding of and identification with ethical striving in the form of EJ activism, which has a reality and agency, and can be seen in its potential beyond the circumstances of the present. It is the open future present in the present.

Environmental Justice is a consciously self-reflective practice (Pellow and Brulle, 2005). It has attracted academics and other researchers who identify with its aims and practice research from within the movement. Critical realism strengthens this approach by its view that social science is always part of either changing or maintaining oppressive structures. According to Bhaskar: 'to criticize a belief is implicitly to criticize any action based on or informed by it' (Bhaskar, 1993, p. 262). Critical realist academics will be predisposed to go into an alliance with activists.

The sheer potential of this relationship challenges DCR to deal with its plethora of new concepts ('jargon') in order to enable its use in combination with a public sociology and citizen science (which are arguably 'jargon-free' approaches underlying EJ intellectual practice), although in practice EJ has also developed its own, very strong neologisms (Martinez-Alier *et al.*, 2014).

Jargon or vocabulary?

Bhaskar's campaign against the exclusion of absence and other thought categories that enable change in philosophy and in practice has led him to invent a range of specific terms that work in tension with hegemonic Western concepts and discourse. These categories carry specific and powerful meanings that allow us to think through situations from a new perspective, and not be inadvertently pulled back into that hegemonic discourse. Many of these are used in this text, including *absence, negativity, positivity, lamination, constellation, open totality* and *subtotality*. Some familiar words are given new meanings, including dialectic (defined against Hegel's understanding of a dialectic resolved in thought and with Marx's implicit but not ever explicitly theorized understanding) (Norrie, 2010).

Precise vocabularies are important. Much of the value of social science depends on the precision of their definition, use and relationships. They enable abstraction to identify and describe objects of research (how reality is 'cut up'), and what operations can be performed on/with them (Sayer, 2000). They also place researchers within ethical/political positions.

However, even within the critical realist tradition, there have been objections to the use of critical realist jargon (Norrie, 2010). An influential critical realist methodology work (Sayer, 1992) manages to avoid much of these terms, and Bhaskar himself emphasizes the 'underlabouring' (Bhaskar, 2013) role of critical realism, meaning that critical realism creates a thinking space (cleared of the undergrowth and obstacles of hegemonic thinking[10]).

This needs to be seen against the perspective that Environmental Justice itself has invented neologisms for a similar, and equally important task: to contest and change existing descriptions and classifications of the world. According to Nowotny (2000, cited in Agyeman, 2005, p. 27):

> framing is, in a very real sense, part of the repertoire of mobilization strategies that are available to a movement, so that the movement filters the problems it is confronting through the history, the beliefs, the language and cultural experiences that are seen by its leaders as most likely to engender widespread sympathy and involvement.

EJ intellectuals emphasize the discursive side of their struggle. Pellow and Brulle (2005, p. 12) for example, argue that:

> changes in social structures are brought about through a redefinition of what constitutes the common sense embodied in the everyday practices of society.... We view this as the central battlefield for the EJ movement – the struggle over the definition of environmental and social reality between social movement groups and the corporate-state structures that produce environmental inequalities.

Environmental Justice theorist Devon Pena specifically defines 'discursive politics' as referring to 'contested encounters between political actors articulating variant ideologies in struggles over "legitimation" of divergent worldviews' (2005, p. 150, footnote 2) and offers the following description of EJ discursive practice:

> Environmental Justice discourses challenged environmental thinkers to reconsider the meaning of basic concepts like nature, environment, ecosystem, wilderness and biodiversity; they forced many to consider the role of race, ethnicity, national origin, class, gender, and culture

> in the framing of environmental history, environmental ethics, and ecological politics.... EJ discourses recentred the problematic of ecological politics in the constellation of cultural differences that construct variant epistemologies of nature.
>
> (2005, p. 131)

Julie Sze in *The environmental justice reader: Politics, poetics and pedagogy* (2002, p. 163) says that 'Environmental Justice is a political movement concerned with public policy issues of environmental racism, as well as a cultural movement interested in issues of ideology and representation.'

It follows then that there is some space for accommodation on both sides, in service of a shared commitment to change. I would propose three guidelines:

1 Keep crucial concepts such as absence, explain them and illustrate their use.
2 Keep the text as close to generally understandable social science use of terms (especially in the Marxist/critical school tradition).
3 From these two moves develop a reasonable vocabulary for DCR analysis of environmental justice case studies that retains the counter-hegemonic thrust of DCR but can function in the tradition of a public sociology.[11]

Conclusions

I conclude this analytical deliberation with what I see as a list of the advantages that DCR holds for EJ.

- The depth ontology of CR enables EJ to talk about reality, construct and claim validity for the most likely, retroduced explanation, and thus challenge reductionist explanations, including denials of the existence of causal links in pollution situations. This is important in contestation about the reality of pollution; specifically in monitoring, mobilizing against pollution, administrative action, court cases, compensation and policy.
- CR is able to deal with both natural and social science and EJ often needs to deal with them together.
- DCR can constellate different knowledges, explanations and theories (within a lamination), deal with the phenomenon of emergence, the influence of the past on the present and the future (thus taking long run and still influential geo-historical trajectories into account).
- CR supports the EJ orientation to take the experiences and knowledges of fenceline communities and activists as a starting point. This validates citizen science and acknowledges their right and their capacity to participate in retroductive (transfactual) explanation building.

- CR allows EJ to escape irrealist dualisms and other 'unserious' thinking.
- CR allows EJ to explore the many harmful absences that drive EJ activism, and understand how to absent them (absent the absences).
- It provides an understanding of the sources of solidarity and the irrepressibility of the human drive to freedom. This encourages activism.
- It allows contradictions to be thought through within a totality, and preserves the idea of holistic approach in an open future.
- It enables envisioning and moving towards a new world, in a strong move sweeping away inherent resistance to change in Western thinking.
- It sees social science as inherently ethical and where it is not, identifies that it is part of oppressive structures. It acknowledges the tight connection between social science and activism, admitting that facts can lead to values.
- Like EJ, it is serious about the possibility of change.

Notes

1 For a parallel argument in Environmental Education, see John Huckle's (2004) excellent chapter.

2 Mainly from the MELD schema in Bhaskar's *Pulse of freedom* and *Plato, etc.*, as mediated by Norrie in his *Dialectics and difference*, with some from earlier work and some from *Reflections on meta-reality*, and from Bhaskar (2010) Chapter 1 on climate change. I will mostly confine the discussion of critical realist concepts to footnotes, in order not to impede the text, but will explain the concepts I use in the text.

3 The first person plural acknowledges the plural nature of producing a text, even a single author text, and asserts the sense of a shared journey via this text.

4 See Abugre in Hallowes (1993) for an example of this thinking at the founding EJNF conference.

5 Their last parliamentary representative was F.W. de Klerk, the nationalist party president who freed Nelson Mandela and initiated the political transition in 1990.

6 The Western Gauteng Services Council.

7 A notable exception to this were black people who only in 1990 had acquired the right to buy land in 'white' areas, and who were sceptical about the reality of the pollution, suspecting that it was a ruse to keep black buyers out.

8 Because countervailing tendencies in open systems resist their operation, for example: although a bridge is subject to gravity, it is held up by the strength of its pillars and does not 'succumb' to gravity.

9 Personal communication (16 May 2006), Dr Murray Coombs, expert witness in Steel Valley court case of 16 applicants, currently director, Elixir Corporate Health Solutions, Pretoria.

10 I use hegemonic thinking in the broad sense of 'Western philosophy from Plato to today' that works in a monovalent fashion, that is, only with the positivity. In this essay, I take that to be the epistemology of reductionism.

11 In the sense that Burawoy (2004) gives to it.

References

Adamson, J., Evans, E. E., and Stein R. (Eds). (2002). *The environmental justice reader: Politics, poetics and pedagogy.* Tucson: University of Arizona Press.

Agyeman, J. (2005). *Sustainable communities and the challenges of environmental justice.* New York: New York University Press.

Bakan, J. (2004). *The corporation: The pathological pursuit of profit and power.* New York: Free Press.

Bhaskar, R. (1993). *Dialectic: The pulse of freedom.* London: Routledge.

Bhaskar, R. (2002). *Reflections on meta-reality: Transcendence, emancipation, and everyday life.* Delhi: Sage.

Bhaskar, R. (2010). Contexts of interdisciplinarity: Interdisciplinarity and climate change. In R. Bhaskar, C. Frank, K. G. Hoyer, P. Naess and J. Parker (Eds), *Interdisciplinarity and climate change: Transforming knowledge and practice for our global future.* London: Routledge.

Bhaskar, R. (2013). Prolegomenon. The consequences of the revindication of philosophical ontology for philosophy and social theory. In M. Archer and A. Maccarini, (Eds), *Engaging the world.* London: Routledge. http://international-critical realism.com/2013/06/10/the-consequences-of-the-revindication-of-philosophical-ontology-for-philosophy-and-social-theory.

Burawoy, M. (2004). American Sociological Association Presidential address: For public sociology. *American Sociological Review,* 2005, *70*(1), 4–28.

Cock, J., and Koch, E. (1991). *Going green: People, politics and the environment in South Africa.* Cape Town: Oxford University Press.

Cock, J., and Munnik, V. (2006). *Throwing stones at a giant: An account of the struggle of the Steel Valley community against pollution from the Vanderbijlpark Steel Works.* Centre for Civil Society, UKZN Report.

Fine, B., and Rustomjee, Z. (1996). *The political economy of South Africa: From Minerals-Energy Complex to industrialisation.* Johannesburg: Witwatersrand University Press.

Fuggle R. F., and Rabie M. A. (Eds). (1983). *Environmental concerns in South Africa: Legal and technical perspectives.* Johannesburg: Juta.

Fuggle R. F. and Rabie, M. A. (Eds). (1992). *Environmental management in South Africa.* Johannesburg: Juta.

Habermas, J. (1996). *The structural transformation of the public sphere.* Massachusetts: MIT Press.

Hallowes D. (Ed.) (1993). *Hidden faces: Environment, development, justice. South Africa and the global context.* Durban: Earthlife Africa.

Hallowes, D., and Butler, M. (2002). Power, poverty, and marginalized environments: A conceptual framework. In D. MacDonald (Ed.), *Environmental Justice in South Africa.* Cape Town: UCT Press.

Hallowes, D., and Munnik, V. (2006). *Poisoned spaces: Manufacturing wealth, producing poverty.* Pietermaritzburg: Groundwork.

Hallowes, D., and Munnik, V. (2007). *Peak poison: The elite energy crisis and Environmental Justice.* Pietermaritzburg: Groundwork.

Harvey, D. (1996). *Justice, nature and the geography of difference.* Oxford: Blackwell.

Henderson, H. (1996). *Creating alternative futures: The end of economics.* West Hartford, CT: Kumarian Press.

Huckle, J. (2004). Critical realism: A philosophical framework for higher education for sustainability. In P. Corcoran and A. Wals (Eds), *Higher education and the challenge of sustainability*. Dordrecht: Kluwer.

ISCOR. (2003). *Executive Report: Environmental Master Plan*.

Jessop, B. (1990). *State theory: Putting capitalist states in their place*. Cambridge: Polity Press.

Kovel, J. (2002). *Enemy of nature: The end of capitalism or the end of the world?* New York: Zed Books.

Luke, T. (1999). Eco-managerialism: Environmental studies as a power/knowledge formation. In F. Fischer and M. Hajer (Eds), *Living with nature: Environmental politics as cultural discourse*. Oxford: Oxford University Press.

Martinez-Alier, J., Anguelovski, I., Bond, P., Del Bene, D., Demaria, F., Gerber, J.-F., Greyl, L., Haas, W., Healy, H., Marín-Burgos, V., Ojo, G., Porto, M., Rijnhout, L., Rodríguez-Labajos, B., Spangenberg, J., Temper, L., Warlenius, R., and Yánez, I. (2014). Between activism and science: Grassroots concepts for sustainability coined by Environmental Justice Organizations. *Journal of Political Ecology, 21*, 19–60.

Mendelsohn, R. (1991). *Sammy Marks: The uncrowned king of the Transvaal*. Cape Town: David Philip.

Munnik, V. (2012). *Discursive power and environmental justice in the new South Africa: The Steel Valley struggle against pollution (1996–2006)*. Unpublished PhD thesis, University of the Witwatersrand, Johannesburg.

Munnik, V., and Wilson, J. (2003). *The world comes to one country: An insider history of the World Summit on Sustainable Development, Johannesburg 2002*. Johannesburg: Heinrich Boell Foundation.

Nixon, R. (2011). *Slow violence and the environmentalism of the poor*. Cambridge: Harvard University Press.

Norrie, A. (2010). *Dialectic and difference: Dialectical critical realism and the grounds of justice*. London: Routledge.

Pellow, D. N., and Brulle, R. J. (Eds). (2005). *Power, justice and the environment: A critical appraisal of the Environmental Justice movement*. Massachusetts: MIT Press.

Pena, D. G. (2005). Autonomy, equity and environmental justice. In D. N. Pellow and R. J. Brulle (Eds), *Power, justice and the environment: A critical appraisal of the Environmental Justice Movement*. Massachusetts: MIT Press.

Price, L. (2014). Mainstream vs. critical realist interdisciplinarity. *Journal of Critical Realism, 13*(1), 52–76.

Proctor, R. N., and Schiebinger, L. L. (Eds). (2008). *Agnotology: The making and unmaking of ignorance*. Stanford: Stanford University Press.

Ramphele, M. (Ed.). (1991). *Restoring the land: Environment and change in post-apartheid South Africa*. London: Panos.

RSA (Republic of South Africa). (1996). *Constitution of the Republic of South Africa*.

RSA (Republic of South Africa). (1998). *National Environmental Management Act* (NEMA).

Sayer, A., (1992). *Method in social science* (2nd edn). London: Routledge.

Sayer, A. (2000). *Realism and social science*. London: Sage.

Schnaiberg, A. (1980). *The environment: From surplus to scarcity*. Oxford: Oxford University Press.

Sze, J. (2002). From environmental justice literature to the literature of environmental justice. In J. Adamson, E. E. Evans, and R. Stein (Eds), *The environmental justice reader: Politics, poetics and pedagogy*. Tucson: University of Arizona Press.

Sze, J., and London, J. (2008). Environmental justice at the crossroads. *Sociology Compass, 2*(4), 1331–1354.

Trapido, S. (1986). Putting a plough to the ground. In W. Beinart, P. Delius, and S. Trapido (Eds), *Putting a plough to the ground: Accumulation and dispossession in rural South Africa 1850–1930.* Johannesburg: Raven Press.

Yudelman, D. (1984). *The emergence of modern South Africa: State, capital and the incorporation of organized labour on the South African Gold Fields, 1902–1939.* Cape Town: David Philip.

17 Absenting absence

Expanding zones of proximal development in environmental learning processes

Heila Lotz-Sisitka

Introduction

This chapter deliberates why and how Bhaskar's concept of absence and dialectical processes of negation can be given greater consideration in the conceptualization of environmental learning and action. I argue that normalized processes of environmental learning can be expanded through engaging not only with real negation and transformative negation (at 1M and 2E), but also with radical (totalizing) negation (at 3L) and linear negation (at 4D) involving self-transformation and transformative praxis conceptualized as *learning processes* in possible zones of proximal development. As stated by Hartwig (2008, p. xiv), 'the ultimate concern of dialectical critical realism is with *thinking being* as incorporating human praxis and reflexivity (emancipatory axiology, or the coherence of theory and practice in practice) at 4D' (my emphasis). The chapter suggests that significant to conceptualization of environmental education as a transformative praxis is the critical realist proposition that the axiology of freedom involves absence and absenting. Hence I found the need to probe this *educationally*, in a rapidly emerging educational arena where change and transformation of society towards a more sustainable, just existence is of central interest.

Absenting as normalized practice in environmental education

Absences as defined in critical realism are constraints on well-being, or ills (Hartwig, 2007; Bhaskar, 2008). In socio-environmental change contexts, absences are ills, or constraints on well-being affecting ourselves,[1] other living beings, and the planet and its immensely complex yet beautiful systems. These may occur at sub-individual psychological levels, individual or biographical levels, or at wider social systemic and/or civilizational/traditional levels (Bhaskar, 2010). Such absences are the object of environmental education processes, as most environmental education processes seek in some way or other to 'address' such absences or ills in society, reflecting 'distinct levels of agency and collectivity with which the social sciences might be concerned' (Bhaskar and Danermark, 2006, p. 289). At a sub-individual psychological level, environmental education programmes may seek to absent an absence of environmental motives or values. At the individual and/or biographical level, environmental education programmes may seek to absent the absence of ecologically oriented identities and practices. At the meso and macro level environmental education programmes exist that seek to absent the absence of sustainability at whole institution, policy or system level.

Overall, it appears that it is the absenting of harmful culture-nature or eco-cultural relations in modernity that are shaping the need for environmental education and learning in society. From this brief introduction (and as also described in a number of chapters in this book), it is possible to see that environmental learning processes can potentially be seen as processes of absenting absences or absenting constraints or ills.

But how might we more explicitly consider this in our work? This chapter argues that in giving in-depth attention to Bhaskar's concept of absence, and the associated processes of dialectical negation (absenting), it may be possible to broaden and enrich what currently seems to be an implicit, normalized[2] practice. To probe this further I use the example of waste management[3] activities in a school context to deliberate *how* we as environmental educators may work with the concept of absence in more explicitly framing and thinking about the implications of 'absence' and 'absenting absence' in our work.

Waste management as an example of absenting an absence

Waste management is a very typical focus of environmental education activities, practised in many schools, colleges and on campuses around the world, as can be seen for example in the global Eco-schools programme or a variety of Green Campus initiatives (UNESCO, 2014). A typical environmental education activity in a school would involve a school teacher encouraging her class to pick up litter or build a recycling station in a school. In picking up the litter, the children are absenting the absence of an effective waste collection system in the school. In building a recycling station, the children are partly absenting the absence of a culture of waste reduction. In these combined activities, children are absenting the absence of a waste management system in their school. While the teacher and children may be successful at absenting the absence of an effective waste management system in their school, or partly successful in absenting the absence of a culture of waste reduction or minimization, it remains difficult for the teachers and the children in the school to fully absent the problem of waste in modern society. This is because the teacher and her learners, like all of us in some way or other, remain locked into the structural realities of waste production in modernity. The waste that they have carefully recycled may just end up in a landfill down the road. Additionally, the underlying generative mechanisms that are producing waste, wasted materials, wasteful cultures and wasteful production systems are often difficult to identify, and even more difficult to dislodge and change, especially at a local level in a school or classroom. The same can be said for absenting other ills at a local level, such as pollution, or food insecurity, or water insecurity, or climate change (see other chapters in this book).

This realization can create challenges for the environmental education teacher and her learners. Even though they are doing what they can with the resources and skills that they have in a school context (or other local context) their local absenting actions can lead to either relativism and/or disillusionment, as is reflected in typical statements such as 'the problems are so big, we can just do our "little bit" ' or 'the problems are so big, what good is our small contribution'? This kind of discourse is increasingly associated with environmental education activities focusing on 'big issues' such as climate change, biodiversity loss, fossil fuel use, food insecurity, unsustainable agribusiness and many other environmental concerns.

Towards a more robust engagement with absenting absences in environmental education

From the example of waste management outlined above (or via an engagement with other examples presented across this book), it is possible to see that absenting absences in environmental education is a challenging affair, as environmental issues have ontological depth; they are shaped by interacting social-ecological generative mechanisms that need to be engaged in and through environmental education praxis, often in combination with other structural interventions. It is possible to suggest that waste management problems are generated by interacting generative mechanisms such as neo-liberal globalization and other forms of capitalist economics with associated tendencies towards planned obsolescence and exclusion of externalities in costing systems, the construction of modern consumer identities and lifestyles, control of resource flows and so forth (Martinez-Alier, 2002, 2012). However, it is not only the analysis of generative mechanisms and stratified reality (at 1M) that is helpful for deepening engagement with complex environmental issues. Bhaskar's (2008) model of dialectical negation which foregrounds absence can further help to deepen ways of thinking through the waste management activity in the school (and by implication other environmental issue contexts) as represented below:

4D the environmental education programme or project seeking to enable agency that absents the ills of nature-culture relations that produce waste that damage environments and societies creating public health risks, ecological and earth system degradation and social injustices; and

3L a waste-free society, closed cycle production, benign nature-culture relations; involving

2La absenting the absent waste management system (e.g. picking up litter); that involves,

2Lb absenting the absence of a culture that values the re-use of waste (e.g. through building a recycling station in the school and re-using materials); and more so

2Lc absenting the absence of a culture that values waste avoidance (i.e. zero-waste), that ultimately requires the bringing into being and;

1M absence of a waste-free society and closed-system production that is based on benign nature-culture relations and social-ecological justice where the flourishing of one is related to the flourishing of all life forms.

From this example, it is possible to see that such a conceptualization can broaden thinking about absenting practices in environmental education activities and it makes more visible the 'challenging affair' of environmental education. The example shows that waste management education ultimately requires absenting the dark sides of modern industrialization and its associated (unsustainable) economic system, and associated production and consumption patterns that reproduce planned obsolescence, that fail to internalize externalities, and that produce consumptive identities and social-ecological injustices (amongst others). Ultimately environmental education seeks to bring into being a new society that is based on benign nature-culture relations and social-ecological justice where the flourishing of one is related to the flourishing of all life forms (Bhaskar, 2008). Environmental education seeks to challenge and transform the foundational narratives of modernity as constructed in the Western Enlightenment tradition, capitalism, the industrial revolution, and the more recent globalization of neo-liberal capitalism (Leff, 2009).

Conceptualizing waste management education via the route of critical realism's perspectives on absence and negation can open normalized practices in environmental education to greater scrutiny and reflexivity. By way of example, it is possible to review the example given above within a broader framing of the dialectical negations embedded in the processes of absenting. If, for example, we were to replace our concern with waste with the concept of love[4] of the world, its people and environments, and with a clear conception of how we are related to all that surrounds us, the absenting process linked to the waste example changes in form. Is waste a sign of loss of love and care for the world around us? If this is the case, what is it that we should be absenting – is it the absence of the absence of a waste management system, the absence of the absence of a culture of unwastefulness (zero-waste), the absence of the absence of harmful eco-cultural production systems, and/or an absence of the lack of love and care for the world and all its beings? And how might these be related? The environmental education project changes form, focus and process. Here it is very interesting to see that in recent times a larger number of environmental education programmes and projects are foregrounding love in their programmatic framing.[5] While this appears to be a positive move, it assumes that the expression of love for nature, rather than *the love involved in the difficult task of absenting of absences*, will result in a caring for the environment, life and all beings. We can love the environment, and implement recycling activities, but we may fail to engage with the need to absent the absence of a culture of zero waste or social-ecological justice, or social and cognitive justice in our societies.

Framing the concerns outlined above in the critical realist language of 'absenting absences' is not to create an obscure or complicated language of referring to what we seem to be doing anyway via our normalized practices in environmental education. The use of the terms 'absenting absence' requires us to frame our love for the world, its beings and people differently, i.e. as 'absenting absences'. This shifts us from a voluntarist concept of agency (where it is assumed that individual agents can do anything at any time without constraints, i.e. 'all we need is love'), to a position that requires us to carefully and lovingly engage with a non-reified notion of structure, and thus a non-voluntaristic, non-deterministic and non-conflationary view of agency. To fully show our love for each other and the world and its beings, we have to stop and ask the question 'Why is X [e.g. a culture of zero-waste[6]] absent, and what needs to be done to absent this absence?' This requires retroductive processes of understanding historically shaped socio-cultural, political-economic, as well as geophysically constituted structural factors and generative mechanisms influencing the problem we are confronting to find possible answers, and potentially to frame more comprehensive education and thoughtfully constituted learning and action possibilities.

Implicitly embedded in our concern for waste is our love for the world. Critical realism therefore helps to underlabour for what is already present but occluded, namely our innate human capacity for loving and caring for others, non-humans and the world around us; it helps to make visible that in our simple absenting actions we are also producing love and care for each other, other beings and the world. It also helps us to see that caring for, and showing love for the world is an act that involves us in developing a deep understanding of the structures and generative mechanisms that may hold powers and the very ills we are seeking to absent in place. We are fortunately not stuck with hyper-critique of the world. We are able to act, and to absent, and through this to realize our love for the world and its systems and beings. While this is the case, we should use our agency judiciously. Collier (2001, p. 309) suggests that 'human emancipation is not about change as such. Rather it is about those changes which absent constraints on the absenting of ills (rather than those changes which build such constraints).' He goes on to say 'that "everything should be changed" was a Nazi slogan', reproduced in today's world as a strategy 'that [often] serves the cause of robbing the poor to give to the rich' (p. 309). Writing from the North, he suggests that today's struggles are to *conserve* institutions of the welfare state that are threatened by neo-liberal changes. Post-colonial and de-colonial theorists in the Global South (e.g. W'Thiongo, 1993; De Sousa Santos 2005, 2007, 2009, 2015; Paraskeva, 2011; Andreotti, 2011) argue that rather than *conserve* existing colonial and neo-colonial institutions there is need to radically reconstitute them.

This brings us to the question of how teachers (or interventionist researchers and concerned citizens) are to 'teach' children and others to love the world and all its beings. From the example outlined above, it would seem helpful for educators, facilitators, interventionist researchers and others to understand the absences and negations necessary to absent waste and other ills in society whilst also understanding that such absenting actions are expressions of love and care for the world around us. This, as shown in the reflective comment on Collier's point above, needs also to be geo-politically and spatio-temporally contextualized.

Importantly, Collier (2001, p. 309) cautions against interpreting such possibilities for agency and change naively, and suggests that there are 'constraints on absenting ills' (e.g. debt and structural adjustment, unemployment, low benefit payments, repressive regimes, colonial intrusion), and that 'The absenting of these is the political task of human emancipation'. Collier (2001, p. 309) also suggests that in privileging negativity or absence, one should not assume that positivity is 'static, passive or constraining'. For example, producing some forms of waste could be potentially helpful in the production of alternative energy resources that might be less damaging to the earth's social-ecological system than carbon emissions, meaning that waste production is not *necessarily* constraining.

Engaging real absences

In the case example above of waste management, it is possible to see that even though it is possible to see our absenting acts as acts of love and care for the world, and through this we can better 'think the world together' relationally, it is still necessary to engage in strong structural critique, and for teachers and their learners to continue to probe the issues critically (i.e. it will not be adequate to switch to love and care building approaches to environmental education only). Even though almost all schools and communities in 'high income' countries are now recycling waste, waste over-production remains on the rise, as does over-consumption and inequality. Bhaskar (2008, p. 206) suggests a conception of social science as involving an 'ethical tetrapolity' which involves being non-neutral against fiduciariness (trustworthiness), which provides content 'for the explanatory critical theory complex', leading to and shaped by 'totalizing depth praxis of emancipatory axiology' and linked to 'freedom qua universal human emancipation'. We need to spell out the substance of the normative order of trust when it comes to waste (i.e. do we trust waste production as being non-harmful to people and life support systems?), as this is the foundation for providing an explanatory critique and identifying absences that need absenting, providing possibilities for 'totalizing depth praxis of emancipatory axiology', meaning our possibilities for action in the world.

Collier's (2001) warning for us not to confuse nominal absences and real absences is helpful here. If we return to our example above, it is possible to suggest that absences such as the absence of love and care for the world and all its beings could be seen to be a real absence when talking about waste production and wasteful squandering of the Earth's resources. Similarly, the absence of non-wasteful production systems that have emerged through the consumptive logic and demands in modern capitalism are similarly real absences, while the absence of an alternative to over-consumptive capitalist forms of development can also be seen to be a real absence in the context of the teachers' waste education programme. Real absences, Collier argues are more difficult to identify as they are related to often invisible but real structural generative mechanisms and tendencies (e.g. global capitalist structures driving consumer cultures). They differ from nominal absences which Collier (ibid.) sees 'all around us' (e.g. the absence of the litter bins in the schoolyard).

Bhaskar (2008) argues that absence is real, not actual (deep or superficial). In an open system, really unhealthy structures can be present and have generative effects even in the absence of empirical evidence. We can have clean schoolyards in all of our schools, yet our cultures of consumption and production remain the same. This differentiation between nominal and real absences is helpful, as it shows us where we can potentially focus our environmental education attention: on creating pedagogies of love and care, and on creating more emancipatory structural formations in and for society (not excluding local actions to pick up litter). Hartwig (2008, p. xiiv) suggests that Bhaskar's work on absenting absence can help us in establishing 'a conatus to ever more complete conceptual formations'. The litter still needs to be picked up, and the recycling system still needs to be in place, but there is need for depth transformative praxis too – for how else will we make our transformative praxis journeys? The important point, however, is to not stop our environmental education activities at the level of litter pick-ups, or making of recycling centres, but to *first or at the same time* engage with deeper negation processes at a structural level. Important for environmental education, too, is that these forms of environmental education are often the most difficult to engage with; it is easy to confuse a clean schoolyard (with all the litter picked up) with 'real' change.

Absenting as process (a): reflexive pedagogy and learning

Bhaskar (2008) argues that absences are not always positively identified, or identifiable. This requires us to be open to indefinite, reflexive forms of pedagogy and learning. As stated in Hartwig (2007:10), absence also connotes 'the *process* of absenting, distancing or *mediating*' (my emphasis). Absenting is 'the hub of *space, time* and *causality*' (Bhaskar, 1994, p. 56) with 'positive biopolars in *emergence* and *creativity*' (ibid.). However, the knowledge and new creative practices that are needed to engage the absenting question may not be readily available, and may need to be 'worked out' in and through emergent forms of praxis. We cannot therefore pre-determine the lesson or the learning process; we need to open it up for discovery, transition and for co-creation, for *mediation*. This does not mean that we cannot have clear starting points, or deep knowledge of the absences or constraints we are seeking to absent, or that we have no idea of what the results of our absenting processes *could* be; it simply means that our pedagogies need to be open-ended enough to allow for further probing, learning and new ways of seeing, doing and becoming as we go along. It comes as no surprise that Bhaskar sees learning processes as being powered by *curiosity*, and *desire to gain knowledge or absent error* in order to meet wants and needs or absent constraints on flourishing (Hartwig, 2007, p. 277). Bhaskar suggests that in so far as ills reflect absences and constraints which 'from the standpoint of practical reason' may be '(i) *unwanted,* (ii) *unnecessary* and (iii) *remediable* or removable, they should be transformatively negated, i.e. absented' (Bhaskar, 2008, p. 260).

Significant to our interest in education, Bhaskar goes on to point to the importance of knowledge, skills and motivation in such a process of transformative negation, and indicates that while 'humans may indeed desire to be free' and to absent ills, they may lack the 'concrete power and/or knowledge to achieve particular freedoms' (ibid.). Thus, as criteria for rational agency, Bhaskar suggests that we would need:

1 knowledge[7] to act on our own real interests (a cognitive requirement);
2 abilities to access skills, resources and opportunities to act (an empowerment dimension); and
3 to have the desire and motivation to act (a motivational requirement).

Bhaskar (ibid., p. 269) suggests that we need only have a 'conatus to knowledge' or a desire for new knowledge to 'get the dialectic of freedom going'. While this is an important educational statement, it should not be interpreted technically (i.e. as transferring knowledge). Knowledge in itself (especially if viewed technically) is not adequate for mobilizing depth transformative praxis, i.e. we cannot assume that framing and transferring bodies of knowledge about environmental issues (as is the tradition in many environmental education programmes) is adequate for absenting ills such as waste. Bhaskar (ibid.) qualifies this as follows:

> to the extent that constraints on knowledge, or more generally on any other aspects of rational autonomy are social ... then only knowledge about constantly developing four planar human nature ... including ... social relations ... and hence the social sciences in sensitized solidarity with the tacit know-how of the subjects$_2$ [oppressed people] can rationally inform emancipatory praxis.

It is with this emancipatory axiology that Bhaskar (1997, p. 141) sees learning being *inherent in the dialectic*, and suggests that:

> Dialectic may be seen as the (experience of {4D}), the process of transformation and dissolution {2E} of stratified (and differentiated) {1M} totalities {3L}. In the human field it constitutes *a general schema for a LEARNING PROCESS* in which absence {2E}, signifying incompleteness, leads to transcendence and greater totality {3L}, in principle reflexively {4D} capable of situating itself and the process whereby it became.

In this citation, we see that Bhaskar sees absenting absence as being at the heart of learning processes, as through absence, it is possible to identify incompleteness, the mobilization of curiosity and the desire to gain knowledge or absent error. Related to our waste example we may not know in advance how to resolve a waste management problem in our community, but it can be 'worked out' along the way once the absence has been identified as an incompleteness, mobilizing our curiosity or desire to gain knowledge or absent the error of waste production and mismanagement. Such learning processes are reflexive, and require teachers and learners to identify the absences at the level of the real, conceptualize alternatives using new knowledge and/or action oriented approaches, and resolve aspects of the absence through transformative praxis activities. For this to be emancipatory, such learning processes should be widely oriented towards the flourishing of all, including all life forms (Bhaskar, 2008). However, such learning processes should not be irrealist or idealized, but should 'sustain a concept of *intentional embodied causally efficacious agency*, that is, of bodily agentive (transformatively negating) agency', involved in absenting absences, 'informed necessarily with real, and potentially good, reasons for action' (p. 277), or 'embodied *agentive* agency' (p. 278). This has implications for how we conceptualize learning processes.

Further insights into absence can provide wider explanations into potential dynamics of the reflexivity and 'embodied *agentive* agency' involved in such learning processes. Importantly, Bhaskar (2008) suggests that what is present and what is absent are not in contradiction, but rather that presence and absence are 'systematically connected' (Bhaskar, 2008). In reflexive learning processes, it is the relational spaces between presence and absence that need to be engaged by learners. In the waste management example, this could, for example, involve undertaking a waste audit to establish what is present, coupled with a retroductive laminated system analysis of the waste problem, followed by thinking of it in terms of absence and what could be possible – i.e. what alternatives might there be, and which of these are viable or feasible for action or to act on in a particular context/situation?

Also interesting for thinking about the possibilities for enabling such reflexive learning processes in our work as environmental educators is the insight that absences may be present somewhere else (e.g. successful re-use systems may be present in Germany, for example, or we may find cultures which show strong love and care for people and all living things in other parts of the world). Absences, as shown by these examples, may also be spatio-temporally distant. For example, the absence of non-wasteful production systems is not present in the school grounds or the school environs, or may only be present on other continents or in other places. Or more benign eco-cultural production systems may be present in indigenous technologies that have lost their social validity under modernity's marginalizations, thus becoming 'absent' through colonial and modernized historical changes in cultures of production and consumption. For our educational work, thinking about the presence of absences elsewhere helps us to *formulate possibilities*, and may also help with *re-imagining what is possible*, or what we may recover (in part) from knowledge that exists in marginalized heritage practices.[8] An example is the practice of *izala*, which is an isiXhosa cultural practice of recycling organic waste at source, destroyed or marginalized by the banning of home composting (*ethutwini*) practices in modern urban legislatively governed settings (O'Donoghue *et al.*, 2013).

Another interesting dynamic for thinking about absence and reflexive learning is the insight from Bhaskar that absences may be non-existent (i.e. they may never be 'anywhere anywhen' (Hartwig, 2007, p. 10)), as is the case with the absence of sustainability in the here and now – it is yet to come into existence and is not (yet) present anywhere. In the past some societies may have been sustainable, but overall planetary sustainability is not in existence in the here and now, hence the ongoing global formulation of documents such as Agenda 21 (UNCED, 1992) which are now culminating in global attempts to frame 'sustainable development goals' to potentially guide absenting actions on a massive scale around the world in a search for sustainability (see sustainabledevelopment.un.org).

If, in our environmental education reflexive learning processes, we accept Bhaskar's argument that non-being has ontological priority over being, we could suggest that the non-presence of sustainability or the absence of a sustainable society has ontological priority over our current state of being, which is riddled with unsustainable practices. Absence therefore becomes a way of beginning to conceptualize eudemonia, or a better society via sustainability thinking. It becomes an opening for mapping, conceptualizing or desiring forms of knowledge and practice that can be considered transformative, without being over-idealistic or instrumentalist. It can provide a means of coming to understand our 'ultimate concerns' which, according to Archer (2003), are significant in our reflexive deliberations, and one might also argue in reflexive learning processes.

It may, however, be too simple a task to suggest that the non-being of sustainability (i.e. its absence) has ontological priority over being, i.e. unsustainable practices – and that this can provide a defensible 'guide' for mobilizing reflexive learning processes. This is because sustainability has been hotly contested as a focus for education, especially in its appropriated form as co-opted into the structural forms that have created the very unsustainable practices it seeks to replace (Jickling, 1992; Bond and Guliwe, 2003). Thus, it may well be too simplistic to say that 'sustainability' as absence has ontological priority over being, unsustainability; and through that suggest that all education should be oriented towards sustainability. There is need to be clearer about the meaning/s of sustainability, and to make a commitment to sustainability as involving practices that result in a healthy environment and flourishing beings.

From this, we can see that absence may well have ontological priority over being, but *what* absence we choose to identify as having ontological priority over being matters. This involves giving attention to democratic process, cognitive justice, deliberation and meaning making in the transitive dimension of reality. Critical realism requires communication and dialogue in knowledge co-production in and for the formation of collective agency (Vandenbergh, 2014) and according to Vandenbergh (2014, p. 24), 'Critical realism cannot and should not neglect hermeneutics.' From a transformative educational perspective, this should take account of the histories of knowledge construction and associated 'epistemicides' (de Sousa Santos, 2005, 2009), and should include a recovery of, or expansion of the varieties of knowledge and languages that could potentially shape transformative meaning making. As Paraskeva (2011) suggests, drawing on De Sousa Santos (2005), this involves an 'itinerant curriculum theory' which involves 'a struggle against epistemological orthodoxy', and which must attempt to 'bring scientific knowledge face-to-face with non-scientific, explicitly local knowledges, knowledges grounded in the experience of the leaders and activists of the social movements studied by social scientists' (De Sousa Santos, 2005, p. xxv).

Absence as process (b): engaging the zone of proximal development

Lev Vygotsky's work on learning and development provides further insight into the meaning making process, as it occurs in what he terms zones of proximal development. Rom Harré (2006) explains that the learning and development theory of Lev Vygotsky is realist in that it combines historical/cultural/instrumental explanations for learning and human development. He says that Vygotsky's conviction was that 'higher mental functions come about through 'an intricate interaction' between biological aspects of *Homo sapiens* and cultural factors that have also evolved over millennia' (Harre, 2006, p. 109). Vygotsky suggests that learning leads development via various mediation processes which involve interactions between subjects (and their biological and psycho-social make-up, histories and cultures), objects (which are also socially and historically imbued), and symbolic tools (also socio-historically, culturally and materially shaped). Such mediation processes, according to Vygotsky, occur for the learner in what he calls the zone of proximal development. What Vygotsky does not address directly is the presence of absences in these zones of proximal development (e.g. the epistemicides referred to by de Sousa Santos (2005), or the real absences referred to by Bhaskar (2008)).

Harre (2006) discusses the links between mediation in the zone of proximal development and pedagogy. He says for pedagogy to be effective,

> Interpsychological functioning [e.g. what and how we see and define absences and their role in environmental and social change/transformation] must be so structured [in our pedagogical practices] so as to enhance the development of intrapsychological functioning. Instruction in the zone of proximal development 'calls to life in the child, awakens and puts in motion an entire series of internal processes of development'.
>
> (Harre, 2006, p. 31, citing Wertsh, 1985)

Pointing to our role as educators, Vygotsky (1978, p. 57) stated:

> Every function in the child's cultural development appears twice: first, on the social level, and later, on the individual level; first between people (interpsychological) and then inside the child (intrapsychological). This applies equally to voluntary attention, to logical memory, and to the formation of concepts. All the higher functions originate as actual relationships between individuals.

Bhaskar in his conceptualization of learning (outlined above) seems to suggest that *a dialectical, reflexive learning process* leads into *transformative praxis* from 1M, via 2E and 3L into 4D. This in some ways is not unlike Vygotsky's perspective on learning, but with further nuance provided as to the ontology of the social level [1M – product], and to the ontological primacy of absence [2E – process]; possibility [3L – process-in-product] (especially from an ethical/transformational perspective) and transformative praxis actualization which must occur also at the intrapsychological level for agency to be mobilized [4 D – product-in-process] (Nunez, 2013). Thus, it is possible to suggest that considering Bhaskar's perspective on learning *in relation to the work of Vygotsky*, provides further nuance to the Vygotskian view that learning leads development, as critical realism begins to qualify *a kind of emancipatory development within the dialectic, that takes account of absences, and which is further qualified in terms of 'flourishing' or the axiology of freedom.* As Nunez (2013, pp. 153–4) states, critical realism situates activity theory ontologically

> in Bhaskar's four-planar social being: a four-dimensional flow consisting not only of the dimensions of our social interactions with other people (already identified in first-generation activity theories) and that of the social sphere or social structure (identified with second- and third-generation activity theory), but also of the dimensions of our transactions with the biosphere or nature and of the stratification of our individual embodied personality. The importance of the four planes of social being is that each dimension is part of our agency, i.e. it is involved in whatever activities we pursue.

This, I would argue provides both curriculum and pedagogical possibilities for environmental education, if both curriculum and pedagogy are understood as being part of Bhaskar's dialectical process that is oriented towards enhancing open, transformative processes of learning, agency, reflexivity, human flourishing and the axiology of freedom on a finite planet. Further consideration of some of the nuanced work on absence could also be helpful for such a view of pedagogical/teaching and learning engagement in the zone of proximal development. Bhaskar suggests that we can absent through:

1 *transformation* (e.g. through creating new types of production e.g. zero waste production systems);
2 *distanciation* (e.g. through not putting lead in petrol, or not packing children's food for school in disposable packaging); or
3 *referential detachment* (i.e. through not associating our work with certain interpretations of sustainable development; or through not accepting modern science as the only epistemology to draw on in our educational work).

This adds further definition for the kinds of teaching and learning engagements that could occur within the zone of proximal development in environmental education, if we are to work more explicitly on the absenting of ills in our environmental education praxis. This more refined view of what types of absenting processes we might engage with also raises the question as to *how we engage in the absenting process in and through our educative acts*, and what kinds of engagements are more interesting/important/appropriate to engage with in the zone of proximal development via our pedagogical practices and choices than others. By way of example, one postcolonial author, Bhambra (2007), comments on the absence of postcolonial thinking in the discipline of sociology, but suggests that to absent this absence there is need for an 'engagement with difference that makes a difference to what was initially thought' (p. 880) (i.e. to the structural foundations of sociology). This is not unlike Paraskeva's (2011) argument (drawing on De Sousa Santos (2005) and Žižek (2006)) that focusing on epistemicides in curriculum provides opportunities for a radical, itinerant curriculum process that can allow an understanding of 'how reality can explode in and change the real' (p. 185). This points to very different types of pedagogical engagements to simply including postcolonial or environmental topics or issues into the existing structural *fabrique* of sociology or curriculum. Similarly, Stephen Sterling (2004) differentiated between what he calls a 'bolt on' model of environmental education (where green issues are 'added in'), and a more transformative model of environmental education where the foundations of existing forms of thinking and praxis are challenged and transformed.

As indicated above, Bhaskar (2008) proposes that absenting processes are crucial to dialectic as the logic of change – to the axiology of freedom; hence often more important than the concept of absence are *the processes of absenting.* I am suggesting in 'Absenting as Process: (a), (b) and (c)' in this chapter (drawing on the work of Vygotsky) that this occurs, or can occur, via reflexive forms of teaching and learning in the zone of proximal development where it is possible to propose processes of transformation, spatio-temporal or other forms of distanciation and/or referential detachment in framing possible environmental education responses. The work of De Sousa Santos (2005) suggests that *radical alterity* may also be a potential form of engaging absences. As outlined above, the process of absenting involves both positive identification of the absence (as discussed in some detail above using the waste education example), but also *transformative elimination,* as hinted at above in the discussion on how critical realism allows us both the moment of social critique, but also real engagement with hope, love, radical alterity, and re-imagining in our search for transformative praxis.

In engaging with absenting processes of transformative elimination, it is also important for us to understand how this may be done, as absences, according to Bhaskar (2008), are multi-faceted and can be of different kinds (examples are for illustrative purposes and relate to the waste management example used across this paper):

1 *Transfactual:* the example here is absenting the ill of waste cannot be solved without recourse to changing global power relations and economic models.

2 *Actual:* absenting the ill of waste can be actual or empirical, and can involve constructing a recycling depot.
3 *In-process:* absenting the ill of waste is always in process as it is a gigantic task and remains incomplete for as long as waste continues to be produced.
4 *Static:* the absence of a dustbin in a schoolyard remains absent until the constraints surrounding the supply of dustbins can be absented.
5 *Internally related in a totality:* absences associated with waste problems are part of a wider constellation of socio-historical, cultural, material and economic practices that make up modern forms of living and ways of being and becoming.
6 *Isolated:* an absence of specialized or effective technology to address waste management problems may be isolated to a specific context or setting.
7 *An inaction or not:* the absence of more benign and sustainable/socially just economic paradigms are known about (at least to some extent) but these appear to be inactive.

These provide further insight into possibilities for pedagogically engaging with transformative elimination practices in the zone of proximal development.

Thinking about absence as possibility for co-defining potential zones of proximal development in environmental education, it is also helpful to consider different forms of negation. Hartwig (2007) suggests that most of the negation processes that critical realists are concerned with are 'real negations' as they are focused on some of the main kinds of absences that critical realism is concerned with such as 'lack of food in a belly', or 'lack of truth in a politician' or, in our case example, 'lack of a zero waste society'. He explains that other forms of negation are also possible, which include 'transformative negation' involving the 'transformation or demise of a pre-existing entity or state of affairs' (p. 11). This involves a 'substantial process'. Achieving a zero-waste culture would reflect the outcome of such a transformative negation process. Bhaskar also refers to 'radical negation' which is a process of absenting that results in 'self transformation', and suggests that ultimately for a transformed society in an interdependent world, all change (transformative negation) must tend towards radical negation (Hartwig, 2007, p. 11). Here we might suggest that absenting the absence of love and care for the world and each other would reflect a 'radical negation' of this kind, and when realized in 'embodied agentive agency' reflects what Bhaskar refers to as 'linear negation' (ibid.).

Absenting as process (c): reflexive learning, agency and action

'Sophia acts, and so absents' (Bhaskar, 2008, p. 43) requires reflexive situating and clarification (Norrie, 2010). Sophia acts in what context, and in relation to what kind of absence (transfactual or actual, static or in-process, isolated or internally related in a totality, inaction or not), and given this, how does she absent? What form/s and possibilities are there for her 'embodied *agentive* agency', and what role for learning in this process? From an environmental education processes perspective, the type of learning interactions that Sophia may engage in as she acts and absents can differ radically; different zones of proximal development are possible. There is a need to engage explicitly with Sophia on that which is emancipatory, and how this might be so. To do this, Bhaskar (2008) suggests that there is need to take into account that agentive possibilities are always shaped by various constraints$_2$ (social constraints) especially and more so amongst subjects$_2$ (oppressed people) (Bhaskar, 2008, p. 251) and there is need to differentiate between power$_1$ (agentive power) and power$_2$ (structural power). Social sciences such as environmental education can only really lead to emancipatory praxis if they operate in 'sensitized solidarity' with the 'tacit know-how of the subjects' (Bhaskar, 2008, p. 251), i.e. if they are based on cognitive justice and solidarity. If Sophia were a subaltern subject (Bhaskar's subject$_2$), she would experience a variety of constraints on her freedom, but would have substantive tacit know-how of waste issues, being at the receiving end of the effects of waste producing societies.

How to co-construct possibilities for learning and agentive action in zones of proximal development becomes a critical question for environmental educators. In discussing an emancipatory axiology of social science, Bhaskar suggests that a transition from fact to value, and from theory to practice-in-practice is necessary. Decolonial theorists such as De Sousa Santos (2005) suggest that co-constructing possibilities for learning in zones of proximal development additionally requires decolonizing the mind (W'Thiongo, 1986; Paraskeva, 2011) and this involves a radical critique of the structuring of existing forms of knowledge, their presence and their construction in education settings, as most have been framed in and through colonial intrusions. How one engages with knowledge, skills and motives, and what knowledge is drawn into the possible absenting engagements in zones of proximal development; which skills are considered or become relevant; and how motives and desires shape what constitutes potential pedagogical engagements in possible zones of proximal development are epistemological questions with ontological roots and axiological implications for learning as a passage through the dialectic. De Sousa Santos (2015, p. 76) suggests that this involves a radical redistributive process that must be inclusive of the 'fulfillment of historical and postcolonial justice,... [which is inclusive of environmental justice, and constitutes] ... a third dimension of justice, alongside socioeconomic and cognitive justice'. Paraskeva (2011) suggests that this work is only just beginning in education.

Of importance is the understanding that absenting (i.e. change) can only occur as a 'tensed spatio-temporalising process'; it happens in the world with people present, in simple or more complex time-space configurations (Bhaskar, 2008; Norrie, 2010). Norrie suggests that such a process understanding is central to avoiding the conflation of change and difference. In this context, Bhaskar (2008, pp. 40–41) highlights the importance of understanding reflexivity in learning as passage through the dialectic. He says:

> Dialectic is at the heart of every learning process.... Unless Sophia sees herself necessarily acting and so absenting, she cannot reflexively situate (and hence detotalizes) herself. That is to say, she in practice alienates and reifies, and hence absents herself and/or her agency, in a way for which she cannot consistently account. Not to admit absence to our ontology (in that very admission) is to commit performative contradiction, the basic form of theory/practice and reflexive inconsistency, and self-referential paradox.

If Sophia were in a school where waste management practices are poor, and where environmental educational activities are used for punishment (e.g. Silo, 2011), she may act to absent the litter in her school grounds in response to school punishment regimes, but may have no motive to do so other than to avoid further oppression, thus her act is an act of alienation, rather than transformation. If Sophia were, however, angry about the waste lying around her school and community, and together with her teachers and fellow learners chose to launch a waste management project in the school and community, she would be more reflexively engaged, her act would be an act of transformation (a power from and a power to[9]), not alienation. From these two examples we can see that the statement 'Sophia acts, and so absents' can *be* different within different educational contexts, where pedagogies are differently deployed in ways that enable situating, meaning making deliberations, and the enabling or constraining of reflexive learning engagements in zones of proximal development. How this is done is key to an emancipatory form of environmental education. Thus, it would seem that *how* education and learning processes are constituted is key in responding to the crisis or urgency surrounding the contemporary social-ecological condition.

As Bhaskar suggests (2008, p. 4), ontic change is not only possible, but is a natural necessity, as 'absenting must occur in a world containing human agency'. In tandem with ontic change, epistemic change is possible, meta-epistemic change is possible, and conceptual change is possible, actual change is possible in events, and real change is possible – in structures and system processes (ibid.). Absenting and the dialectic is, as Bhaskar (2008, p. 40) says, the 'great loosener'. It permits an empirical 'open texture', a 'structural fluidity and interconnectedness', which we need to understand as possibility for environmental education teaching, research and learning. As argued in this chapter, such possibilities occur in zones of proximal development that are/can be constructed by teachers and learners in relation to what is at 1M, and what is possible via 2E, 3L and 4D agency. Deepening our understanding of 2E is essential for the dialectic to emerge between 1M and in relation to 3L and 4D; it is essential for an emancipatory axiology and for learning as passage through the dialectic.

Education, as explained by Vygotsky (1978), is critical to the ontogenesis of culture. His work highlighted that education, in this case environmental education, can potentially enable someone to become what they are not yet (i.e. becoming). Neocosmos (2015, p. 20, my emphasis), writing about decolonization politics in the global South, suggests that:

> If any teaching is to be done it should emphasize *the struggles for an equal and just society and a dignified life*.... One cannot search for emancipatory inspiration in past or present idealized cultures, but *only in the exceeding of culture through the contradictions which it itself engenders.*

In this sense, the zone of proximal development is important to consider since it is a dynamic region of sensitivity in which the transition from interpsychological functioning to intrapsychological functioning can be made. Educational researchers suggest that engaging contradictions in zones of proximal development allows for generativity that potentially leads to transformed human activity via learning that leads also to transformative agency (Engeström and Sannino, 2010). This requires an in-depth understanding of the historically specific context (ibid.). Griffin and Cole (1984) suggest that in order to understand interaction in the zone of proximal development, one must identify the 'leading activities' that characterize various phases of ontogenesis. They qualify this as follows: 'as an alternative to internal, individual stage approaches to development, *leading activities provide for a notion of societally provided progressions,* the sort of context-selection mechanisms that we have considered important for understanding development' (1984, p. 51, my emphasis). The types of learning activities that Griffin and Cole have in mind are play, formal learning and work. The types of learning activities suggested in this chapter also include learning in communities and via everyday practices (see also Chapters 11, 12 and 13 in this book).

Conclusion

I conclude this chapter by suggesting that working carefully, critically, democratically and dialogically (with a commitment to radical alterity) with absenting absences in environmental education is key to reflexive learning as passage through the dialectic, change and the axiology of freedom. It can potentially enhance thinking on the possibilities for co-constructing zones of proximal development and thus the way in which learning can potentially lead transformative praxis or 'development' in Vygotskian terms. As Wertsch (1985, p. 75) suggested, 'the zone of proximal development provides a point where the ontogentic and sociohistorical domains may be examined in interaction'. As shown here, there is much work to be done to consider the possibilities for this interaction, more so to do this in a way that is not technically deterministic, but dialogically and reflexively co-constructed *with* learners in ways that are not irrealist, and that recognize possibilities for 'embodied agentive agency' in emerging zones of proximal development. Educational processes oriented to the absenting of absences via processes of embodied, agentive agency would seem to be required. Bhaskar (2008, p. 261) describes this in terms of a 'theory and practice of transition', that includes 'the sensitized solidarity' referred to above. In doing this, he suggests a need to differentiate in our work between instrumental, critical, explanatory critical, depth explanatory critical, totalizing depth explanatory critical, and dialectical rationality processes, appended by 'geo-historical directionality', indicating the value implicational character of social science, and by implication environmental education.

As discussed in this chapter, absenting absence is a tensed spatio-temporalizing participatory process that recognizes that becoming involves engaging in a process of change as 'being in motion', involving continuing tensed processes of reflection-action-reflection, expanding knowledge and possibilities. Change-oriented or transformative environmental learning in this sense is seen as relational reflexive movements of *transformative elimination* of absences or ills in and through learning processes.

Perhaps critical realism and a focus on absenting absences has the power to offer environmental education teachers, learners and researchers ways of thinking the world together, rather than 'thinking the world apart', and through this, see their absenting practices not as just a small action that is hard to validate or as inconsequential, but rather as part of a much bigger story in which all absenting actions are potentially processes of radical alterity that also enable love and care for others, non-humans and the wider world.

Acknowledgements

I would like to thank Leigh Price for her careful and thoughtful rounds of comment on this chapter as it emerged over the period of a year or more.

Notes

1 I use 'ourselves', 'we' and 'us' in this chapter to indicate that we are all implicated in the social-ecological crises of our times, as has been shown by recent scientific studies such as the IPCC (2014) reports on climate change.

2 I use the concept of normalized as used by Foucault (1975) in *Discipline and punish: The birth of the prison.*

3 Almost any other environmental concern (e.g. biodiversity loss, climate change, water pollution etc.) could be used as a case example, this one was chosen for its ubiquity, and practical resonance across countries, educational institutions and cultural contexts. Waste is also one of the 'iconic' concerns of the global environmental movement and examples of environmental education practice focusing on waste management and reduction can be found in countries around the world – from Botswana to China, India, Columbia, USA, Nigeria, Australia.

4 Bhaskar (2008, p. 249) suggests that an emancipatory 'normative order would be based on trust, solidarity and care, if not indeed love'.

5 For example, see the IUCN campaign 'Love, Not Loss'. www.iucn.org/about/union/commissions/cec/cec_how_we_work/love__not_loss_/ (accessed 9 May 2015).

6 Or a culture of social and cognitive justice, or food security, or water security, equity etc.

7 Bhaskar says that this knowledge could be tacit, propositional, practical and/or discursive (ibid.).

8 De Souza Santos (2005) describes such practices as 'epistemicides', when he states that 'in the name of modern science, many alternative knowledges and sciences have been destroyed, and the groups that used these systems to support their own autonomous path of development have been humiliated' (p. xviii).

9 Bhaskar (2008, p. 260) suggests that absenting constraints involves 'freedom *from* and freedom *to*'.

References

Archer, M. (2003). *Structure, agency and the internal conversation.* Cambridge: Cambridge University Press.

Andreotti, V. (2011). *Actionable postcolonial theory in education.* New York: Palgrave Macmillan.

Bhambra, G. (2007). Sociology and postcolonialism: Another 'missing' revolution? *Sociology, 41*(5), Special Issue on Sociology and its Public Face(s), 871–884.

Bhaskar, R. (1994). *Plato, etc.: The problems of philosophy and their resolution.* London: Verso.

Bhaskar, R. (1997). On the ontological status of ideas. *Journal for the Theory of Social Behaviour, 27*(2/3), 135–147.

Bhaskar, R. (2008). *Dialectic: The pulse of freedom.* London: Routledge.

Bhaskar, R. (2010). Contexts of interdisciplinarity. In C. Frank, K. G. Høyer, P. Naess and J. Parker (Eds), *Interdisciplinarity and climate change: Transforming knowledge and practice for our global future* (pp. 9–10). London: Routledge.

Bhaskar, R., and Danermark, B. (2006). Metatheory, interdisciplinarity and disability research: A critical realist perspective. *Scandinavian Journal of Disability Research, 8*(4), 278–297.

Bond, P., and Guliwe, T. (2003). Contesting sustainable development: South African civil society critiques and advocacy. In G. Mhone and O. Edigheji (Eds), *Governance in the new South Africa: The challenges of globalisation* (pp. 313–345). Cape Town: University of Cape Town Press.

Collier, A. (2001). On real and nominal absences. In J. Lopez and G. Potter (Eds), *After postmodernism: An introduction to critical realism* (pp. 299–310). London: Athlone Press.

De Sousa Santos, B. (2005). *Democratising democracy: Beyond the liberal democratic canon.* London: Verso.

De Sousa Santos, B. (Ed.). (2007). *Another knowledge is possible.* London: Verso.

De Sousa Santos, B. (2009) [2014]. *Epistemilogicas do Sul.* Coimbra: Almendina [Translated into English as *Epistemologies of the South: Justice against Epistemicide*]. London: Verso.

De Sousa Santos, B. (2015). *If God were a human rights activist.* Stanford: Stanford University Press.

Engeström, Y., and Sannino, A. (2010). Studies of expansive learning: Foundations, findings and future challenges. *Educational Research Review.* doi:10.1016/j.edurv.200212.002.

Griffin, P., and Cole, M. (1984). Current activity for the future: The zo-ped. In B. Rogoff and J. V. Wertsch (Eds), *New directions for child development: No. 23. Children's learning in the 'zone of proximal development'.* San Francisco: Jossey Bass.

Harré, R. (2006). *Key thinkers in psychology.* London: Sage.

Hartwig, M. (Ed.). (2007). *Dictionary of critical realism.* London: Routledge.

Hartwig, M. (2008). Introduction. In R. Bhaskar, *Dialectic: The pulse of freedom.* London: Routledge.

IPCC (Intergovernmental Panel on Climate Change). (2014). *Climate Change 2014: Impacts, adaptation, vulnerability.* Cambridge: Cambridge University Press.

Jickling, B. (1992). Why I don't want my children to be educated for sustainable development. *Journal of Environmental Education, 23*(4), 5–8.

Foucault, M. (1975). *Discipline and punish: The birth of the prison.* New York: Random House.

Leff, E. (2009). *Latin American environmental thought.* Paper presented at the 6th Latin American Congress of Environmental Education. San Clemente de Tuyu, Argentina, 19 September 2009.

Martínez-Alier, J. (2002). *The environmentalism of the poor: A study of ecological conflicts and valuation.* Cheltenham: Edward Elgar Publishing.

Martínez-Alier, J. (2012). Environmental justice and economic degrowth: An alliance between two movements. *Capitalism Nature Socialism, 23*(1), 51–73.

Neocosmos, M. (2015). *Thinking African politics in an era of increasing violence.* Paper presented at UHURU Seminar, Rhodes University, 15 April 2015.

Norrie, A. (2010). *Dialectic and difference: Dialectical critical realism and the grounds of justice.* New York: Routledge.

Nunez, I. (2013). Transcending the dualisms of activity theory. *Journal of Critical Realism, 12*(2),141–165.

O'Donoghue, R., Shava, S., and Zazu, C. (2013). *African heritage knowledge in the context of social innovation: Learning contributions of the Regional Centres of Expertise on Education for Sustainable Development.* Tokyo: UNU Institute of Advanced Studies.

Paraskeva, J. (2011). *Conflicts in curriculum theory: Challenging hegemonic epistemologies.* New York: Palgrave MacMillan.

Silo, N. (2011). *Exploring opportunities for action competence development through learners' participation in waste management activities in selected primary schools in Botswana.* Unpublished doctoral study, Rhodes University, Grahamstown.

Sterling, S. (2004). Higher education, sustainability and the role of systemic learning. In P. Blaze Corcoran and A. E. J. Wals (Eds), *Higher education and the challenge of sustainability: Problematics, promise and practice* (pp. 49–70). Dordrecht: Kluwer.

UNESCO (United Nations Educational, Scientific and Cultural Organization). (2014). *Shaping the future we want: UN Decade of Education for Sustainable Development (2005–2014). Final Report.* Paris.

Vandenbergh, F. (2014). *What's critical about critical realism? Essays in reconstructive theory*. London: Routledge.

Vygotsky, L. S. (1978). *Mind in society*. Cambridge, MA: Harvard University Press.

Wertsch, J. V. (1985). *Vygotsky and the social formation of mind.* Cambridge, MA: Harvard University Press.

W'Thiongo, N. (1986). *Decolonising the mind: The politics of language in African literature.* London: James Currey.

W'Thiongo, N. (1993). *Moving the centre: The struggle for cultural freedoms.* London: Heinemann.

18 Some implications of metaReality for environmental educators

Leigh Price

Introduction

In this chapter I explore how Bhaskar's metaReality might be useful for environmental educators because it allows us, without contradiction and despite our religion or lack of religion, to make use of the insights and ethics of certain spiritual traditions associated with the environmental movement, such as indigenous knowledge, ecofeminism and the transcendentalism (often inspired by Christianity) of the early American environmental conservationists, such as Henry David Thoreau (1817–1862), Mary Austin (1868–1934) and Aldo Leopold (1887–1948).[1] Such meta-theoretical unity will, I believe, improve our environmental education practices and potentially lead to a world of flourishing beings.

Bhaskar's critical realism and 'faith'

A common theme found in all three phases of Roy Bhaskar's work[2] is that we have 'faith', where faith is defined as believing in things unseen (Hebrews 11:1) (Bhaskar, 2008, p. 73). This is because, for Bhaskar, reality is layered and only a small part of it can be measured (is empirical). We can only know about the non-empirical aspects of reality by supposition. Philosophically speaking, such faith is known as fallibilism. Take the example of quarks. As Christine Davies, a theorist at the University of Glasgow in the United Kingdom put it: 'You can't just put a quark on a scale and say, "This is what its mass is".' Scientists have thus long been aware that not everything that exists can be directly measured. However, Bhaskar's critical realism goes further than this because it suggests that some things are also real even if they are not measurable because they are immaterial or as yet unmanifest. Therefore absences are real, such as the (presently absent) past, risks and unexpressed tendencies. Relationships, which in society take the form of structures and mechanisms, are real. Indeed, they would be real even if they were unexpressed. For example, there may be a real tendency towards racism in a society, but the racism may be unexpressed due to legal constraints and careful policing. Empiricist science's materialism and lack of an understanding of the layered nature of reality has some unfortunate consequences. One consequence is that scientists who choose to study aspects of reality which fall outside the realms of the empirical can be marginalized. For example, historians, psychoanalysts and social scientists have often been belittled as pseudoscientists. Tesh (1988, p. 169) subscribed to this marginalization for social scientists when she stated:

> [R]esearch that takes the social structure as the unit of analysis gets pushed to the periphery of science. At most it is a lesser kind of science – social science. At worst it is not science at all but a pseudoscience contaminated with politics.

Similarly, many academics feel that Bhaskar's 'metaReality' should be avoided, since it is associated with faith not only in social science, psychoanalysis and history but also in the world's spiritual traditions (Creaven, 2014). It seems likely that these academics are atheists.[3] However, as an atheist myself, I have not found metaReality to require me to 'suspend my disbelief'[4] in order to engage with it. Bhaskar's position, whilst it is not exclusive of theism, also leaves space for atheists by advocating an immanent spirituality.[5] Indeed, possibly to ensure that he could include all religious beliefs, including atheism, Bhaskar's later work does not insist on the existence of a godhead or any kind of humanoid transcendent. He leaves it up to the reader to decide whether they want to add this possibly extra aspect to their understanding. In Bhaskar's first book of his spiritual turn, he referred to God with a capital 'G' (Bhaskar, 2000). However, in his later work, Bhaskar's preferred word for this aspect of reality is not 'God' but 'god' or even 'the cosmic envelop'. In one example, he describes the realm of non-duality (oneness) as normally being thought of as 'a sort of thing which is to do with god and ultimata or bliss or peak experiences, such as one might have during a moment of scientific or artistic creativity or while listening to music or in prayer or worship, etc.' (Bhaskar, 2002, p. xii). Thus Bhaskar's spirituality is based on an awe-filled awareness of the interconnectedness of all things and is associated with creativity. If we manage to achieve awareness of our unity then Bhaskar would say that we have achieved 'god-realisation'. Nevertheless, he admits that 'This does however raise a problem of how we conceptualise the goal of unity, in one's ground-state, with the totality or the cosmic whole, or its ground, which I have called "god-realisation".' Note his quotation marks. A rose by any other name is still a rose. Awareness of our interconnectedness, whether we call this 'god-realisation' or not, is still awareness of our interconnectedness. Or alternately, 'God-realisation, whether we call it "awareness of interconnectedness or not", is still God-realisation'.

Ironically, it is not unusual for atheists to hold such transcendental views. For example, the famous atheist Daniel Dennett (2007, p. 245) declared that 'The world is sacred.' Even Richard Dawkins (2007, p. 14) explained that he was not attacking the pantheistic reverence for the cosmos which he admitted he shared with such distinguished exponents as Albert Einstein. For Dawkins (ibid., p. 40), the pantheist's God is a poetic synonym for the laws of the universe and 'Pantheism is sexed-up atheism.' I would suggest that it might be helpful to say that 'Panentheism is transcendentalized atheism', that is, it is a kind of atheism that assumes the existence of the level of the real and not just the actual and the empirical. Note that I prefer to talk about panentheism, not pantheism. Pantheism makes the rather large claim that everything in the universe is a full expression of God/the cosmic envelope. Panentheism suggests that all beings are an expression of, some might say God/the cosmic envelope, but do not exhaust it (Agar, 2013).

One might ask why Bhaskar chose to refer to god when it was not necessary for his argument and it has threatened his reputation in academia. Perhaps there was method in his madness. First, the message that Bhaskar wanted to share cannot but sound spiritual. To purposively make it sound non-spiritual would ultimately makes it seem contrived. The ancient mystics and many of our non-Western ethnic religions arrived at spiritual tenets after intense study of human nature and our relationship with the universe. For Bhaskar to make his insights sound original would potentially have left him open to accusations of plagiarism. However, this is not to detract from Bhaskar's original contribution, which was to formulate a useful philosophical ontological and epistemological language for better understanding what has perhaps been called god. Thus, one can perhaps justify why it was necessary for Bhaskar to take the path that he did, when a less controversial packaging of his work would have been easier for some to accept.

Another reason Bhaskar perhaps began exploring the world's great spiritual traditions could also be a practical one. If I am wandering along, trying to solve a problem and start to get an inkling of the solution, but I find that someone else has already spent a long time working on that solution and can teach me about it, it seems wise to save myself time and effort by taking advantage of the work already done. If Bhaskar had tried to keep his work separate from the spiritual traditions, it would have been inefficient. In the end, his engagement with transcendental realism and its expression in the spiritual traditions has potentially benefited us all and made the wisdom of the mystical, metaphysical traditions accessible to everyone, even atheists such as myself.

Why might Bhaskar's metaReality be useful?

A major reason Bhaskar's metaReality may be useful is because it seems likely that metaphysical thinking is in fact both unavoidable and vital to human survival. Research into how the brain functions suggests that human beings are hard-wired for transcendental thinking due to its evolutionary advantages (Boyer, 2002). A key advantage of metaphysical thinking includes, amongst other things, its ability to detect environmental patterns, which critical realists would say is a retroductive[6] step in the process of detecting non-empirical structures and mechanisms that explain events, the critical realist level of the transcendentally real. Thus the same kind of thinking that we associate with religion is also involved in detection and representation of animacy and agency, social exchange, moral intuitions, precaution against natural hazards and understanding of misfortune (Boyer, 2003, p. 119). Essentially, metaphysical thinking is present even in atheists.

However, transcendental thought has developed a bad reputation in the academy. Indeed, Jürgen Habermas (1988) wrote a book entitled *Post-metaphysical Thinking: Philosophical Essays.* The stand against metaphysical or transcendental thinking makes sense if we lack a concept of an emergent and layered ontology, since it requires a strict and damaging dichotomy between body and the transcendent mind. In order to avoid this untenable dichotomy, many philosophers tried to reduce truth to either the empirical body (such as David Hume)[7] or to the transcendent mind (such as Emmanuel Kant). Critical realism does not require this dichotomy since it assumes that the transcendent levels are emergent from the empirical levels. In other words, structures and mechanisms are emergent from empirical events; the present is emergent from the absent past; and the mind is emergent from the body. However, to achieve this, critical realism has had to change our scientific standards by insisting that the non-empirical can also be real. Indeed, critical realism was originally called transcendental realism (Bhaskar, 1975).

Being an advocate of such transcendentalism can be frustrating in that it results in being misunderstood by contemporary mainstream scientists. The environmentalist Thoreau (5 March 1853, p. 469) was frustrated in a similar way:

> I felt that it would be to make myself the laughing-stock of the scientific community to describe to them that branch of science which specially interests me, inasmuch as they do not believe in a science which deals with the higher law. So I was obliged to speak to their condition and describe to them that poor part of me which alone they can understand. The fact is I am a mystic, a transcendentalist, and a natural philosopher to boot. Now that I think of it, I should have told them at once that I was a transcendentalist. That would have been the shortest way of telling them that they would not understand my explanations.

However, if Thoreau had had access to critical realism he would not have needed to assume the impossibility of cross-disciplinary understanding. For critical realists, higher laws (at the level of the real) are provided by a metaphysics which is 'a conceptual science'. It simply clarifies ontological presumptions within the sciences and connects the real with the empirical via the process of emergence. It nevertheless does not support metaphysics as 'a dialogue-stopping claim to have a privileged access to Being' (Cruickshank in Hartwig, 2007, p. 181) and thus it does not return us to original versions of metaphysics which were de-coupled from the actual and material. Another environmentalist, Leopold (2001 [1949]), provides us with an example of a spirituality that appears to be based on conceptual science which allows us to better understand the structures and mechanisms behind empirical events. For example, in *A Sand County Almanac* (p. 29), he explained: 'There are two spiritual dangers in not owning a farm. One is the danger of supposing that breakfast comes from the grocery, and the other that heat comes from a furnace.' Here we see him equating a lack of spirituality with a lack of understanding the real structures and mechanisms (the scientific concepts) behind food and energy production. The environmentalist Mary Austen (1903) also discussed the connection between the successful study of scientific principles, and spiritual sensibilities. She wrote:

> The first effect of cloud study is a sense of presence and intention in storm processes. Weather does not happen. It is the visible manifestation of the Spirit moving itself in the void. It gathers itself together under the heavens; rains, snows, yearns mightily in wind, smiles; and the Weather Bureau, situated advantageously for that very business, taps the record on his instruments and going out on the streets denies his God, not having gathered the sense of what he has seen. Hardly anybody takes account of the fact that John Muir, who knows more of mountain storms than any other, is a devout man.
>
> (pp. 246–247)

Austen here suggests that stopping simply at the level of the empirical (the records on one's instruments) provides a shallow understanding of events such as weather. Thus we need an in-depth understanding; I would say that we need an understanding of structures and mechanisms leading to the weather at the level of the real. Referring back to my earlier passage about the evolutionary usefulness of transcendental thinking, and interpreting Austen's words somewhat poetically, I argue that Austen's 'weather' exists at the level of the empirical, and is the manifestation of transcendental structures and mechanisms (her 'storm processes') which exist at the level of the real (which she calls 'the void' – note her concept of the void is that of a causally efficacious absent thing).[8] Given our contemporary understanding of transcendental thinking as being both that associated with religion and that associated with detecting patterns, it is not surprising that she links being devout with seeing these patterns. Furthermore, some research has shown that meditation (which we might associate with being devout) has been shown to increase our ability to be intuitive through increasing the brain's conscious access to subliminal messages and understandings (Zeidan *et al.*, 2010; Strick *et al.*, 2012).

The metaphysical and indigenous axiology

Nevertheless, despite its common-sense appeal, a pervasive lack of understanding about the metaphysical, which the environmentalist Thoreau feared would make him a laughing stock is, I argue, a reason for the glass ceiling that limits the full utilization of indigenous knowledge in an environmental education context (Price, 2005). Spirituality is also one of the ways that ecofeminism was critiqued and possibly explains, at least in part, the unfortunate demise of ecofeminism[9] as a respectable academic discipline and a leader of ecological activist movements.[10]

In terms of indigenous knowledge, not only is this glass ceiling effectively racist in that it ultimately assumes the superiority of Western knowledge, but it also prevents us from giving credence to, and thus making use of, indigenous ethics or ways of interacting with each other and the environment (O'Donoghue, Chapter 9, this book). As an environmental educator, I have wished for justifiable axiological access to the Earth ethic that we find amongst many indigenous and spiritual people. It seemed like a sad irony that the more ethically scientific I was, the less access that I had to an environmental ethic (like Austen's Weather Bureau that denies God, but loses touch with natural phenomena at the same time). It appeared to be impossible to translate the environmental ethic of spiritual people into an Earth ethic for non-believing Westerners. This was because it seemed to require suspension of disbelief, that is, it seemed to require some kind of acceptance of the existence of a transcendental God (of the conversation-stopping sort), or gods, or ancestral spirits or pagan magic.

However, equipped with my new commitment to immanent spirituality, I revisited the indigenous knowledge stories. I realized that the epistemology of Western scientists was differently applicable to different levels of reality as described or explained by indigenous knowledge. Essentially, indigenous knowledge that relates to transcendental structures and mechanisms relevant to natural science is fairly amenable to acceptance by Western scientists, although even here there has been a battle for acceptance (O'Donoghue, Chapter 9, this book; Sabai, Chapter 10, this book). So, when a traditional person places the seeds for next year's crop above the wood stove, storing them there to keep them under the care of the hearth gods, Western scientists might understand that the protection for the seeds comes from the transcendental mechanism by which smoke keeps away weevils. However, indigenous knowledge that relates to transcendental structures and mechanisms relevant to social science are generally ignored. This is linked to what I pointed out earlier, which is that mainstream scientists are fairly happy to accept mechanisms and structures which relate to natural science (which we can or at least may one day be able to measure), but not so happy to accept the existence of structures and mechanisms that we will never be able to measure empirically, such as absences or social structures. Nevertheless, it is the non-empirical, psycho/socio/spiritual aspects of indigenous knowledge that is behind the motivation of adherents' ethic of care for the environment, a motivation that elsewhere seems elusive and whose absence is often assumed to result in the well-acknowledged, problematic gap between knowledge and action in environmental education (Kollmuss and Agyeman, 2002).

In other words, for explanations of actions relating to physical mechanisms, there was need for only one layer of suspension of disbelief amongst non-believing Westerners. Once the mainstream scientists stopped dismissing the indigenous knowledge outright as mere superstition, they could assume that only the words describing the referent were different, but that the referent itself (for example, the physical mechanism that makes smoke repel weevils) was mutually present for both indigenous and non-indigenous parties. However, for the understanding of actions pertaining to society, such as ethical action, that is the indigenous people's axiology, there was a double requirement for the suspension of disbelief, since mainstream science denies both (a) the words describing the referent, that is the spiritual explanations given for actions to promote social structures leading to a flourishing society, as well as (b) the referent itself, that is the very existence of social structures. Thus the transcendental truths about society – such as that human beings are likely to be motivated to protect the environment out of the need to care for fellow beings based on an awareness of our connectedness (God-realization) – are not easily assimilated by mainstream science. Currently, it is as if we cannot 'see' these structures because we cannot measure them directly. These structures and mechanisms are the same ones that indigenous knowledge and many of the spiritual traditions refer to when teaching us how to look after our society and the environment through various forms of compassion and reciprocity. I am not saying that critical realism will get both Westerners and non-Westerners speaking the same words about referents, but I am saying that when it comes to social structures, it can make it possible that they agree that a referent exists. Critical realism is thus able to return the Western scientist to just one layer of disbelief when it comes to questions of action to improve the functioning of society. Furthermore, critical realism reduces the seeming gulf between the different ways of explaining the referent (perhaps changes the necessity for disbelief at all) by acknowledging epistemic relativism. Thus the referent is assumed to exist but there are always different ways of describing the referent. Nevertheless we can choose between better and worse explanations of that referent depending on what we want to achieve. Furthermore, it is not always necessary to have exactly the same words to understand a phenomenon to arrive at similar axiological conclusions. This is similar to the argument that I make about how it is not that important whether we use the term 'god realisation' or 'awareness of our connection' in reference to 'whatever it is that makes us speak of it in these terms' (i.e. the referent): what is important is that our words lead us to act in ways that are commensurable with (thus relatively true to) the referent and thus cause us to act in accordance with the natural laws associated with that referent. To act out of line with those laws would be detrimental to our well-being. For example, if someone convinced me that hearth gods were imaginary, and therefore I stopped putting my seeds above the hearth in winter, come summer, I would wake to find my seeds filled with weevils.

It gets a bit complicated here because it might seem that I am arguing for a kind of instrumentalism, whereby we treat human beings like non-living things as I am contending that scientists treat the transcendental structures and mechanisms of natural science the same way as the structures and mechanisms of social science, and for the same purpose – to achieve a sound way of determining action (or in the case of humans, ethical action). However, my position is light years away from a heartless, meaningless instrumentalism that is based on a Humean view of the world as disjointed and disconnected, where empirical constant conjunctions are the only measure of causality and there is no place for non-empirical reality, such as values. Instead, I am suggesting that through the existence of a depth ontology, the world is intrinsically meaningful and the way to knowing our connectedness is through our compassion and empathy,[11] facilitated by our reason, which provides us with transcendental understanding. So, when a plant releases oxygen, it *means* that certain transcendental structures and mechanisms called photosynthesis are likely to have occurred. When humans act to protect the environment, it *means* that certain socio/spiritual/psychological structures and mechanisms, such as compassion towards future generations, are likely to have occurred. If I want oxygen, I need to provide the conditions for photosynthesis to occur; if I want a healthy environment, I need to provide the conditions for such things as compassion for future generations to occur.

Spiritual awareness as a basis for axiology in environmental education

I am not the only environmental educator writing about spiritual awareness as a basis for axiology. In fact, it seems that I may be part of a majority (see also Lupele, Chapter 4, this book). Fraser *et al.* (2014) identified five general discourses in environmental educator language relating to the purpose of environmental education. The most popular discourse, elements of which were to be found often co-existing with other discourses, was one that they labelled 'spiritual instrumentalism'. They placed the following sort of statements into this category (ibid., p. 7):

- It's using nature to reconnect people to their spiritual centre.
- Teaching that nature has the same rights as people.
- Promotes the understanding that everything is connected to everything in nature.
- Nature education is the path to help us appreciate God through His work.

Thus Fraser *et al.* (2014) take a similar position to the one that I take in this chapter, which is that not all spiritual statements associated with the purpose of environmental education invoke God with a capital 'g'. Whilst I am loathe to use the word instrumentalism, because I associate it with some of the worst abuses of epistemologically-won power, I nevertheless concede that ultimately environmental educators aim to act to transform the world. My guess is that this approach, which acknowledges spirituality or awareness of interconnectedness, will result in greater appreciation of the need for emotional attachments to people and places to motivate action, resulting in something of a turning back to early environmental education practices which valued activities that developed a personal relationship with nature. Such nature-experience environmental education (providing opportunities for becoming aware of our interconnectedness – or 'God-realization') was somewhat discredited (Russell, 1999) as not paying enough attention to the other forces of the environmental crisis, such as socio-economic factors. However, research has shown that those people who have had significant nature experiences especially in their formative years are more likely to exhibit environmentally sound behaviour (Palmer and Suggate, 1996; Bögeholz, 2006). Possibly, it is exactly this 'awareness of our interconnectedness' that allows us to fully understand how socio-economics are connected to environmental issues.

Conclusion

Based on the above, this chapter suggests that environmental education can benefit from engaging with critical realist metaReality in a number of ways. First, by advocating an immanent spirituality, it allows even atheists to have access to transcendental ethics, where that transcendentalism is based on a conception of reality as being layered such that real structures are emergent from empirical structures (for example, mind is emergent from body or society is emergent from individuals). This does not mean a return to naïve transcendentalisms that assumed a damaging mind/body split and are associated with fundamentalist religious stands. Rather it is simply a transcendentalism that assumes that we are all interconnected at the level of the transcendental real. Second, metaReality allows academic environmental educators (who, statistically speaking, are likely to be atheists) to engage more honestly with ethical positions of religious/spiritual people, for example indigenous peoples, ecofeminists, adherents of any of the major world religions and the early environmentalists of the USA. This is because it allows us to assume a common referent (its ontological reality) even if there are differences in the language that we use to describe that referent (epistemological relativism). This position is not going to prevent disagreements, but at least it will allow conversations to take place since we can agree that we are talking about the same thing. This is similar to the position taken by Peircean scholar, Chiasson (1999), who suggests that perhaps there is simply a language-based misunderstanding between what one person might call a hopeful vision that must be connected to a belief in God, but which another might be willing to agree is a form of long-term optimism. Finally, metaReality may be able to encourage greater creative theorizing in terms of explanations for our situations and thus greater efficaciousness in dealing with them. This is because this version of spirituality is similar to the version invoked by the environmentalists Leopold and Austen, which is associated with the way that the brain understands complex causal mechanisms at the level of the real, not just the empirical.

Notes

1 I refer to these religious or transcendental views because I have a little knowledge of them. No doubt similar positions are to be found in other religions too.
2 That is, common to basic critical realism, dialectical critical realism and metaReality.
3 According to Ecklund and Scheitle (2007, p. 287), 52 per cent of US academics admit to having no religion compared to 14 per cent of the general population.
4 The term 'willing suspension of disbelief' was coined by Samuel Coleridge (1817) and refers to the way that, in order to engage with a story, it is sometimes necessary to suspend one's judgement concerning the implausibility of the narrative.
5 One can without contradiction be a critical realist and an adherent of a religion. For example, Matthew Wilkinson (2013) has argued for a critical realist Islam, the former Archbishop of Canterbury on occasion quotes Roy Bhaskar (see Williams, 23 November 2006) and in 2014 well-known critical realist Professor Lady Margaret Archer was named by Pope Francis as President of the Pontifical Academy of Social Sciences (Lenartowick, 2014).
6 Charles Sanders Peirce coined the term 'retroduction' and defined it as follows: 'Retroduction, or Hypothetic Inference,... depends on our hope, sooner or later, to guess at the conditions under which a given kind of phenomenon will present itself' (Peirce, 1997 [1913], pp. 385–388). There are strong parallels between critical realism and the pragmaticism of Peirce. Peirce (1908) also connected retroductive thinking to spirituality.
7 Empiricists such as Hume were often motivated by a reaction against the political tyranny associated with disembodied (non-emergent) transcendentalism, since its understanding of 'truth' was unfettered by the need to be accountable to empirical evidence.
8 Bhaskar (2008 [1993], p. 39) talks about 'causally efficacious absent things'.
9 Another reason for the demise of ecofeminism is the accusation that it was essentialist (Gaard, 2011); critical realism potentially redeems ecofeminism

from this charge, too, with its understanding of the concrete universal and concrete singular. Provided one's theoretical understanding is balanced by the presence of the unique, concrete singular, strongly influenced by specific mediations, it is possible to talk in general terms without being essentialist. Therefore, at the level of the empirical, each woman is a unique being, powerfully influenced by the specificities of her life history, yet at the level of the real she is part of, and shares the characteristics of, the category 'women' (although if certain of these characteristics are not expressed in her this does not demonstrate the lack of truth of the category 'women' – it rather demonstrates that the woman exists in an open system context).

10 According to Google Trends, in 2014 the term 'ecofeminism' was searched 92 per cent less often than in 2004.

11 I would argue that the scientists who first discovered/described photosynthesis would have been exercising empathy as they tried to understand the activity of the relevant molecules – the intellectual activity that goes into asking why the water molecule joined up with the carbon dioxide molecule is not much different to asking why Mary got together with Tony; in both situations we put ourselves in the position of the person or the molecule and imagine what might get them to act as they did. Anecdotally, scientists often describe their love for their objects of study. Bhaskar (2002, p. 176) has explained that the emotions are squeezed out between the terms of the mind-body dualism. He also explained that 'The emotions mediate all agency-action on the world and all perception-reaction to the world'. Negatively, emotions can block awareness, experience and agency but positively they facilitate and accompany self-expansion. 'Nothing can be done or felt or experienced without emotions' (Bhaskar, 2002, p. 176).

References

Agar, J. (2013). *Post-secularism, realism and utopia: Transcendence and immanence from Hegel to Bloch.* London: Routledge.

Austin, M. (1903). *The land of little rain.* Boston: Houghton, Mifflin. Retrieved 17 May 2015, from http://memory.loc.gov/cgi-bin/query/r?ammem/calbkbib:@field%28NUMBER+@band%28calbk+vg17%29%29.

Bhaskar, R. (1975). *A realist theory of science.* Leeds: Leeds Books.

Bhaskar, R. (2000). *From East to West: Odyssey of a soul.* London and New York: Routledge.

Bhaskar, R. (2002). *Meta-reality: The philosophy of meta-reality. Volume 1: Creativity, love and freedom.* London: Sage.

Bhaskar, R. (2008) [1993]. *Dialectic: The pulse of freedom.* London and New York: Routledge.

Bloom, H. (2007). *Henry David Thoreau.* New York: Infobase Publishing.

Bögeholz, S. (2006). Nature experience and its importance for environmental knowledge, values and action: Recent German empirical contributions. *Environmental Education Research, 12*(1), 65–84.

Boyer, P. (2002). *Religion explained: The evolutionary origins of religious thought* (Reprint edition). New York: Basic Books.

Chiasson, P. (1999). *Revisiting a neglected argument for the reality of God.* Retrieved 17 May 2015, from www.iupui.edu/~arisbe/menu/library/aboutcsp/chiasson/revisit.htm.

Coleridge, S. T. (1817). *Biographia literaria; or, Biographical sketches of my literary life and opinions*. London: R. Fenner. Retrieved 17 May 2015, from http://archive.org/details/biographialitera00colerich.

Creaven, S. (2014). *Against the spiritual turn: Marxism, realism, and critical theory*. London: Routledge.

Dawkins, R. (2007). *The God delusion* (New edn with additions). London: Black Swan.

Dennett, D. C. (2007). *Breaking the spell: Religion as a natural phenomenon* (Reprint edition). London: Penguin.

Ecklund, E. H., and Scheitle, C. P. (2007). Religion among academic scientists: Distinctions, disciplines, and demographics. *Social Problems, 54*(2), 289–307.

Fraser, J., Gupta, R., and Krasny, M. E. (2014). Practitioners' perspectives on the purpose of environmental education. *Environmental Education Research, 20*(1), 1–24.

Gaard, G. (2011). Ecofeminism revisited: Rejecting essentialism and re-placing species in a material feminist environmentalism. *Feminist Formations, 23*(2), 26–53.

Habermas, J. (1988). *Postmetaphysical thinking: Philosophical essays*. Massachusetts: MIT Press.

Hartwig, M. (2007). *Dictionary of critical realism*. London: Routledge.

Kollmuss, A., and Agyeman, J. (2002). Mind the gap: Why do people act environmentally and what are the barriers to pro-environmental behavior? *Environmental Education Research, 8*(3), 239–260.

Lenartowick, K. (2014). British woman appointed by Pope to head Pontifical Academy. *Catholic News Agency*. Retrieved 17 May 2015, from www.catholicnewsagency.com/news/british-woman-appointed-by-pope-to-head-pontifical-academy.

Leopold, A. (2001) [1949]. *A Sand County almanac*. Oxford: Oxford University Press.

Palmer, J. A., and Suggate, J. (1996). Influences and experiences affecting the pro-environmental behaviour of educators. *Environmental Education Research, 2*(1), 109–121.

Pierce, C. S. (1908). A Neglected Argument for the Reality of God. *Hibbert Journal, 7*, 90–112.

Peirce, C. S. (1997) [1913]. *Collected papers of Charles Sanders Peirce* (C. Hartshorne, P. Weiss, and A. W. Burks, Eds) (New edition). Bristol: Thoemmes Continuum.

Price, L. (2005). Playing musement games: Retroduction in social research, with particular reference to indigenous knowledge in environmental and health education. *Southern African Journal of Environmental Education, 22*, 87–96.

Russell, C. L. (1999). Problematizing Nature Experience in Environmental Education: The Interrelationship of Experience and Story. *Journal of Experiential Education, 22*(3), 123.

Strick, M., van Noorden, T. H. J., Ritskes, R. R., de Ruiter, J. R., and Dijksterhuis, A. (2012). Zen meditation and access to information in the unconscious. *Consciousness and Cognition, 21*(3), 1476–1481.

Tesh, S. N. (1988). *Hidden arguments: Political ideology and disease prevention policy*. New Brunswick: Rutgers University Press.

Thoreau, H. D. (1981) [1853]. *Journal: Volume 5 1852–1853* (J. C. Broderick and R. Sattelmeyer, Eds). Princeton: Princeton University Press.

Wilkinson, M. L. (2013). Introducing Islamic critical realism. *Journal of Critical Realism, 12*(4), 419–442.

Williams, R. (2006). Rome Lecture: *Secularism, faith and freedom*, 23 November 2006. Retrieved 17 May 2015, from http://rowanwilliams.archbishopofcanterbury.org/articles.php/1175/rome-lecture-secularism-faith-and-freedom.

Zeidan, F., Johnson, S. K., Diamond, B. J., David, Z., and Goolkasian, P. (2010). Mindfulness meditation improves cognition: Evidence of brief mental training. *Consciousness and Cognition, 19*(2), 597–605.

Index

Page numbers in *italics* denote tables, those in **bold** denote figures.

For Product Safety Concerns and Information please contact our EU representative GPSR@taylorandfrancis.com
Taylor & Francis Verlag GmbH, Kaufingerstraße 24, 80331 München, Germany

www.ingramcontent.com/pod-product-compliance
Ingram Content Group UK Ltd.
Pitfield, Milton Keynes, MK11 3LW, UK
UKHW022007180926
475701UK00003B/245
* 9 7 8 0 3 6 7 5 9 7 6 8 9 *